SUSTAINABLE RETROFITTING OF COMMERCIAL BUILDINGS

Despite recent improvements in energy efficiency being made in new buildings, it is important that similar improvements are made in existing commercial buildings and to take action to meet emission reduction targets. The objectives and challenges of such action will reduce the risk of the commercial building sector becoming obsolete due to high energy use and poor environmental performance.

Sustainable Retrofitting of Commercial Buildings presents a theory-based, practice-support methodology to deal with sustainable retrofitting opportunities for existing commercial buildings in warm climates using bioclimatic design as the basis. The book has four main parts, focusing on eco-design and renovation, bioclimatic retrofitting, technological and behavioural change, and case studies of retrofitting exemplars. In Part I, the context of climate change effects on design and renovation at the city scale are discussed. In Part II, we consider bioclimatic retrofitting as a 'design guide' for existing buildings highlighting the significance of architectural design and engineering systems for energy performance. The technological and behavioural contexts of the existing building sector, with policies, modelling, monitoring and trend analysis in respect to energy and environmental performance, are covered in Part III. The final part gives some case studies showing the effectiveness of strategies suggested for effective environmental performance. This book provides compelling information for all involved in the design and engineering of retrofitting projects in warm climates.

Richard Hyde is Professor of Architectural Science at the University of Sydney, Australia. He is a registered architect engaged in the field of sustainable architectural design and research for buildings. He is currently Associate Dean Research, Editor-in-Chief of *Architectural Science Review* and Coordinator of the Sustainable Design Program, Faculty of Architecture Design and Planning, the University of Sydney.

Nathan Groenhout is a consultant providing ecologically sustainable design, thermal and energy modelling, and research and development in building innovation. He is currently Adjunct Associate Professor in the Faculty of Architecture Design and Planning at the University of Sydney, Australia. In January 2002, he was a Visiting Research Scholar at the Institute of Fluid Science, Tohoku University, Japan whilst completing his PhD in numerical and experimental investigations of advanced solar water heaters.

Francis Barram is an author and energy analyst with extensive experience in the design of sustainable building services and industrial processes. He is a member of the Australian Standards Energy Audit Standard Committee and founder and Managing Director of Ensight, an award winning national consultancy.

Ken Yeang is an architect, planner and ecologist. He is the Chairman and Design Director of Llewelyn Davies Yeang (UK) and principal of T. R. Hamzah & Yeang (Malaysia). He is the Distinguished Plym Professor at the University of Illinois, USA, and has received interna
masterplans.

SUSTAINABLE RETROFITTING OF COMMERCIAL BUILDINGS

WARM CLIMATES

EDITED BY RICHARD HYDE, NATHAN GROENHOUT,
FRANCIS BARRAM AND KEN YEANG

LONDON AND NEW YORK

First published 2013
by Routledge
2 Park Square, Milton Park, Abingdon, Oxon, OX14 4RN

Simultaneously published in the USA and Canada
by Routledge
52 Vanderbilt Avenue, New York, NY 10017

First issued in paperback 2020

Routledge is an imprint of the Taylor & Francis Group, an informa business

British Library Cataloguing in Publication Data
A catalogue record for this book is available from the British Library

Library of Congress Cataloging-in-Publication Data
Sustainable retrofitting of commercial buildings : warm climates /
edited by Richard Hyde ... [et al.].
p. cm.
Includes bibliographical references and index.
1. Commercial buildings–Energy conservation. 2. Buildings–Energy
conservation. I. Hyde, Richard, 1949–
TJ163.5.B84S87 2012
690.028'6–dc23
2012006900

ISBN 13: 978-0-367-57667-7 (pbk)
ISBN 13: 978-1-84971-291-0 (hbk)

Typeset in Univers by
Keystroke, Station Road, Codsall, Wolverhampton

CONTENTS

LIST OF FIGURES

LIST OF TABLES

NOTES ON CONTRIBUTORS

Francis Barram is an electrical technologist and economist with extensive experience in the design of sustainable building services and power generation. He has designed efficient services for several groundbreaking projects in South East Queensland. He has served as a member of the National Steering Committee of the Technical Reference Group for the Greenhouse Rating Scheme and as a consultant to the Australian Building Code's Energy Efficiency Technical Advisory.

Brett Beeson is an ESD consultant and mechanical services engineer. His expertise is in building physics, computer simulations (energy, day-lighting, fluid dynamics) and low-energy mechanical systems. He sees Australia's widely varying climate and unique culture as a challenge to produce technically sound and people-focused designs.

Christina Candido is an architect by training and a researcher by curiosity. She graduated and completed her MPhil in Architecture and Urban Planning at the Federal University of Alagoas (Brazil) and recently finished her PhD as a double degree in Civil Engineering at Federal University of Santa Catarina (Brazil) and Environmental Science at Macquarie University (Australia). Her research focuses on indoor environmental quality bioclimatic design.

John Cole is Professor and Director of the Australian Centre for Sustainable Business and Development, University of Southern Queensland.

Nathan Groenhout is a consultant providing ecologically sustainable design, thermal and energy modelling, and research and development in building innovation. He has extensive experience as a professional engineer conducting research, both experimentally and using computer simulation tools and in managing research and product development projects. He has also been involved in a number of research projects using CFD and FEA modelling to analyse and provide solutions for real-world engineering problems, such as vehicle composite structures and complex heat transfer and fluid flow. In January 2002, he was a Visiting Research Scholar at the Institute of Fluid Science, Tohoku University, in Japan.

Edward Halawa is Senior Research Fellow at the Centre for Renewable Energy, Research Institute for the Environment and Livelihoods, Charles Darwin University, Darwin, Australia. After completing his PhD at the University of South Australia in 2005 he worked as a Postdoctoral Researcher there until March 2011 during which he was involved in the CSIRO funded Project: *Intelligent Grid in a New Housing Development*. From May 2011 to June 2012 he was Senior Lecturer in the Mechanical Engineering Department, Universiti Teknologi PETRONAS in Tronoh, Malaysia. In his earlier research carrier he was a UNDP Fellow on Solar and Renewable Energy at SOGESTA, Urbino, and ENEA-Casaccia, Rome, Italy. His research competence and expertise include thermal energy storage, thermal comfort, building energy modelling and simulation using TRNSYS and solar energy for cooling, drying, space and water heating.

Kat Healey is an engineer and PhD candidate with the Centre for Sustainable Design within the School of Architecture at the University of Queensland, undertaking research focusing on qualitative aspects of thermal comfort. Her industry experience includes sustainability leadership, mechanical services design, renewable energy systems and building simulation.

Richard Hyde is Professor of Architectural Science at the University of Sydney. He is a registered architect who has been working in the field of sustainable architectural design and research for buildings. He is currently Head of the Architectural and Design Science discipline and Coordinator of the Sustainable Design programme at the University of Sydney. He has a national and international reputation in the area of environmentally sustainable design.

Rosemary Kennedy is Director of the Centre for Subtropical Design, an interdisciplinary research collaborative, at the Queensland University of Technology School of Design. She also teaches architecture in both the undergraduate and Master's programmes. Rosemary convened the ground-breaking international *Subtropical Cities* conferences in Brisbane in 2006 and 2008 and was co-chair at the third conference held in Fort Lauderdale, USA, 2011.

David Leifer is both a Registered Architect and an Incorporated Engineer. He has been a university academic since 1986 at the Universities of Glasgow, Queensland, Auckland and Sydney, and is currently Dean of the newly established Asian School of Architecture & Design, Cochin, India.

Margaret Liu is a PhD student in the Faculty of Architecture, Design and Planning at the University of Sydney and a registered architect in New South Wales. She has four years of experience as a project manager in capital works at the same university, where she managed an extensive range of retrofit projects. She also has over five years of architectural practice experience in the private sector. Her research interests are bioclimatic retrofit of existing buildings, low energy architecture, sustainable planning and design, and the benchmarking and assessment of sustainable architecture.

Susan Loh is a Lecturer at the School of Design, Queensland University of Technology, Brisbane. Susan has worked in architectural firms in Australia and Canada for ten years in aged care, commercial buildings and residential projects. Her main areas of academic research and teaching involve sustainable design and environmentally responsive buildings.

Mark B. Luther, Associate Professor, is the Consortium Director of MABEL (the Mobile Architecture and Built Environment Laboratory) at Deakin University, Geelong, Australia. He is a registered architect in Michigan

(USA) and Victoria (Australia). He lectures in the School of Architecture and Building at Deakin University, in graduate and undergraduate courses on lighting, acoustics and building services as well as a course in concepts Sustainable Future. Through the MABEL programme, he has been involved with measuring the Indoor Environmental Quality (IEQ) of over 35 buildings throughout Australia. Projects have included schools, offices, airports, houses and hospitals. MABEL's projects have partnered with the Australian Greenhouse Office (now DEWHA), Sustainability Victoria, the Department of Human Services, the CSIRO and the Building Commission (Victoria).

Caimin McKabe has over 20 years' consulting experience in environmentally responsive building design both in Australia and Europe. Key to his success is his multi-disciplinary, open approach to design and understanding on how to improve the sustainability of buildings, in particular existing buildings, as well as the ability to communicate complex analysis and results in under-standable ways, enabling quick and effective decisions to be made at the early stages of design. He also has considerable experience in the analysis and assessment of ESD strategies in buildings which has allowed his ideas to be incorporated into designs with engineering assurance. Included in his experience are a number of developments, which enjoy both national and international praise for innovative design, energy efficiency, occupant com-fort and reduced environmental impact, including the award-winning Monash Science Centre and 60L Green Building Project in Australia, and the Commerzbank Headquarters in Frankfurt.

Alexandra McKenna is responsible for delivering The Warren Centre's programmes in line with the strategic plan and ensuring the success of innovative projects including the Low Energy High Rise Project. The Warren Centre for Advanced Engineering is a not-for-profit company which aims to foster excellence and innovation in advanced engineering throughout Australia. This project, working in collaboration with the property sector, seeks to identify low cost strategies for operating energy-efficient high-rise office buildings. Prior to joining The Warren Centre, Alex was responsible for delivering the sustainability platform for the DEXUS Property Group including a $41 million energy improvement upgrade plan to deliver a 4.5 star NABERS Energy average for the office portfolio. Previously, Alex held the role of Manager, Risk & Sustainability at Colonial First State Global Asset Management. Alex holds a Master of Environmental Law, Business and Management, a Bachelor of Science in Agriculture and a Graduate Diploma in Applied Finance and Investment.

Wendy Miller is a Senior Research Fellow in the School of Physics, Chemistry and Mechanical Engineering, Queensland University of Technology, Brisbane. Wendy has worked with communities, industry and research in the fields of energy services, energy efficiency and renewable energy. The focus of her academic research is on interdisciplinary systems approaches to the provision of energy services to meet occupants' needs.

Michelle Nurman is a practising architect and currently working in the NSW Government Architect's Office. She has nine years' experience in all facets of architectural design and documentation, particularly in public buildings and commercial fit-outs. She is a Green Star accredited professional. Her Master's Honour's thesis was on the topic of bioclimatic principles for retrofitting commercial buildings in a temperate climate zone. Her area of interest is in retrofitting projects to bring buildings in line with current sustainable best practice. She has been involved in upgrades to make NSW government buildings meet the NSW government Sustainability Policy and in initiatives to make organisations carbon neutral.

Alan Obrart is the principal of Obrart & Co. and senior lecturer and coordinator of the graduate programme in building services at the University of Sydney.

Lester Partridge is a Technical Director of AECOM's Building Engineering Business Line and is their Global Director of the Applied Research and Advanced Design Technical Practice. He has over 25 years of involvement in all facets of building services consultancy. With project experience gained in Australia, Britain, Singapore, Hong Kong, China, New Zealand and the United States of America, he provides expertise in master-planning, system concept design, detailed design and project administration. Lester provides specialist expertise in thermal and energy modelling and sustainable 'Green' building design. Lester is a qualified mechanical engineer and a Fellow of the Australian Institute of Engineers. Lester's expertise includes design and modelling for passive ventilation and daylighting along with building energy simulation and comfort analysis modelling.

Brett Pollard is an architect and landscape architect, and Head of Knowledge and Sustainability at HASSELL. He has a Master of Design Science (Sustainable Design) and is a LEED Accredited Professional of the United States Green Building Council as well as a Green Star Accredited Professional of the Green Building Council of Australia. Brett has over 25 years' experience gained in public and private practice in Australia and Europe, and spent five months researching at Vancouver's University of British Columbia in 2007. He has experience and expertise in a wide range of sustainable design areas, including renewable energy systems, building energy performance, high performance façades and green building rating systems. Brett is a regular presenter at conferences, universities and workshops, and has researched and written many research papers and articles on sustainable design issues. He regularly provides strategic input and high-level sustainable design advice for a broad range of projects throughout all of HASSELL's studios and disciplines.

Bruce Precious is National Sustainability Manager – Operations, for the GPT Group. GPT's business is focused on the ownership, management and development of real estate. GPT's mission is to 'create and sustain environments that enrich people's lives'. Bruce is a Mechanical Engineer with a keen interest in making things work the way they were designed to. He is also Chair of the City of Sydney's Better Building Partnership and Chair of the Green Star Performance – Indoor Environment Quality Working Group.

Indrika Rajapaksha is a Senior Lecturer at the University of Moratuwa in Sri Lanka, and has been involved in research in London and in Australia at the University of Sydney. She received her Dr. Eng., Environmental Science in Japan and has a Master of Science, Architecture, from the University of Moratuwa. Her specialisation is in computer simulation of buildings.

Upendra Rajapaksha is a Senior Lecturer in architectural design and science in the Department of Architecture, University of Moratuwa in Sri Lanka. He earned his doctorate in architecture from the University of Queensland, Australia, and since then he has provided research and practice expertise in green buildings, low energy architecture and environmental sustainability in the built environment. He has over 16 years of involvement in all fields of architectural practice and education in Sri Lanka. He is a qualified chartered architect, an associate of the Sri Lanka Institute of Architects and an overseas member of the Royal Institute of British Architects.

Craig Roussac directs the Investa Sustainability Institute's programme of action research for sustainability in the built environment through collaborations with leading universities and research institutions. He also works for Investa Property Group, one of Australia's largest and most respected owners of commercial property where he has responsibility for the sustainability, safety and environmental management platforms. Craig has a longstanding interest in energy and buildings and is a PhD candidate at the University of Sydney in the Architectural Science discipline. His primary research interest is in how technologies and non-technological factors can combine to achieve significant greenhouse gas emission reductions from the operation of buildings.

Sattar Sattary has studied and worked in the environmental architecture area (construction process impact, carbon metric of construction sites, optimum embodied energy) and sustainability area since 2002 in Australia. He is an accredited professional member of Green Globe 2004, Green Building Council since 2008, participating in developing green tools for Green Globe, Water Theme Park, Memorial Park, Aquatic Centre and most recently working with Australian Green Infrastructure Council (AGIC). Prior to this he held an academic post at the National University in Iran.

Leena Thomas is a Senior Lecturer in the School of Architecture at the University of Technology, Sydney. Leena has served as Course Director, Director of Teaching and Learning and Acting Head of School of Architecture, and currently heads the Environmental Studies strand at the school where she has introduced a strong emphasis on integrated environmental design across design studio and architectural science subjects in the School. Leena's research focuses on interrogating and transforming contemporary design practices to be responsive to global concerns for climate change, zero carbon development and high quality living/work environments. Her research and consultancy work on design process, building environmental rating and performance, user studies and post-occupancy evaluation includes a number of iconic green buildings in Australia and India. These outcomes have been used by enabled funding partners and industry stakeholders to validate design decisions, and improve the development and management of buildings. Leena regularly participates as an expert panellist for awards, technical committees and policy reviews for the government, or professional and industry bodies such as the Australian Institute of Architects, the Green Building Council of Australia and the National Australian Building Environmental Rating Scheme.

Mark Thompson is an Architect and ESD Consultant who became the Corporate Sustainability Principal for the Australian Schiavello Group of Companies in February 2010. Prior to this, he was a Director of Brisbane-based Architects, The TVS Partnership, for 18 years. Since 2007 he has been President of the Australian Green Development Forum, where he actively pursues practical solutions to ESD implementation and education.

Marci Webster-Mannison is known for the design of ecologically inspired architecture as a practising architect, speaker and writer. Marci is the Director of the Centre for Sustainable Design at the University of Queensland.

Ken Yeang is an architect and Director in an international architecture and planning practice. His buildings are cited in many texts as exemplars of bioclimatic design. He has been appointed to a number of positions at universities worldwide and has received international awards for architecture on the basis of his design research. Dr Yeang's major contribution in the field is the development of propositions for bioclimatic skyscrapers as an alternative built form for tall buildings, which he is generally attributed as having pioneered. He is committed to research on the design and post-occupancy evaluation of his buildings and to working in an integrated manner with all members of the design team.

FOREWORD

After many years at a university heading a research group and managing a Swiss national programme 'Solar Architecture', I decided to try life in free enterprise and establish a private company. Almost immediately realtors got wind of my search for office space and offered me the usual modern facilities. Instead, I chose a building from the 1920s. The office space was previously the home of an ancient accountant who happily added numbers for half a century, never noticing that the plaster walls and ceiling might collapse at any moment. The opportunities for renovation were exciting, to say the least! The building owner gave me a budget and so came the challenge to practise what I teach and write. The basics were right: most windows faced north (good for daylight and summer comfort north of the equator), there was a quiet roof terrace for coffee breaks and it had a good location (first rule of real estate wisdom) close to the train station 12 minutes from the centre of Zürich. In addition to the priority I gave to effective lighting, I had the kitchenette, toilet room and corridor demolished and completely rebuilt. These spaces, although not places of hour-long occupancy, have a disproportionate effect on the perception of the quality of the work environment.

In contemporary commercial buildings it would seem that comfort, regardless of the climate or architecture, has been possible since the advent of air conditioning in the 1950s. The reality is different than it seems in buildings where 'design' was more important than occupants, and depletion of non-renewable energy and climate change were of no concern. As a result, thermal and visual comfort leave much to be desired.

The catchy term, 'bioclimatic design' suggests an alternative design approach: adapting architecture to a climate so that biological life is nurtured. The *bio* prefix infers that the impact of a building on its environment (locally and globally), as well as its interior are considered. Given comfort and energy performance shortcomings of so much of the building stock, *retrofitting* is an urgent topic. *Commercial buildings* are an important sector because in an ever increasing urban, hi-tech civilisation much of the workforce is engaged within buildings. Quality of life, and connected with this, productivity, are directly affected by the indoor climate created within four walls, windows and a roof.

Achieving indoor comfort in *warm climates* is a challenge. In temperate and cold climates heat from passive solar gains and office equipment helps provide a warm environment. If all else fails, occupants can dress warmer. In a warm climate, solar gains and appliance heat are detrimental and there is only so much clothing you may take off. Commercial buildings poorly designed for their climate can be truly awful places to work. Yet, historically in many warm and humid as well as warm and dry regions of the world, there was wisdom in using natural climatic forces to improve comfort. Such historic solutions may be inadequate today, given our higher comfort expectations, the size of buildings and types of activities pursued. However, such concepts may provide adequate comfort at least part of the year and in any case can reduce the loads on energy consumption of mechanical systems.

In this context I met the co-author and editor of this book, Richard Hyde. He led a working group on exactly this topic in the framework of a programme on sustainable housing I was running for the International Energy Agency (IEA-SHC Task 28). The team he assembled from the Middle East, South-east Asia and Australia brought knowhow about indigenous design and a richness in building cultures. From a visit to Richard in Brisbane, I learned how important comfort can be in a hot humid climate!

A paradigm shift from 'architecture' for architecture journals to bioclimatic design can be accelerated by a few key steps. First, public awareness of the benefits of this approach is essential. This should be reflected in changes in building codes and policy, but more importantly in the priorities of real estate investors and demands of clients. A re-look at the economics, including assigning value to less tangible benefits can help sell concepts. Once clients are motivated and enthusiastic, designers need usable analytical tools to develop skills. Unlike the brute force available from a 50 ton chiller, bioclimatic design strategies work in subtle and less predictable ways. It is better to fail in a computer model than in a built project! After a sensible concept has been drawn up, the next key step is articulating it in the design brief so the right contractors are attracted (and your lawyer can back you). Lastly, when a renovation has been completed, monitoring and fine tuning are important to optimise performance.

This book aims to help change designer and client thinking. For this purpose, the large section on exemplary retrofit projects is the proof of concept. The illustrated buildings allow the reader to form an opinion on how holistic the renovation is: its completeness in all aspects. Included here for me are of course energy and comfort, but also less definable qualities. Is there a sense that as a result of the renovation, people will feel good, even joy in the building? There needs to be a computer model to assess the 'joy factor'. In the office space I renovated, I felt joy to come to work. A commercial building retrofitted to work better in the climate where it is located is intuitively a more relevant and satisfying architecture.

Robert Hastings, Donau University

PREFACE

Despite recent improvements in energy efficiency being made in new build it is important that the existing commercial building sector also take action to meet emission reduction targets. The objectives and challenges of such action will reduce the risk of the sector becoming obsolete due to high energy use and poor environmental performance.

This book presents a theory-based, practice-support methodology to deal with sustainable retrofitting opportunities for existing commercial buildings in warm climates using bioclimatic design as the basis.

The book is divided into four main parts focusing on eco-design and renovation bioclimatic retrofitting, technological and behavioural change and case studies of retrofitting exemplars. In Part I, the context of climate change effects on design and renovation at the city scale are discussed. In Part II we consider bioclimatic retrofitting as a 'design guide' for existing buildings highlighting the significance of architectural design and engineering systems for energy performance. The technological and behavioural contexts of the existing building sector in which policies, modelling, monitoring and trend analysis in respect to energy and environmental performance are covered in Part III. The final part gives some case studies showing the effectiveness of strategies suggested for effective environmental performance. This book provides compelling information for all involved in the design and engineering of retrofitting projects in warm climates.

This book addresses a number of questions around the renovation of commercial building though retrofitting with 'green' technologies in the face of increasing pollution affecting many facets of the earth's climate and ecological systems. It is the culmination of a six-year research project involving a wide range of contributors who address particular questions for research and practice in the area of bioclimatic design for retrofitting commercial buildings in warm climates. Each of the contributors has a particular story about retrofitting for climate change, which collectively is diverse and complex. The object of this volume is to draw together some of the directions and findings from these stories into some collective wisdom for design practice. Each chapter addresses a particular question posed from research into sustainable retrofitting, given as the subtitle of the chapter.

ACKNOWLEDGEMENTS

The project has been supported by an Australia Research Council (ARC) Linkage grant LP0669628, 'Exploring Synergies with Innovative Green Technologies: Redefining Bioclimatic Principles for Multi Residential Buildings and Offices in Hot and Moderate Climates', initially from 2006 at the University of Queensland and subsequently at the University of Sydney from 2008 to the conclusion of this project in 2012.

During this time the research work has been actively carried out in collaboration with international partners Llewellyn Davies Yeang and T.R. Hamzath & Yeang, through Dr Ken Yeang, and AECOM through Dr Nathan Groenhout and ENSIGHT through Francis Barram. The industry partners were joined in 2009 by INVESTA through Craig Roussac.

The project was born at the Centre for Sustainable Design at the University of Queensland. This joint Centre between Architecture and Engineering provided the interdisciplinary foundation to the project. Without the support from this Centre, the involvement of joint Director Professor Peter Dux and a number of research personnel, this project would not have been possible.

I am sincerely indebted to Edward Halawa who played an important role in the preparation of the proposal for this ARC Linkage project; his work provided the initial draft, which formed a substantial basis for the submission of the ARC Linkage proposal. The proposal was used to form the basis for a paper presented in AIRAH's Pre-Loved Building Conference in 2007 in Brisbane and I would like to acknowledge his co-authorship on that paper.

The early work while in Queensland involved a very successful initial workshop, joint hosted with the EPA Queensland, Architecture programmes of Queensland University of Technology and the University of Queensland. This included involvement of Kate Healey and Marci Webster-Mannison. This set the initial stages of the project. In particular, I would like to thank Professor Deo Prasad for his involvement in the early work and his presentation at the workshop.

My move to the University of Sydney in 2008 provided a marked stepping up of the project; the empirical research work got off the ground and became embedded in both the teaching and research culture of the Faculty of Architecture Design and Planning. The support from the Discipline of Architectural and Design Science is acknowledged and central to the success of this project. The support from the then Dean of Faculty Warren Julian and Craig Symes from SYDNOVATE helped to bed the project down in its new university milieu.

We attracted new research staff. I am indebted to Dr Indrika Rajapaksha and Dr Upendra Rajapaksha who joined the project from the Department of Architecture, University of Moratuwa in Sri Lanka and did a substantial amount of the research involving the design and experimental research work using computer simulation. We changed the methodology thanks to the offer of a case study building from INVESTA. The property and facilities staff involved in the project provided access to the building and to the energy and other data to provide thorough ongoing monitoring. We are indebted to the INVEST organisation for this facility and to them for their support. They may find the results in Part II useful. Importantly, thanks to Craig Roussac, we managed to obtain further support for the second workshop help at the University of Sydney in 2009.

The workshop involved Sydney University staff, a number of postgraduate students and industry experts. I am indebted to them for the workshops and meetings in which they brought together their own particular skills and knowledge. In particular, Dr David Leifer provided assistance alongside Mr Alan Obrart. It should be noted that their help and the championing of the project across the University of Sydney were important to this project and have helped in the work of research students, Michelle Nurman, Margaret Liu and others who have contributed to this project.

A third workshop was held in 2010 and would not have been possible without the help of Dr David Leifer and Karen Sharp from AECOM. The International Islamic University of Malaysia hosted this workshop. I am indebted to the students from Sydney and from Malaysia who joined forces for an International Research Colloquium in collaboration with Professor Shireen Kassim from IIUM and Dr Ken Yeang and Andy Chong, Director, T.R. Hamzath & Yeang. I am indebted to them for hosting this conference and design workshop. The first day was a research colloquium with the presentation of papers followed on the second day with a tour of the buildings by T.R. Hamzath & Yeang. The final day involved a Design Charrette which led to new ideas for ecological retrofitting. This was an enriching experience for the students, staff and practitioners. The outcomes of the three workshops are available on the Faculty of Architecture, Design and Planning website, the University of Sydney. 'Exploring Synergies with Innovative Green Technologies for Advanced Renovation', http://sydney.edu.au/architecture/research/archdessci_advanced_rennovation_project.shtml.

In the later stages, Dr Christina Candido assisted with pulling the book together with Nicki Dennis and Alice Aldous from Earthscan, Routledge and Taylor & Francis.

In conclusion, I would like to thank Professor Robert Hastings for the Foreword to the book and the involvement of the International Energy Agency, Solar Heating and Cooling Program for their support. We are continuing to diffuse the outcomes of the research through Task 47, which is instrumental in bringing together research and practice of sustainable retrofitting.

Richard Hyde
Professor: Chair of Architectural Science,
The University of Sydney

LIST OF ABBREVIATIONS

AGBR	Australian Greenhouse Building Rating
AGP	Accredited Green Power
AUD	Australian Dollar
Base building energy	Energy consumed in supplying building central services
BEEM	Building Environmental and Energy Modelling
BESTEST	Building Energy Simulation Test
BIM	Building Information Modelling
BMS	Building Management System
CBD	Commercial Building Disclosure scheme
CFD	Computational Fluid Dynamic
CO_2e	Carbon dioxide equivalent
ETS	Emissions Trading Scheme
FiT	Feed in Tariff
GBF	Green Building Fund
GHG	Greenhouse Gas (emissions)
HVAC	Heating Ventilation and Air Conditioning
IRR	Internal Rate of Return
IWEC	International Weather file for Energy Consumption
$kgCO_2$	kilogram Carbon Dioxide
$kgCO_2/m^2$	kilogram Carbon Dioxide on a per square metre net floor area basis
KPI	key performance indicator
kW	Kilowatt
kWh	Kilowatt hours
kWh/m^2	Kilowatt-hour unit of measurement of energy on a per square metre net floor area basis
MJ	Mega Joule unit of measurement of energy
MJ/m^2	Mega Joule unit of measurement of energy on a per square metre net floor area basis
NABERS	National Australia Built Environment Rating System
NLA	Net Lettable Area
NPV	Net Present Value
PV	Photovoltaic
REC	Renewable Energy Certificates
SCS	Solar Credit Scheme
SEDA	Sustainable Energy Development Authority
TRY	Typical Reference Year
VAV	Variable Air Volume
W	Watt – unit of energy

GENERAL INTRODUCTION

Richard Hyde

This book arose from a project to explore the use of 'green technologies' in novel and innovative combinations for renovation of larger-scale buildings to reduce energy consumption and greenhouse gas emissions. The project started in 2006 with a mission to provide design guidance to assist the transformation of the property industry to a more sustainable future. The design guidance includes redefinition of bioclimatic design principles, demonstrates opportunities for synergies in technologies to reduce environmental impacts, and provides cost benefit information focused on the renovation of non-residential buildings in warm climates. This general discussion introduces the original mission. Parts of this mission reappear as themes in the subsequent chapters.

First, the significant drivers to this work in 2006 concerned international and national policy. At that time, according to the Productivity Commission, the need for energy conservation was a national imperative to be enforced in due course by legislation in large-scale buildings, being Class 5-9 as defined by the Building Code of Australia (The Australian Government Productivity Commission, 2005). The Property Council of Australia also argued that little research and training, to enable the industry to meet the energy targets in this proposed legislation, had been carried out (The Property Council of Australia, 2005).

Second, research was initiated by the International Energy Agency which started investigating how to harness the processes of renovation to improve environmental performance as a critical pathway for improving the sustainability of our cities (International Energy Agency, 2005). This led to similar research to the IEA task work beginning in Australia. The Property Council of Australia argued that renovation is a neglected area with new buildings representing only 1 or 2 per cent of the building stock; the remaining stock will need to be upgraded to achieve the new energy targets (Australian Greenhouse Office, 1999). Principles, strategies and cost benefit information are needed to enable this transformation (The Property Council of Australia, 2005; Australian Greenhouse Office, 1999).

Third, research into combinations of new and existing technologies could be exploited for this process of transformation to more sustainable buildings, hence sustainable renovation, and later we called this sustainable retrofitting since it required the replacement of existing technology with new high performance systems. These systems are examined in terms of achieving synergies between systems – between passive (fabric) and active systems (mechanics) – advancing knowledge in the key field of Passive Low Energy Architecture (PLEA, 2007).

For the purposes of this work, a clear distinction is drawn between refurbishment, renovation and retrofitting in the context of building obsolescence. Refurbishment may be defined as returning the building, or its systems, to their original condition, addressing the forces of physical obsolescence. Renovation takes this a step further and may incorporate changes to the physical parameters of the building, while retrofitting refers to the replacement and upgrading of systems and technology to address technological or environmental obsolescence. The overall goal was to develop new bioclimatic principles, strategies and techniques for large-scale buildings that have some form of obsolescence impacting them. While these principles, strategies and techniques are well established in the new build space, they were at that time lacking in the existing building domain.

BIOCLIMATIC DESIGN

Bioclimatic issues in architecture were identified by Olgyay in the 1950s and developed into a process of design in the 1960s (Olgyay, 1963). This design process brings together the disciplines of human physiology, climatology and building physics; it has been integrated within the building design professions in the context of regionalism in architecture, and in recent years has been seen as a cornerstone for achieving more sustainable buildings (Hyde, 2000). PLEA (Passive and Low Energy Architecture) is 'a commitment to the development, documentation and diffusion of the principles of bioclimatic design and the application of natural and innovative techniques for heating, cooling and lighting' (http://www.plea-arch.org/). Hence research into bioclimatic issues mainly takes the form of passive low energy architecture research (PLEA-r), a well-developed field carried out worldwide (see PLEA conferences, http://www.plea-arch.org/).

This research has led to the development of bioclimatic design principles, which are used by design professionals as a starting point for designing with climate in mind. So far, they have been developed primarily for low and medium-scale buildings, since these types of buildings are relatively easy to make bioclimatically interactive, with the form and fabric of the buildings matched to human and climate factors to optimise climate response (Rajapaksha and Hyde, 2002).

Large-scale buildings have largely escaped attention due to issues of programme complexity, the dense urban context in which these buildings are usually located, and the availability of cheap energy for cooling and providing comfort. The design principles used have largely ignored bioclimatic influences, achieving an adequate internal environment by using mechanical energetic systems to add back comfort. Exceptions to this approach are increasing, and can be found in a number of buildings, such as the pioneering high-rise buildings of Yeang and the large floor plate buildings of Bligh Voller Nield (Yeang, 1999).

The existence of a set of bioclimatic principles, strategies and best practice solutions for larger-scale buildings has not yet been acknowledged or fully researched within the field (Pedrini, 2003). Whilst a number of case studies have been written on the bioclimatic design of large-scale buildings, the extent to which general principles can be adduced from these examples is limited. Jones (1998) has developed some bioclimatic principles for large buildings, concerned with: the form of energy used, renewable or non-renewable; the efficient use of energy and its conservation; and human well-being, comfort and amenity. However, these are largely unrelated to building strategies in a building science context, or fully demonstrated as best practice solutions for large-scale buildings.

Furthermore, design professionals remain sceptical of such approaches, due to the lack of workable models of large-scale bioclimatic buildings and the cost of additional design work such as simulation modelling to demonstrate proof of concept, cost effectiveness and comfort of these types of building (Jones, 1998). Improving energy efficiency has centred on designing more efficient mechanical systems rather than examining factors, such as the passive elements of the building, in order to engage in synergies that lead to an integrated solution.

Importance of passive elements: the building's microclimate, form and fabric

Jones suggests that the bioclimatic office building will use five to six times less energy over its life than a conventional office building (Yeang, 1998). This is achieved primarily through the use of the building's microclimate, form and fabric, rather than by using more efficient mechanical equipment. For example, in warm climates where cooling is needed most of the year, 34 per cent of the energy used for cooling is addressed at the mitigation of solar radiation into the building. Cooling is normally achieved through air conditioning and so incurs large environmental penalties – high energy use, high greenhouse gas emissions. Bioclimatic design refocuses on providing high quality passive design of the form and fabric of the building envelope and on the use of new technologies. Pioneering work by Yeang has defined a range of passive biophysical elements, which can be used to achieve 'net zero-energy buildings', but more emphasis on research in this area is needed (Gilijamse, 1995).

DEVELOPING ENVIRONMENTAL TARGETS: TOWARDS THE 'NET ZERO-ENERGY BUILDING'

The 'net zero-energy building' is an ideal concept in which no fossil fuels are used and sufficient electricity is generated from natural sources to meet the service needs of the occupants. This idea sets the optimal design target for a building in terms of minimising the environmental effect and energy cost of the dwelling. The appealing aspect of this idea is the availability of 'free' energy derived from nature such as solar and wind energy, and of energy sinks such as air, water and earth. Ideally, this 'free' energy can be used to achieve the zero-energy target by employing both passive and active solar technologies. Since the technologies come in various forms and states of development, it is crucial to explore how they can be synergised to achieve the goal (Wittchen, 1993).

EMERGING AND PROVEN TECHNOLOGIES

Theoretically, the zero-energy dwelling concept can be achieved by simultaneously:

- reducing the energy demand of the dwelling by increasing energy efficiency, through various energy conservation measures; and
- utilising the solar energy incident on the walls, roof and ground surfaces surrounding the house for electricity generation and to satisfy the dwelling's heating requirement.

Implementing these measures entails careful assessment of their economic feasibility, environmental benefits and effects on human comfort. Various technologies for energy conservation and solar energy utilisation are already available or are in development. Energy conservation measures available to reduce energy demand include proper insulation of the building envelope and the selection of high efficiency (high energy rating) appliances, among others. Solar technologies can be divided into passive and active technologies. Passive solar technologies include passive solar heating, natural ventilation, daylighting, thermal mass storage, ground cooling, etc. Passive solar heating and cooling employ the structural elements of a building to collect, store and distribute solar energy without mechanical equipment or with minimal use. The Trombe wall (or bioclimatic wall) is one popular means for collecting solar energy. The solar energy transmitted through a transparent cover is absorbed by the outer surface of the wall; it is either conducted through the wall, reaching the inner surface several hours later, or picked up by air flowing through the space between the cover and the wall outer surface (Manz and Egolf, 1995). Research studies by Manz and Egolf and Stritih and Novak explored the use of phase change materials (PCMs) as the thermal storage in the Trombe wall, a technique that reduces energy losses and energy requirements (Hyde, 2000; Stritih and Novak, 2002). In cooler climates the Trombe wall has transformed into a bioclimatic wall that combines a range of environmental functions such as shading, rain protection, ventilation, light diffusion and glare control (Clarke *et al.*, 1997; Yeang, 1999).

Building-integrated photovoltaic systems have been the subject of intensive investigation in recent years. In this concept, the roof or façade of a building is designed to accommodate solar panels (or cells), thereby minimising the cost of their support structures (Halawa, 2005). A similar concept has been applied to solar thermal systems, where steel roofing is used as a thermal absorber. To improve efficiency, PCM thermal storage can be incorporated into the system (AIRAH, 2003).

The technologies mentioned above are only a few of those emerging related to the goal of creating buildings with extremely low energy demand. Given these technologies, the concept of zero-house buildings can theoretically be achieved. Realisation, however, depends on whether the three criteria previously mentioned can be satisfied, namely:

- environmental benefits;
- improvement in human comfort; and
- economic feasibility.

The first and second criteria can easily be delivered by solar technologies: a fact that has been used by solar advocates backing their 'switch to solar' movement. The third criterion, economic feasibility, is equally important and should not be treated separately from the other criteria.

Until now, there have been no significant research studies exploring the synergies of these emerging green technologies in order to bring about new and proven solutions to the environmental and energy problems of residential buildings. Research efforts so far have been fragmented, with researchers and/or research groups focusing on particular technologies without looking at the whole problem and exploring the best solution.

WHY RENOVATE?

A central issue for design professionals is to assist the building industry to transform the design and construction of buildings to improve their environmental performance. The importance of this has been identified by the private sector in order to meet pressure to improve environmental performance. The evidence found in studies by the Property Council of Australia (The Property Council of Australia, 2005) and a member survey of the Australian Institute of Refrigeration, Air-conditioning and Heating (AIRAH) in 2003 (Haycox, 2003) supports this view. A way to achieve this is to identify opportunities and mechanisms to support change within the industry.

Renovation as a path to improve environmental performance and achieve sustainability

One opportunity is arising due to the inherent entropy in the build stock. Building obsolescence and rapid expunction of sunk cost in buildings – payback periods of four to five years are common – mean there is scope for rebuilding and renovation (Hyde, 2007). Current estimates for Melbourne are that 10 per cent of the city's building stock is undergoing renovation at any given time, hence within ten years, the city will have, in theory, fully upgraded its building stock. This provides the opportunity to improve building stock to meet new and emerging needs for sustainability, linked to emerging international and national initiatives through specific agencies and organisations. The International Energy Agency's Solar Heating and Cooling Program (IEA SHC) (www. http://www.iea-shc.org/) is one such scheme, and the Green Building Council of Australia (GBCA) is another (http://www. gbcaus.org/). These two are part of the building industry's efforts to self-regulate in the move to sustainable ecological development; both bring together experts in the building field to work on transforming the industry. The GBCA's programme is applied, working to develop environmental standards through its rating tools and advocacy, while the IEA is a more research-based initiative, and is therefore relevant to this research project.

Provide initiatives to steer the industry towards sustainable ecological development

The International Energy Agency, through its Solar Heating and Cooling Program Tasks, has brought together experts from many countries to investigate industry transformation related to particular problem areas and building types. Its Task 28/36 recognised a significant opportunity for transformation in the growing trend to ecological housing, with extremely low purchased energy usage for heating and cooling, and minimal CO_2 emission (see IEA SHC Task 28/38 on Solar Sustainable Housing (http://www.iea-shc.org/task 28/index.html)). This involves the identification of the most promising and viable solar and/or green technologies and exploration of their synergies, and can be expected to directly minimise reliance on fossil fuels and to improve the environmental and thermal performance of residential buildings and/or non-residential buildings.

IEA SHC Task 47 Solar Renovation of Non-Residential Buildings points out:

> Buildings are responsible for up to 35 per cent of the total energy consumption in many of the IEA participating countries. The EU Parliament approved in April 2009 a recommendation that member states have to set intermediate goals for existing buildings to a fixed minimum percentage of buildings to be net zero energy by 2015 and 2020.
>
> (http://www.iea-shc.org/task47/)

DON'T DETONATE, INNOVATE

There are a number of areas where innovation has been investigated. These are as follows.

Review of Green Passive Technologies (see Part I)

A range of passive elements using a range of biophysical elements were examined, as follows:

- thermo hydronic: thermal mass and water sinks, thermo syphons, chimneys;
- kinetic: adaptive thermal defences;
- organics: heat sinks;
- aerodynamics: adaptive wind defences;
- materials: phase change, heat storage, insulation, radiant and evaporative defences;
- ground effects: heat storage.

(Yeang, 1999; Law, 2001; Szokolay 2004)

Furthermore, research has examined the limits of these passive systems in terms of their heat and/or humidity modification for warm climates. Part I deals with this area of innovation. We realised that we started with a very narrow view of green technology, where green technology is a subset of man-made technology. By 2010, we had moved to a broader ecological definition of green technology. To this end the design challenge is to further integrate this definition into both the design of our buildings and extend to the planning of our cities. The eco-city as yet does not exist. Piecemeal landscaping and ecological refuges are islands in a sea of man-made technology. Innovation at many levels is required to harness the power of green technology. The scale of the sustainable initiatives concerning the eco-city are defined at the planning level for the city. These decisions impact on a scale of multiple buildings and urban environments. Bioclimatic design impacts at the individual building scale for the purposes of this research. The bulk of this research is focused on the building scale. A start is with the existing systems and innovation in the systems that are used. It is argued this can be achieved through bioclimatic retrofitting. Part II examines the application of green technologies, particularly at the building scale.

BIOCLIMATIC RETROFITTING: SYSTEM INNOVATION AND SOLUTION SETS (SEE PART II)

The industry is currently searching for new technologies that can potentially be integrated into the renovation stage of residential buildings and/or non-residential buildings and generally have the following characteristics:

- are aesthetically acceptable and unobtrusive;
- can be easily integrated into the building;
- are reliable and durable;
- are more environmentally friendly than conventional systems;
- are socially and economically acceptable.

The 'solution set' methodology was developed and tested through the IEA SHC Task 28/38 programme. A solution set is a descriptive model of combinations of passive and active systems that can be used to achieve energy efficiency (International Energy Agency, 2005). Solution sets are tested in terms of meeting social, environmental and economic targets for performance. The research considered case studies of both theoretical and existing non-residential buildings. The theoretical studies lay the foundation for identification and formulation of new solution sets that can be tested through a range of tools using computer modelling, while the case studies of existing buildings allowed a widening of the design parameters and demonstrated the viability of the new solution. This approach follows the IEA SHC Task 28/38 methodology.

Selection of technologies and energy targets

The research explored and identified available solar and energy conservation technologies and non-technical systems that can be used to improve environmental and thermal performances of as-built non-residential buildings.

The NABERS Rating tool can be used to set targets for environmental performance and to assess the performance improvements of the application of the solution sets (The Australian Government Department of the Environment and Heritage (DEH), 2005). This tool was developed for the purpose of improving environmental performance of office buildings by collecting data on their operational performance and comparing this against a prescribed performance standard. Data from the buildings can be collected to assess current performance. The as-built buildings can then act as reference buildings to compare improved performance of the retrofitted systems (Stephenson and Mitalas, 1967).

Selection of solution sets

The next step leads to the development of several configurations of passive and active systems applicable to non-residential buildings. Further analysis of these configurations leads to new solution sets that can easily be integrated into the renovation and retrofitting process of non-residential buildings.

Simulation, monitoring and cost benefit modelling

The next step was aimed at testing the influence of the solution sets on the performance of a representative group of reference buildings in a range of climatic zones. This involved testing the retrofitting solution sets through modelling of the reference buildings with the changes incorporated and then comparing the results against the benchmark reference. However, this proved to be problematic. The development of comprehensive typologies for office design proved impractical, hence a critical case study approach of an existing building was used which had many of the characteristics common to the kind of non-residential buildings requiring renovation at this time. Hence although generalisations about the application of the solution sets more broadly are difficult, the change in approach provided more depth and allowed both simulation and monitoring information to be utilised in the design approach.

Modelling and simulation of the critical case building with alternative energy system configurations were carried out to show how these selected configurations contribute to the reduction of energy demand. The implications of the introduction of these systems were investigated (Langstone *et al.*, 2001). A built-in multi-zone building model based on the transfer function method developed by Stephenson and Mitalas (1967) was utilised, and is recommended by ASHRAE (1989) (Cellura *et al.*, 2003).

This integrated methodology evolved into a whole building calibrated simulation approach, which, when harnessed with the NABERS' performance standard, provided a powerful tool for assessing the renovation and retrofitting options which have the potential to provide valid data to building owners wishing to consider retrofitting options. This tool was complemented with cost benefits studies using a Life Cycle Costing methodology (Barram, 2007). This approach utilised a predictive energy-modelling tool *Energy Analyser*(tm) (Barram 2011b).

EXPLORING THE BUILDING TYPOLOGY IN THE CONTEXT OF TECHNOLOGICAL AND BEHAVIOUR CHANGE FOR PERFORMANCE IMPROVEMENTS (SEE PART III)

Part III examines some of the drivers around what needs changing in terms of technology and behaviour to achieve performance improvements.

A review of the typology of building forms is carried out. It is suggested that underlying this typology it is recognised that a disparity exists between the environmental and social performance of a building and the corresponding valuation of that property. Both in Australia and internationally, the linkage between sustainability and financial worth is hindered by a lack of information and conclusive evidence relating to costs and benefits (Prasad, 2007). There is, nevertheless, a growing perception that green buildings represent added value through lower vacancy risk due to improved marketability, lower outgoing costs, increased rental growth and future opportunities in carbon trading. There is a clear need for research to quantify costs and benefits and to develop key performance indicators that will aid meaningful communication between people within the property industry (Thomson, 2007).

Sustainable building design has moved beyond passive solar principles – it now takes in wide-ranging considerations such as energy, water and waste efficiency and occupant productivity (Cole, 2007). Ultimately, the way the interdependencies between people, architecture, technology and the environment work will define the success of a building. A refurbishment needs to start from the perspective of a new building design rather than just attempting to overhaul the existing design. This could occur, for example, by mapping paths for occupant circulation, natural ventilation and daylight penetration, and using these paths to guide the design (Loughnane, 2007). At the same time, the refurbishment should attempt to retain or reuse materials where possible, and make use of simple, existing strategies such as insulation, materials, shading and glazing choices (Luther, 2007).

Developing the link between sustainability and property value will require methods to measure a building's performance in areas such as energy and water efficiency, acoustics, ventilation, thermal comfort and lighting. This can be done through computer simulation of parameters such as thermal comfort, energy systems or daylighting performance (Miller, 2007). Alternatively, direct measurement using systems such as MABEL can be used to quantify the performance of an existing building using measures such as air change effectiveness or predicted mean vote (Luther, 2007). Occupant satisfaction can also be quantified through surveys, which can glean more subjective, behavioural or operational insights (Loughnane, 2007). These techniques are increasingly being practised in the industry, and are helping to develop a knowledge base and bridge the gap between design expectations and as-built performance.

PRINCIPLES, CONCEPTS AND BEST PRACTICE (SEE PART IV)

The final section includes case studies of some recent Australian projects. This is focused on types of retrofitting such as adaptive reuse in the PMM building in Brisbane where the existing building was upgraded with new functions to improve both its economic and environmental sustainability.

The renovation and retrofitting of passive and active systems of 388 George Street in Sydney involved the replacement of chillers and boilers and a lighting upgrade (Loughnane, 2007). Variable speed drives were introduced for the cooling towers, ventilation fans and pumps to reduce energy use at part load. New controls systems are being installed that let the ventilation system use an economy cycle and allow intelligent control of the lighting. Simulation of daylighting, thermal performance and energy use predicted that the energy performance could improve its NABERS rating from 2.8 stars (average) to 4.5 stars (excellent). A black water system was also considered but not installed which predicted a saving of 12.5 ML per year.

55 St Andrews Place in Melbourne is another renovation project that focused on remedying existing problems that were due to poorly performing or outdated services and improving the energy, water and waste efficiency of the building. An ESD building improvement plan, involving the architects, services consultants, owner's representative, environmental manager and a dedicated ESD consultant, was created as a first step and considered the site, form, fabric, services and resources. This led to a re-think of the glazing systems and air intakes and the introduction of better insulation and shading as well as new paths for daylight penetration. A typical retrofitting approach would have focused on replacing the services, fixtures and fittings and refitting the interior, and would have failed to deal with the building's inherent problems.

Lessons from Part IV stress the importance of benchmarking groups of buildings rather than working on single buildings and trying to improve the building stock incrementally. Unfortunately, in practice this is rarely possible due to a lack of data relating to the existing building stock within the public domain. This necessitates the use of calibrated simulation models of the existing building being used as the base case and a test bed for assessing improvements to the building fabric and systems.

Advanced renovation includes both renovation and retrofitting where existing systems are replaced when they become physically obsolete, with new technology to improve performance. Unlike new build, sustainable retrofitting has at its core a different design conception, methodology and process and is expected to make a significant contribution to the transformation of the building industry towards ecological sustainability through the provision of information and training. This will assist organisations to move from the status quo to transforming organisations that adopt environmental thinking, strategies and standards as part of their core business. We will need to use renovation as the context for future research as this is seen as a new and important opportunity for sustainable and ecological development, given that the buildings that exist now will continue to make up the vast majority of Australia's and the world's building stock in the years to come.

ACKNOWLEDGEMENTS

AECOM (formerly Bassett Applied Research)
Australian Research Council
Integrated Energy Services
T.R. Hamzah and K. Yeang SdN. Bhd, Llewelyn Davies Yeang UK

REFERENCES

AIRAH (2003) *Member Survey: Energy Efficient Design in the Commercial Building Sector*, report prepared by The Australian Institute of Refrigeration, Air Conditioning and Heating (AIRAH). Available at: http://mail.airah.org.au/downloads/AIRAHEnergy Results.pdf (accessed 4 May 2012).

Australian Greenhouse Office (1999) *Australian Residential Building Sector Greenhouse Gas Emissions 1990–2010*, Dept. of the Environment and Heritage. Available at: http://www.climatechange.gov.au/what-you-need-to-know/buildings/publications/greenhouse-gas-emissions/~/media/publications/energy-efficiency/buildings/commbuild.pdf.

Barram, F. (2007) 'Financial modelling and property valuation', in Retrofitting Using Bioclimatic Principles: Looking For Value Adding', mini-conference, The Royal Australian Institute of Architects, Queensland Chapter, Continuing Professional Development Course, May, 2007.

Barram, F. (2011a) *Exploding the Carbon Myths; Secret Carbon Response Strategies to Beat the Carbon Tax*, forthcoming, Global Publishing Group, Australia.

Barram, F. (2011b) *Energy Analyser, Predictive Precinct and Building Modelling Tool Manual*, Australia: ECA Training.

Cellura, M., Giarre, L., Brano, V.L. and Orioli, A. (2003) 'Thermal dynamic models using Z-transform coefficients: an algorithm to improve the reliability of simulations', in Proceedings of the Eighth International IBPSA Conference, August 11–14, the Netherlands, pp. 139–46.

Clarke, J.A., Johnstone, C., Kelly, N. and Strachan, P.A. (1997) 'The simulation of photovoltaic-integrated building façades', *Proceedings of Building Simulation '97*, Volume 2, pp. 189–95.

Cole, J. (2007) 'Introduction', in Retrofitting using bioclimatic principles: looking for value adding, mini-conference, The Royal Australian Institute of Architects, Queensland Chapter, Continuing Professional Development Course, May, 2007.

Gilijamse, W. (1995) 'Zero-energy houses in The Netherlands', *Proceedings of Building Simulation '95*, pp. 276–83.

Halawa, E. (2005) 'Modelling and thermal performance: evaluation of roof integrated heating system with PCM thermal storage', PhD thesis, University of South Australia.

Haycox, M. (2003) 'Government Policy and Procurement,' in the *Proceedings of the Inaugural Green Building Council*, Sydney, Australia, 13–15 October,

Hyde, R.A. (2000) *Climate Responsive Design: A Study of Buildings in Moderate and Hot Humid Climates*, New York: E and F N Spon.

Hyde, R.A. (ed.) (2007) *Bioclimatic Housing: Learning from Innovative Projects in Warm Climates*, London: Earthscan.

International Energy Agency (2005) *Task 37 Advanced Renovation of Housing*, Solar Heating and Cooling Group. Available at: http://www.iea-shc.org/task37/ (accessed 8 Aug. 2005).

Jones, D.L. (1998) *Architecture and the Environment: Bioclimatic Building Design*, London: Laurence King.

Langston, C. and Ding, G. (2001) *Sustainable Practices in the Built Environment*, 2nd edn, Oxford: Butterworth-Heinemann.

Law, J.H.Y. (2001) 'The bioclimatic approach to high-rise building design: an evaluation of Ken Yeang's bioclimatic principles and responses in practice to energy saving and human well-being', BArch thesis, The University of Queensland, St. Lucia, Australia.

Lord, R. (2007) 'St Andrews Place: turning a sparrow into a peacock', in Retrofitting using bioclimatic principles: looking for value adding, mini-conference, The Royal Australian Institute of Architects, Queensland Chapter, Continuing Professional Development Course, May, 2007.

Loughnane, E. (2007) 'Simulation modelling for optimisation of existing buildings: case study', in Retrofitting using bioclimatic principles: looking for value adding, mini-conference, The Royal Australian Institute of Architects, Queensland Chapter, Continuing Professional Development Course, May, 2007.

Luther, M. (2007) 'Mobile architecture and built environment laboratory,' in Retrofitting using bioclimatic principles: looking for value adding, mini-conference, The Royal

Australian Institute of Architects, Queensland Chapter, Continuing Professional Development Course, May, 2007.

Manz, H.P. and Egolf. W. (1995) Simulation of radiation induced melting and solidification in the bulk of a translucent building façade', *Proceedings of Building Simulation*, 95: 252–8.

Miller, W. (2007) 'Same latitude – new attitude: pilot research project for reducing greenhouse gas emissions', in Retrofitting using bioclimatic principles: looking for value adding, mini-conference, The Royal Australian Institute of Architects, Queensland Chapter, Continuing Professional Development Course, May, 2007.

Olgyay, V. (1963) *Design with Climate: Bioclimatic Approach to Architectural Regionalism*, Princeton, NJ: Princeton University Press.

Pedrini, A. (2003) 'Integration of low energy strategies to the early stages of design process of office buildings in warm climate', PhD thesis, The University of Queensland, St. Lucia, Australia.

PLEA, *Passive Low Energy Architecture*, available at: http://www.plea-arch.org/ (accessed 8 Aug. 2007).

Prasad, D. (2007) 'Active solar systems and integration', in Retrofitting using bioclimatic principles: looking for value adding, mini-conference, The Royal Australian Institute of Architects, Queensland Chapter, Continuing Professional Development Course, May, 2007.

Rajapaksha, U. and Hyde, R.A. (2002) 'Passive modification of air temperature for thermal comfort in a courtyard building for Queensland', in the *Proceedings of the International Conference Indoor Air*, Monterey, June 30–July 5, USA.

Stephenson, D.G. and Mitalas, G.P. (1967) 'Cooling load calculations by thermal response factor', *ASHRAE Transactions*, 73(1).

Stritih, U. and Novak, P. (2002) 'Thermal storage of solar energy in the wall for building ventilation', *IEA, ECES IA Annex 17, Advanced Thermal Energy Storage Techniques*. Feasibility studies and demonstration projects, 2nd Workshop, 3–5 April, Ljubljana, Slovenia.

Szokolay, S.V. (2004) *Introduction to Architectural Science: The Basis of Sustainable Design*, London: Architectural Press, UK.

The Australian Government Department of the Environment and Heritage (DEH) (2005) NABERS. Available at: http://www.deh.gov.au/settlements/industry/construction/nabers/#what (accessed 11 Nov. 2005.)

The Australian Government Productivity Commission (2005) *The Private Cost Effectiveness of Improving Energy Efficiency*, Productivity Commission Inquiry Report No. 36, Camberra, Australia.

The Property Council of Australia (2005) *Response to the Regulatory Impact Statement*, Submission 2: RD 2004-01 Proposal for Class 5-9 Buildings, May, 2005.

Thomson, M. (2007) 'Retro-fitting sustainability', in Retrofitting using bioclimatic principles: looking for value adding, mini-conference, The Royal Australian Institute of Architects, Queensland Chapter, Continuing Professional Development Course, May, 2007.

Wittchen, K. (1993) 'A solar wall simulation module', in *Proceedings of Building Simulation*, 93: 377–83.

Yeang, K. (1998) *T.R. Hamzah and Yeang: Selected Works*, Mulgrave, Vic.: Images Publishing.

Yeang, K. (1999) *The Green Skyscraper: The Basis for Designing Sustainable Intensive Buildings*, Munich: Prestel.

PART I
ECO-DESIGN AND RENOVATION

1.1
INTRODUCTION
Richard Hyde

Part I examines the argument for developing an eco-design approach to the built environment and, by extension, how eco-design can be integrated into building retrofitting.

Chapter 1.2 by Ken Yeang examines the question, why should buildings be designed to integrate with the natural environment and climate for indoor comfort in environmentally sustainable efforts in the face of changing climate conditions? Green design, it is argued, has to go beyond conventional rating systems such as LEED or BREEAM. While they are tools which are useful indexes for providing a common basis for comparing the greenness of building designs, they are, however, not totally effective design tools. They do not provide a sufficiently comprehensive approach to the issues of environmental design at the local, regional and global levels. The set of eco-design principles outlined here provides the fundamental basis to this approach but remains only partially tested.

Chapter 1.3 by Richard Hyde aims to address the issue of how eco-design can be applied to the renovation of office buildings. Further testing of the use of eco-design for retrofitting sustainability into buildings was carried out. To be successful, it requires a broad approach at the city scale in order that strategies can operate at the building level. In eco-design, the use of nature as a metaphor for design provides a powerful tool to conceptualise eco-retrofitting projects. However, we have assessed the process of building obsolescence into the physical: buildings that are now 20 years old need to have their systems, façade and service systems renovated. The temptation is to replace systems with the existing specifications, so we argue that retrofitting is needed to replace buildings with new green technologies or face further obsolescence from social, economic and environmental drivers. This provides an emerging context for change and hence we argue that eco-design can provide a framework for sustainable retrofitting which can operate across the levels of city and building obsolescence.

Chapter 1.4 closes Part I with a summary by Richard Hyde.

1.2
STRATEGIES FOR DESIGNING OUR GREEN BUILT ENVIRONMENT

Why should buildings be designed to integrate with the natural environment in the face of changing climate conditions?

Ken Yeang

INTRODUCTION

We are all only too aware of the numerous pressing global social issues that need to be addressed. These include issues such as addressing abject poverty, providing clean water, adequate food and shelter, proper sanitation, and so forth. But ultimately if we do not have a clean environment, such as clean air, clean water, and clean land, all those other pressing global social issues become even more difficult and costly to resolve. Thus, saving our environment has to be the most vital issue that humankind must address today, feeding into our fears that this millennium may be our last.

For the designer, the compelling question is: how do we design for a sustainable future? Globally, businesses and industries face similar concerns of seeking to understand the environmental consequences of their functions and processes, to envision what these might be if they were sustainable, and to take action to realise this vision with comprehensive ecologically benign strategies, with new business models, new production systems, materials and processes. More than these, our human society has to change to a sustainable way of life; we need to change how we live, behave, work, make, eat, learn, and move about.

We would be mistaken if we regarded green design as simply about eco-engineering. These engineering systems are indeed an important part of green design (see the section on 'grey eco-infrastructure'), giving us an acceptable level of comfort that is sustainable, while such technologies continue to rapidly develop and advance towards greener and cleaner engineering solutions for our built environment. However, it must be clear that eco-engineering is not exclusively the only consideration in green design.

Neither is green design just about rating systems (such as LEED (Leadership in Energy and Environmental Design), BREEAM (Building Research Establishment Environmental Assessment Method), carbon profiling, etc.). These are certainly useful checklists and guidelines but they are not comprehensive. They are useful as a partial tick list of reminders of some of the key items to consider in green design or for comparing buildings and masterplans using a common standard. They have also been useful in proselytising green design to a wider audience. But as they are not comprehensive and ecologically holistic (an aspect crucial in eco-design), many designers, once they have achieved the highest level of rating (such as platinum) are asking – What next? Where do we go from here?

Clearly, green design has now entered the mainstream of architecture. Ask any architect today about green design and you will find the same pitch – use of renewable energy systems (such as photovoltaics, wind generators, etc.), compliance with accreditation systems, carbon profiling, planning as new urbanism, etc. We need to ask whether this is all there is to green design.

The contention here is that achieving effective green design is much more than the above and that green design is not as easy as it once was claimed to be. It is complex. While still incomplete, there are a number of design strategies that can be adopted in combination to get as close as we can to the goal of achieving a state of stasis of our built environment with the natural environment.

FOUR STRANDS OF ECO-INFRASTRUCTURE

The first design strategy is to view green design in terms of weaving four strands of eco-infrastructure, colour-coded here as follows:

1   the 'green' (the green eco-infrastructure or nature's own utilities which must be linked);
2   the 'grey' (the engineering infrastructure being the eco-sustainable clean-tech engineering systems and utilities);
3   the 'blue' (water management and closing of the water cycle by design with sustainable drainage);
4   the 'red' (our man-made systems, spaces, hardscapes, society, legislative and regulatory systems).

Green design is the seamless and the benign blending of all these four sets of eco-infrastructures into a system. This concept provides a platform for green design. Like the factors in DNA which reduce a complex concept into four simple sets of instructions, these four sets of eco-infrastructures and their integration provide the integrative bases for green design and planning – the blending of all these four sets of infrastructures into a system.

The green eco-infrastructure

The green eco-infrastructure is vital to every design and masterplan. It parallels the usual grey urban infrastructure of roads, drainage systems and utilities. This green eco-infrastructure is nature's utilities. These are the interconnected network of natural areas and other open green spaces within the biome that conserve natural eco-system values and clean air and water. This network also enables the area to flourish as a natural habitat for a wide range of wildlife as well as delivering a wide array of benefits to humans and the natural world alike, such as providing habitats linked across the landscape that permit fauna (such as birds and animals) to move freely. This eco-infrastructure is nature's functioning infrastructure (equivalent to our man-made engineering infrastructures, designated here as grey, blue and red eco-infrastuctures), and in addition to providing cleaner water and enhancing water supplies, it can also result in some, if not all, of the following outcomes: cleaner air; a reduction of the heat-island effect in urban areas; moderation of the impact of climate change; increased energy efficiency; and the protection of source water.

Incorporating an eco-infrastructure is thus vital to any eco-master-planning endeavour. Without it, no matter how clever or advanced the eco-engineering systems are, the design or masterplan remains simply a work of engineering, and can in no way be called an ecological masterplan, nor in the case of larger developments, an eco-city.

These linear flora and fauna corridors connect existing green spaces and larger green areas within the locality and link to the landscape of the hinterland, and can create new larger habitats in their own right, or may be in the form of newly linked existing woodland belts or wetlands, or existing landscape features (such as overgrown railway lines, hedges and water-ways). Any new green infrastructure must clearly also complement and enhance the natural functions of what is already there in the landscape.

In the masterplanning process, the designer identifies existing green corridors, routes and green areas, and possible new routes and link-ages for creating new connections in the landscape. It is at this point that additional green functional landscape elements or zones can also be inte-grated, such as linking to existing waterways that also provide ecological services, such as drainage to attenuate flooding.

This eco-infrastructure takes precedence over the engineering eco-infrastructure in the masterplan. By creating, improving and rehabilitating the ecological connectivity of the immediate environment, the eco-infrastructure turns human intervention in the landscape from a negative into a positive. Its environmental benefits and values are as a green armature and framework for natural systems and functions that are ecologically fundamental to the viability of the locality's plant and animal species and their habitats, such as healthy soils, clean water and clean air. It reverses the fragmentation of natural habitats (as a consequence of urban sprawl and transportation routes, etc.) and encourages an increase in biodiversity to restore functioning eco-systems, while providing the fabric for sustainable living, and safeguarding and enhancing natural features.

This endeavour by design to connect the landscape must extend to the built form, both horizontally and vertically. An obvious demonstration of horizontal connectivity is the provision of ecological corridors and links in regional and local planning that are crucial in making urban patterns more biologically viable. Connectivity over impervious surfaces and roads can be achieved by using ecological bridges, undercrofts and ramps. Besides improved horizontal connectivity and ecological nexus, vertical connectivity with buildings is also necessary since most buildings are not single-storey but multi-storey. Design must extend the ecological corridors vertically upwards, with the eco-infrastructure traversing a building from the foundations and landscape at the ground level to create habitats on the walls, terraces and rooftops.

The grey eco-infrastructure

The grey infrastructure is the usual urban engineering infrastructure such as roads, drains, sewerage, water reticulation, telecommunications, and energy and electric power distribution systems. We need not be prescriptive of any specific engineering system, but require that these systems be clean technologies, of low embodied energy and be carbon-neutral as much as possible, and at the same time be integral with the green infrastructure.

The blue eco-infrastructure

Parallel to the green ecological infrastructure is the water infrastructure (the blue eco-infrastructure) where the water cycle should be managed to close the loop, although this is not always possible in locations with low rainfall. Rainfall needs to be harvested and water use must be recycled. The surface water from rain needs to be retained within the site and be returned back to the land for the recharging of groundwater and aquifers by means of filtration beds, pervious roadways and built surfaces, retention ponds and bioswales. Water used in the built environment (both grey and black water) needs to be reused sustainably as much as possible.

Site planning must take into consideration the site's natural drainage patterns and provide surface water management so that rainfall remains within the locality and is not drained away into water bodies, which means the water is then lost. Combined with the green eco-infrastructure, stormwater management uses the natural processes to infiltrate, evapotranspire, or engineer the capture and use of stormwater on or near the site where it falls, while potentially generating other environmental benefits.

Waterways should not be culverted or deculverted as engineered waterways, but should be replaced by the introduction of wetlands and buffer strips of ecologically functional meadows and woodland habitats. Sealed surfaces can reduce soil moisture and leave low-lying areas susceptible to flooding from excessive run-off. Wetland greenways need to be designed as sustainable drainage systems to provide ecological services. Green buffers can be used, together with linear green spaces, to maximise their habitat potential.

Eco-design must create sustainable urban drainage systems that can function as wetland habitats. This is not only to alleviate flooding, but also to create buffer strips for habitat creation. While the widths of the buffer may be constrained by existing land use, their integration through linear green spaces can allow for wider corridors. Surface water management maximises the habitat potential. Intermittent waterway tributaries can be linked up using swales.

The red (or human) eco-infrastructure

This red or human eco-infrastructure is our human community, its built enclosures (buildings, houses, etc.), hardscapes and regulatory systems (laws, regulations, legislation, ethics. etc.). This is the social and human dimension that is often missing from the work of many green designers. It is evident that our present profligate lifestyles, our economies and industries, our mobility, our diet and food production, etc., need to be changed to be sustainable.

SEAMLESS AND BENIGN BIO-INTEGRATION

The second design strategy is to regard green design as bio-integration – as the seamless and benign environmental bio-integration of the synthetic and the artificial (the man-made) with the natural environment. It is the failure to successfully bio-integrate that is the root cause of all our environmental problems. In effect, if we are able to bio-integrate all our business and industrial processes and functions, all our built systems and essentially everything that we do or make in our built environment (which, by definition, includes our buildings, facilities, infrastructure, products, refrigerators, toys, etc.) with the natural environment in a seamless and benign way, there will, in principle, be no environmental problems whatsoever. Successfully achieving this is, of course, easier said than done, but herein lies our challenge for the future.

We can draw an analogy here between eco-design and prosthetic design in surgery. A medical prosthetic has to integrate with its organic host, i.e. the human body. Failure to integrate successfully results in dislocation in one or in both. By analogy, this is what eco-design should achieve: a total physical, systemic and temporal integration of our man-made built environment and our activities with our organic host, i.e. the natural environment, in a benign and positive way. Eco-design is thus design that successfully bio-integrates our artificial systems both mechanically and organically, with its host system being the eco-systems in the biosphere. Our designing for bio-integration can be considered through three aspects: physically, systemically and temporally.

Physical and systemic integration requires discernment of the ecology of the locality. Any activity arising from our design or our business/industries must physically integrate benignly with the eco-systems. To achieve this, we must first understand the locality's eco-system before imposing any human activity or built system upon it. Every site has an ecology with a limited capacity to withstand stresses imposed upon it. If stressed beyond this capacity, the site will be irrevocably damaged. Consequences of this can range from minimal localised impact (such as the clearing of a small land area for access), to the total devastation of the entire land area (such as the clearing of all trees and vegetation, levelling the topography, diversion of existing waterways, etc.).

We need to ascertain the locality's eco-system structure, energy flow, its species diversity and other ecological properties and processes. Then we must identify which parts of the site (if any) can permit different types of structures and activities, and which parts are particularly sensitive. Finally, we must consider the likely impacts of the intended construction and use over time.

This is, of course, a major undertaking. It needs to be done diurnally over the year and in some instances over several years. To reduce this lengthy effort, landscape architects have developed the 'sieve-mapping' technique for landscaping mapping. We must be aware that this method is an abbreviated approach, and generally treats the site's eco-system statically and may ignore the dynamic forces taking place between the layers and within an eco-system. Between each of these layers are complex interactions.

Another major design issue is the systemic integration of our built forms and their operational systems and internal processes with the eco-systems in nature. This integration is crucial because if our built systems and processes do not integrate with the natural systems in nature, then they will remain disparate, artificial items and potentially polluting and destructive to the ecology of the locality. Their eventual integration after their manufacture and use can only be through bio-degradation. Often, this requires a long-term natural process of decomposition.

Temporal integration involves the conservation of both renewable and non-renewable resources to ensure that these are sustainable for future generations. This includes designing low-energy built systems that are less or are not dependent on the use of non-renewable energy resources.

ECO-MIMESIS

The third design strategy is to regard green design as 'eco-mimesis', as imitating the attributes and properties of eco-systems – their processes, structure, features and functions. This is one of the fundamental premises for eco-design. Our built environment must imitate eco-systems in all respects, e.g. recycling, using energy from the sun through photosynthesis, using systems that increase energy efficiency, and achieve a holistic balance of biotic and abiotic constituents, etc.

Nature without humans exists in stasis. Can our businesses and our built environment imitate nature's processes, structure, and functions, particularly its eco-systems? For instance, eco-systems produce no waste. Everything is recycled within the system. Thus, by imitating this, our built environment will produce no waste. All emissions and products are continuously reused, recycled within the system and eventually reintegrated with the natural environment, in tandem with efficient uses of energy and material resources.

Eco-systems in a biosphere are definable units containing both biotic and abiotic constituents acting together as a whole. From this concept, our businesses/industries and built environment should be designed analogously to the eco-system's physical content, composition and processes. For instance, besides regarding our architecture as merely art objects or as engineering-serviced enclosures, we should regard architecture as an artefact that needs to be operationally and eventually integrated with nature.

As is self-evident, the material composition of our built environment is almost entirely inorganic, whereas eco-systems contain a complement of both biotic and abiotic constituents, or of inorganic and organic components. We need to reverse our building systems to make them more organic.

Our myriad of construction, manufacturing and other activities are, in effect, making the biosphere more and more inorganic, artificial and increasingly biologically simplified. To continue without balancing the biotic content means simply adding to the biosphere's artificiality, thereby making it increasingly inorganic and synthetic. This results in the biological simplification of the biosphere and the reduction of its complexity and diversity. We must reverse this trend and balance our built environment with greater integral levels of biomass, ameliorating biodiversity and ecological connectivity in the built forms.

Eco-design also requires the designer to use green materials and assemblies of materials, and components that facilitate reuse, recycling and reintegration for temporal integration with the ecological systems. We need to be eco-mimetic in our use of materials in the built environment. In eco-systems, all living organisms feed on continual flows of matter and energy from their environment to stay alive, and all living organisms continually produce outputs. Here, an eco-system generates no waste, the waste from one species being the food for another species. Thus, matter cycles and recycles continually through the web of life. It is this closing of the loop in reuse and recycling that our man-made environment must imitate.

ECO-DESIGN TO RESTORE IMPAIRED ENVIRONMENTS

Fourth, eco-design can be regarded not only as the creation of new artificial 'living' urban eco-systems or the rehabilitation of existing built environments and cities, but also as one of restoring existing impaired and devastated eco-systems regionally within the wider landscape following our designed system. Eco-design must look beyond the limitations of the project site, and at the larger context of the locality. Where necessary, we should improve the ecological links between our designed systems and our business processes with the surrounding landscape and hardscapes, not just horizontally but also vertically.

Achieving these linkages ensures a wider level of species connectivity, interaction, mobility and greater sharing of resources across boundaries. Such real improvements in ecological nexus enhance biodiversity and further increase habitat resilience and species survival. Providing new ecological corridors and linkages in regional planning is crucial in making urban patterns more biologically viable.

Crucially, we need to apply these concepts to retrofit our existing cities and urban developments. We must bio-integrate the existing inorganic aspects of our built environment and its processes with the landscape so that they become mutually eco-systemic. We must create 'man-made eco-systems' compatible with the eco-systems in nature. By doing so, we enhance the ability of the man-made eco-systems to sustain life in the biosphere.

ECO-DESIGN AS A SELF-MONITORING SYSTEM

The fifth strategy for eco-design is to regard our designed system as a series of interdependent environmental interactions, whose constant global and local monitoring (e.g. through GPS and biosensors, etc.) is necessary to ensure global environmental stasis, enabling an anticipatory approach to and the immediate repair and restoration of environmental devastation by humans, natural disasters, and the inadvertent negative impacts of our man-made environment, activities and industries. These sets of environmental interactions need to be monitored for appropriate corrective action to be taken immediately to maintain global ecological stability.

The above are strategies that can be adopted singly or in combination to approach green design. Green design has to go beyond conventional rating systems such as LEED or BREEAM, etc. While these processes are indeed useful indexes for providing a common basis for comparing the greenness of building designs, they are, however, not totally effective design tools. They are not comprehensive enough in approaching the issues of environmental design at the local, regional and global levels.

In general, ecological design is still very much in its infancy. The totally green building or completely green eco-city does not yet exist. There is still much more theoretical work, technical research and invention, environmental studies and design interpretation to be done and tested before we can have a truly green built environment. We all need to continue this great pursuit for the sake of the environment and all its inhabitants.

1.3
ECO-DESIGN FOR RETROFITTING

How can eco-design work apply to retrofitting buildings for climate change?

Richard Hyde

INTRODUCTION

Ecological design is seen as a critical way to improve the sustainability of our cities. In this chapter eco-design is harnessed to the process of renovation to improve the environmental performance of buildings. We have called this *ecological retrofitting* for a number of reasons. We now see this process as an extension of bioclimatic retrofitting, as a central strategy in the move towards both the mitigation of the pollution which is arguably responsible for climate change and adaptation to it. Retrofitting should be differentiated from renovation. Renovation attacks the effects of obsolescence in buildings while retrofitting is the addition of new technologies to old systems for the purpose of improved efficiency.

A combination of the effects of obsolescence and those of climate change has occurred. This state of play requires mitigation of the causes of climate change and the development of adaptation strategies through renovation and retrofitting. Hence this gives a leverage point in sustainable design to address two forces at work in the building stock. The first section of this chapter examines some of these effects and argues the case for ecological retrofitting, as a further extension to the bioclimatic approach to design. The second section provides a study of how eco-design can be used for retrofitting. Finally, the feasibility of this approach is discussed in terms of the inherent characteristics of the buildings to be retrofitted and the process of obsolescence.

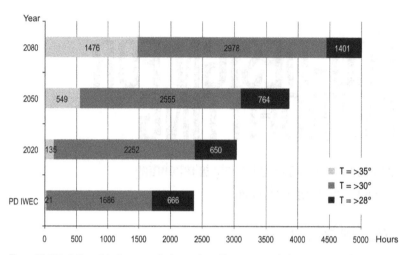

Figure 1.3.1 Modelling of the increases in the number of hours external air temperatures for years 2020, 2050 and 2080. Three stages are shown: (1) hours which exceed 35°C; (2) hours exceeding 30°C; and (3) hours exceeding 28°C as modelled by simulation, for Malaysia

Note: T = temperature

Source: Indrika Rajapaksha.

CLIMATE CHANGE AND BIOCLIMATIC DESIGN

Warm climates

Concern about the effect of human activities and the processes of indus-trialisation on anthropogenic heat and carbon pollution is increasing. While there is a clear emphasis on carbon pollution and the reduction of this at source, the issues associated with waste heat are often ignored. The work by Barry and Chorley on climate and atmosphere reports that solar radiation in winter averages around 25 W/m² in Europe. This is similar to the heat production from large cities (Barry and Chorley, 2003: 381). Furthermore, waste heat, sometimes called secondary heat or low-grade heat, comes from heat produced by machines, electrical equipment and industrial processes in our cities. No useful application can be found for this heat. Hence the challenge is to find useful strategies for accommodating waste heat. Some even argue that we should no longer use air conditioning to reduce these heat loads in buildings. Figure 1.3.1 shows the simulation of climate effects for the next 70 years or so. This problem is facing existing buildings and retrofitting entails dealing with several imperatives.

The adaptation imperative

The first imperative for retrofitting is adaptation to the effects of climate change. Evidence of climate change is an increase in outdoor temperatures in warm climates, suggesting an increase in the environmental heat loads on buildings. Failure to address such negative impacts on existing buildings through appropriate design interventions, and the degrading of active systems that already are in use in buildings, can lead to an increase in energy use.

Further, it is evident that much of the existing building stock performs very inefficiently in terms of energy at present (Wilkinson and Reed, 2008), and, therefore, a critical analysis of the climatic behaviours of macro, meso and microclimates and their negative effects on the design of buildings in order to assess the performance of buildings is required. The CC World Weather Gen Tool from the University of Southampton can model such predicted climate change effects on local macroclimate patterns. These predictions are based on data from the HadCM3 (Hadley Centre Coupled Model Version 3) and the International Panel on Climate Change (IPCC) Third Assessment Report (TAR) on climate change.

The mitigation imperative

The second imperative is mitigation of the reported causes of climate change, which are greenhouse gases (GHG) emissions. It is argued that the solution of major retrofitting of existing buildings can be a major plank in the world's struggle to halt the increase in GHG emissions. In terms of emerging 'futures', namely, climate change predictions, there is great potential for technological and behavioural change to provide the necessary mitigation and adaptive conditions to address GHG emissions reductions.

According to the 2007 IPCC Report, buildings offer the largest share of cost-effective opportunities for GHG mitigation of all industry sectors, and have the greatest potential for decreasing CO_2 emissions. Achieving a lower carbon future will require significant attempts to enhance programmes and policies for energy efficiency in buildings well beyond what is happening today (IPCC, 2007: 306). These acts are focused on a technical approach to the architectural and engineering fields. Several technical and engineering improvements for buildings have been proposed and four major areas of improvements have been identified: (1) reduce energy for heating, cooling and lighting loads; (2) improve the use of insulation, and lower the thermal mass of the building; (3) increase the efficiency of appliances and the HVAC systems; and (4) increase the efficiency of the lighting systems.

In addition to the technical factors to improve the carbon footprint of buildings, there is agreement but limited evidence to show that the potential reduction through non-technological options is important; however, this has rarely been assessed. The potential of policies to improve these determinants, such as occupancy behaviour, culture, consumer choice and use of technology, is poorly understood (IPCC, 2007: Chapter 6). It is argued that this approach is not only poorly understood but also poorly conceptualised and hence the IPCC is unlikely to find evidence for this unless it looks more broadly into completing frameworks. A response to these issues is to develop retrofitting which uses bioclimatic principles.

Based on this thinking it is now thought necessary to move beyond efficiency to the zero-energy building concept. This can be achieved by the simultaneous actions of: (1) reducing the energy demand and thereby increasing the energy efficiency of buildings through various energy conservation measures; and (2) utilising the solar energy incident on the wall surface, roof and ground surrounding the house for electricity generation and to satisfy the building's heating requirements. The implementation of these actions entails the careful assessment of its economic feasibility, and benefits for the environment and human comfort.

A central initiative in retrofitting is therefore the inclusion of the many emerging 'green technologies' relating to the goal of realising extremely low energy demand in buildings and generating green energy from renewable sources. This raises the question whether the concept of zero-energy (fossil fuel) buildings is theoretically possible. The realisation of this concept, however, depends on whether it can satisfy three criteria: (1) environmental benefits; (2) human comfort improvement; and (3) economic feasibility.

Research is needed which aims at exploring the synergies of these emerging green technologies to bring about new and valid sets of solutions to the environmental problems and energy issues relating to existing buildings. Research efforts so far have been fragmented, where researchers focus on a particular technology without looking at the whole problem and exploring the best solution.

The main measures have been mainly focused on new construction rather than existing buildings. It is recommended that more efficient equipment for heating and cooling, advanced or 'intelligent' building envelopes as well as more effective insulation need to be effectively applied in newly constructed buildings. Furthermore, IPCC suggest, 'occupant behaviour, culture and consumer choice and use of technologies are also major determinants of energy use in buildings and play a fundamental role in determining CO_2 emissions' (IPCC, 2007). Hence, the application of both technical and non-technical strategies for new building can be applied to existing buildings through the process of retrofitting. However, we would argue that the selection and integration of these processes should be framed in a bioclimatic context using eco-design principles.

PROGRESS ON BIOCLIMATIC RETROFITTING AND ITS BROADER ECOLOGICAL INTEGRATION

Bioclimatic design

Bioclimatic issues in architecture were identified by Olgyay in the 1950s and developed as a process of design in the 1960s (Olgyay, 1963). Figure 1.3.2 shows bioclimatic design and its interlocking fields of biology, climatology, technology and architecture. It is argued that by interlocking these fields we can achieve a stasis often called balance between humans and the climate. The bioclimatic design process brings together the disciplines of human physiology, climatology and building physics. It has been integrated into the

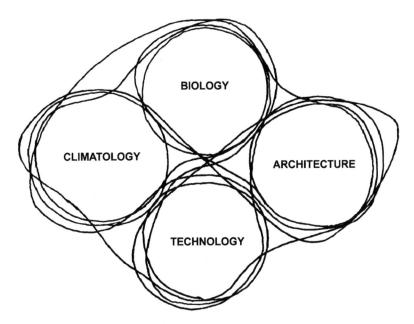

Figure 1.3.2 Bioclimatic design and its interlocking fields to achieve climate balance

Source: Olgyay (1963).

building design professions in terms of regionalism in architecture and in recent years has been seen as a cornerstone to achieve more sustainable buildings (Szokolay, 2008). This research has led to the development of bioclimatic design principles, which are used by design professionals as a starting point for designing with the climate in mind. These have been developed primarily for low and medium-sized buildings, and the rationale for this is that these types of buildings are relatively easy to make climatically interactive. This means the form and fabric of the building can be matched to human and climate factors to optimise climate response (Hyde, 2000).

Non-domestic buildings have largely been ignored because of pro-gramme complexity, the density of the urban context where these buildings are located, and the availability of cheap energy for cooling and providing comfort. The design principles have largely excluded bioclimatic influences and achieve an adequate internal environment by adding comfort through mechanical energetic systems (Rajapaksha and Hyde, 2002). A comparison between the characteristics of bioclimatic and conventional buildings is shown in Table 1.3.1.

The goal of bioclimatic retrofitting is to assess conventional build-ings to see if they can become more like bioclimatic buildings. Very often one hears of retrofitting buildings with more air conditioning; rarely is the con-verse carried out. Work to address this apparently simple goal is yet to be undertaken.

This is evident by the lack of a set of principles, strategies and best practice solutions for bioclimatic retrofitting of larger-scale buildings, which is

Table 1.3.1 Comparison of the characteristics of bioclimatic buildings and conventional buildings

Bioclimatic buildings	Conventional buildings
Narrow plan	Deep plan
Optimally orientated	Glass façades
Use 'fresh' air	Mechanically lit
Daylight	Fully air conditioned
Solar-heated	Reliant on fossil fuel energy
Naturally ventilated	
Highly shaded	
Well insulated	
Reliant on the natural environment	

yet to be fully researched and acknowledged in the field (Yeang, 1999). While a number of case studies on the bioclimatic design of non-domestic buildings have been written, the extent to which general principles can be advanced from these examples is limited. Jones developed some bioclimatic principles for large buildings. These are concerned with the form of energy used – renewable or non-renewable – energy efficiency and conservation and human well-being, comfort and amenity. These elements are largely unconnected to building strategies in a building science context, or fully demonstrated in best practice solutions of non-domestic buildings. Furthermore, the design professionals remain sceptical of such an approach. This is due to the lack of workable models of bioclimatic retrofitting of large-scale buildings and the cost of additional design work such as simulation modelling to demonstrate the proof of concept, cost effectiveness and comfort of these types of buildings (Pedrini, 2003).

Importance of the passive elements: the building's microclimate, form and fabric

Jones suggests that the bioclimatic office building will use five or six times less energy than a conventional office building over its life (Jones, 1998: 45). This is achieved primarily through the use of the building's microclimate, form and fabric rather than through the use of efficient mechanical equipment. For example, in warm climates where cooling is needed most of the year, 34 per cent of the energy for cooling addresses mitigation of solar radiation into the building (Parlour, 1997). This is normally achieved through air conditioning and, as a consequence, incurs large environmental penalties – high energy use, high greenhouse gas emissions. Bioclimatic design refocuses on providing high quality passive design of buildings though new technologies in the building envelope and in its form and fabric. Pioneering work by Yeang and others has defined a number of passive strategies using a range of biophysical elements, as shown in Table 1.3.2.

Clearly there is a relationship between climate change and the bioclimatic design of buildings. Table 1.3.2 presents these technologies in an

Table 1.3.2 Passive strategies for cooling and heating non-domestic buildings

Strategies	Example
Thermo hydronic – thermal water sinks	Mewah-Oils Headquarters, Port Klang (Westport), Selangor, West Malaysia (Figure 1.3.3)
Kinetic – adaptive thermal defences	Guthrie Pavilion, Shah Alam, Selangor, Malaysia (Figure 1.3.4)
Organics – heat sinks	DIGI Technology Operation Centre, Malaysia (Figure 1.3.5)
Aerodynamics – adaptive wind defences	Menara UMNO, Pulau Pinang (Figure 1.3.6)
Materials – phase change, heat storage, radiant defences	Menara Mesiniaga Subang Jaya, Selangor, Malaysia (Figure 1.3.7)
Ground effects – heat storage	VADS Plaza, Malaysia (Figure 1.3.8)

Source: Yeang (1999, 2002), Hamzah and Yeang (1994), Law (2001).

ecological context. It addresses an important question of how buildings can integrate with the natural environment and climate for indoor comfort in environmentally sustainable efforts in the face of changing climate conditions. However, the difficulty is that while the building attributes can quite easily be identified, the influence on the performance of the buildings is less measurable.

Notwithstanding this lack of evidence of the performance improvements, many of the green technological strategies found in the examples shown in Table 1.3.2 can apply to the retrofitting of existing buildings. For example, Figure 1.3.3 shows the Mewah-Oils Headquarters, in Port Klang (Westport), Selangor, West Malaysia, using thermo hydronic features to provide thermal water heat sinks in and around the building. Bioclimatic strategies can influence the microclimate of the building to reduce local temperatures.

The Guthrie Pavilion in Figure 1.3.4 shows the application of a large parasol roof, which is used to mitigate the heat gain to the roof and provide additional open space above and around the building.

The DGI building shown in Figure 1.3.5 is noted for the use of plants as heat sinks in this data centre in Kuala Lumpur, Malaysia. The vertical planting is attached to the building in steel trays containing soil. The plants are held in place by a porous fabric. Water is reticulated from an eco-cell pumping station on the ground. This type of green technology can be retrofitted to existing buildings, creating a new skin or building envelope. The concept for this data centre is to wrap the ecological elements up the side of the building to create this green wall. Local species were collected for this building. The climatic effects of this on the heat load of the building are hard to gauge. Data centres like this produce large amounts of waste heat and an exercise in heat evacuation is needed. However, this approach provides a

Figure 1.3.3 Mewah-Oils Headquarters, Port Klang (Westport), Selangor, West Malaysia, features the use of thermo hydronic features to provide thermal water heat sinks around the building

Source: Hamzah and Yeang (1994).

Figure 1.3.4 Guthrie Pavilion, Shah Alam, Selangor, Malaysia. Kinetic – adaptive thermal defences used to protect the building from climatic effects and to provide additional open space above and around the building.

Source: R. Hyde.

new perspective for renovation, reintroducing an ecological approach to retrofitting.

Menara UMNO, Pulau Penang, shown in Figure 1.3.6, is noted for the carved form of the floor plate in a V shape which aims to influence the aerodynamics created by building form to adapt the ambient wind flow and to enhance ventilation. However, this building has a large deep plan floor plate and the effectiveness of the daylight penetration is limited, requiring permanent artificial lighting to the interior (Hyde, 2008).

Figure 1.3.7 shows the Menara Mesiniaga project, started in 1989 and completed in 1992. This building is noted for integrating the principles of the bioclimatic approach into tall buildings. Notable features include the way the ground-based materials and landscaping have been introduced into the building façade and at high levels form the sky courts. The landscaping includes an earth berm at the ground level, which slopes up to the first floor. Landscaping is integrated in the façade and on the sky courts. Passive low-energy features are used with shading on all the window areas, particularly facing the area where the low angle sun strikes the building to the east and west orientations. The building service cores are naturally ventilated and are sunlit with views to the outside. The roof space provides future flexibility to retrofit solar panels.

Figure 1.3.5 DGI building (architect: Ken Yeang): use of landscaping as heat sinks in this data centre in Kuala Lumpur, Malaysia

Source: R. Hyde.

The VADS Plaza, built in the 1970s, shown in Figure 1.3.8, provides another example of how to integrate bioclimatic principles in tall buildings. In this case, it makes use of the influence of ground effects such as landscaping and the evaporative effects of cooling associated with plants to reduce thermal loads on the building. However, unlike the previous example in Figure 1.3.7, it does not have the integration of the sky court and greater integration of ground features.

Figure 1.3.6 Menara UMNO, Pulau Penang (architect: Ken Yeang): the carved form of the floor plate in a V shape aims to influence the aerodynamics created by the building form to adapt the ambient wind flow and to enhance ventilation

Source: Hamzah and Yeang (1994).

Figure 1.3.7 Menara Mesiniaga Subang Jaya, Selangor, Malaysia (architect: Ken Yeang): this building uses strategies for radiant defences through landscaping to effect the integration of bioclimatic principles in tall buildings

Source: Hamzah and Yeang (1994).

Many of these buildings are now approaching 20 years of age and we see many of the aspects of building obsolescence now emerging, and this now presents an opportunity to retrofit these buildings following the principles for ecological design. The next section introduces design-led research into how eco-design principles can be applied to both renovation and retrofitting.

Figure 1.3.8 VADS Plaza (architect: Ken Yeang): external elevation showing the shading systems and landscaping element that runs from the ground to the roof

Source: R. Hyde.

APPLYING ECO-DESIGN TO RETROFITTING: VADS PLAZA

The eco-design process

The eco-design principles discussed in Chapter 1.2 were applied to the Vads Plaza project. A one-day charette examined the application of these eco-design principles to the VADS Plaza. Notions of the ecological imperative are found in the work of Ken Yeang and these principles are operationalised into a manifesto for green design, involving the following (see Chapter 1.2):

- Eco-infrastructure, broken down into subsystems:
  - green: supporting natural systems;
  - grey: supporting engineering physical systems;
  - blue: supporting water systems;
  - red: supporting human systems.
- Bio-integration:
  - biotic (living);
  - abiotic (non-living).
- Eco-mimesis:
  - eco-systems mimicking and translation.
- Eco-design to restore impaired environments:
  - biodiversity regeneration.

- Eco-design as self-monitoring:
  - life-cycle measurement.

Many of these principles can be applied to renovation involving retrofitting based on ecological thinking. This is a new approach to retrofitting, which has not yet been fully tested. Currently the building is undergoing renovation and retrofitting so the theoretical work provided here may provide some input into the design.

Project description

VADS Plaza (formerly Plaza IBM until July 2006) is located in Kuala Lumpur. At 3.8 degrees north, it lies close to the equator in a hot humid climate. Powell reports the building has:

> [A] hybrid form. It has a crisp, sharp geometry; each of the upper floors projecting over the floor below in a configuration like a slightly displaced pack of cards. This is a development of Ken Yeang's 'filter' ideas, the broken irregular surface, the shading of windows, and the reaction against the flat-wall and thin-skin aesthetic of Modernist architecture. What is evident in this building is the departure from the tenets of Modernism. Here, the aesthetic is derived from the simultaneous resolution of a number of climatic regulatory devices.
> (Powell and Kurokawa, 1989)

To renovate this building, one is obliged to follow the original intentions of the project (Figure 1.3.9). Hence the aim was to interweave the basic strategies of bioclimatic deign in this project with an increased focus on the application of the ecological principles espoused by Yeang in his later work. Ecological renovation means a new perspective which involves a larger view of the site. The current thinking about site boundaries in terms of property and ownership are less important, given that the forces of nature tend to permeate through and breach these boundaries with little regard for man-made regulations.

Hence the line of enquiry for this project was to start with the urban ecology, the bio-region and begin to assist nature in breaching the boundaries of the site and to establish opportunities within the building to increase the contribution from the indigenous biota. The fundamental principle for eco-design, as a renovation and retrofitting process, is to start with the ecology of the city.

Eco-design to restore impaired environments: biodiversity regeneration

Restoring impaired environments is a complex task; it is bound up with largely determining the causes of the ecological impairment and providing some advice primarily to repair the environmental damage (US Environmental Protection Agency, 2008, 2011). The main aim is to provide clear guidance to determine the causes of biological impairment; to develop methods to

(a)

(b)

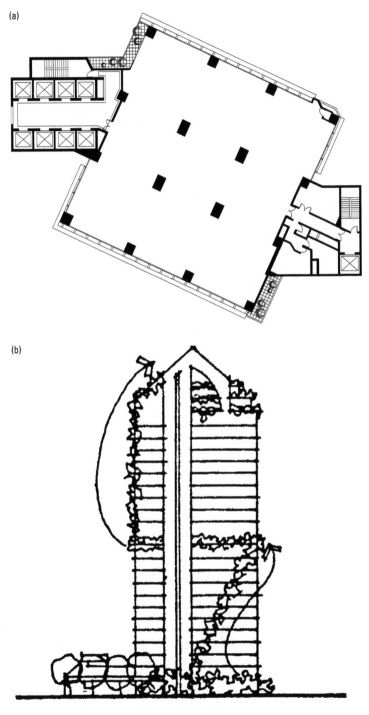

Figure 1.3.9 VADS Plaza: (a) floor plate; (b) elevations

Source: Powell (1989).

repair such impairments as excess sediment, nutrient enrichment, habitat and flow alteration and chemical contamination; to test and illustrate causal analysis through stakeholder case studies; to ensure that many of the stakeholders have experience of determining the causes of biological impairment; and to develop and provide ready access to tools and databases that make causal analysis easier and faster. In an ecological context it is a matter of looking not only at the relationship between biological systems but also at the relationships between systems, and in the case of architectural design, examining the relationship between the man-made and the natural environment. The notion of restoring impairment is normally seen as part of a retrofitting process. Hence goals such as enhancing bio-integration on the site and restoring impaired environments through biodiversity regeneration are developed. The work of Zarsky is useful in this regard. She suggests this imperative provides 'a way in which humans should interact with the biosphere to maintain its life support function'. She advocates five principles:

1   *Biodiversity:* all species and habitats should be conserved, maintaining 'natural' evolution.
2   *Eco-systems conservation:* the natural stock of ecological resources – soil, ground and surface water, land biomass, water biomass – has regeneration limits.
3   *Interconnectedness:* improvements in one area of a country should not be at the expense of others.
4   *Aversion of risk:* precautionary principles, unknown thresholds to incremental change which could have significant systematic consequences.
5   *Scale of impact:* human minimisation of energy and material flows into ecosystems.

(Zarsky, 1990)

These principles require a level of analysis at the bio-region level. From this the availability of existing eco-systems with which to connect or which can be connected to the building can be assessed. Mapping of existing species of flora and fauna in local nature reserves and areas, which preserve wildlife habitats, can enhance this kind of approach. The problem is largely spatial in nature, how to find space within the city to apply the principles of ecology.

The first step in the VADS Plaza retrofitting was to identify the areas of the city where, as Zarsky notes, 'all species and habitats should be conserved, maintaining "natural" evolution'. The areas in Kuala Lumpur where Nature Reserves exist were identified. However, it was argued that an increase in the amount of the green space on the site of the VADS Plaza would be less effective without connection to the local Nature Reserves. Eco-systems' conservation and hence retrofitting mean increasing the natural stock of ecological resources – soil, ground and surface water, land biomass, water biomass – which in turn can lead to the regeneration of flora and fauna. The initial mapping work that was done to establish this kind of approach is seen in Figure 1.3.10.

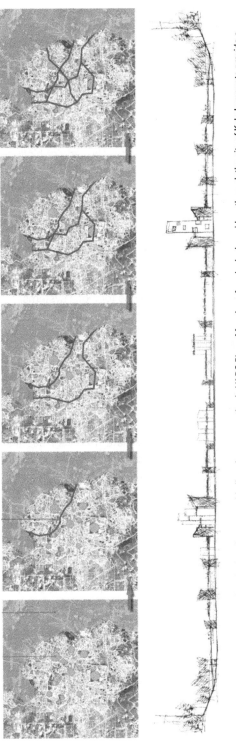

Figure 1.3.10 Eco-design to restore impaired environments: biodiversity regeneration in VADS Plaza. Mapping of ecological corridors through the city of Kula Lumpur to provide a connection to the forest reserves where the process of evolution is still continuing.

Source: Danny Si and Yibie Liu.

The benefit of this work is seen terms of the scale and inter-connectedness that are created, a web of green biotic corridors through the city, interconnecting buildings and the Nature Reserves. Hence it is argued that this increase in biotic links in the city may act to mitigate the heat island effects. However, the scale of this kind of intervention is outside the scope of the project.

Bio-integration of biotic (living) and abiotic (non-living) elements

Urban areas are congested with non-living elements. The ratio of biotic to abiotic is very small. An examination of the biotic to abiotic ratio of the current VADS Plaza site was calculated. Of the current area (site and building floor area), it was found to have approximately 0.8 per cent usage by biotic and the reminder abiotic. Hence the task of renovation and retrofitting was to increase the biotic percentage. It was found that by replanning the spatial zoning of the building, it was possible to achieve an improvement of 40 per cent biotic usage, as seen in Figure 1.3.11. The retrofitting design proposal worked on the assumption that planning and urban design changes to the city infrastructure would provide better ecological access for biotic systems.

Eco-mimesis: eco-systems mimicking and translation

Replanning of the building worked on the basis of eco-mimesis – eco-systems mimicking and translation. This created a hierarchy in the building similar to the tropical rainforest. This was seen as complementary to the original form, matching the intent of the building. Powell reports:

> In Yeang's approach to the typical floor plate he recognises that there are two geometries (and axes): one being the sun's geometry on its east–west path, and the other the geometry of the site in relation to the roads . . . The typical floor plan is orientated towards the north–south axis in relation to the sun's path. However, the service cores (lifts, stairs and toilets) are located on the hot sides of the tower (that is, the east and west sides). They provide shading and respond to the geometry of the site. All the lift cores and staircases are located on the outside of the building (as against central-core positions) so that they receive natural light and ventilation. By this disposition of the building, the configuration of the built forms responds to the Malaysian climate.
>
> (Powell and Kurokawa, 1989)

In response to this retrofitting, a range of replanning and rezoning of all levels of the ground plane is included. Again the problem is spatial, that is, providing space for ecological systems. The aim is to keep the same functional area but increase the allocation of space for bioclimatic elements. This means making better use of the available space, displacing elements of the building, which

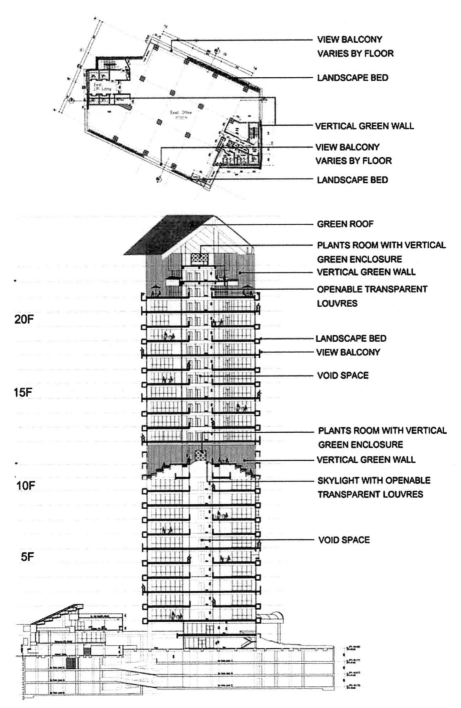

VIEW BALCONY
VARIES BY FLOOR

LANDSCAPE BED

VERTICAL GREEN WALL

VIEW BALCONY
VARIES BY FLOOR

LANDSCAPE BED

GREEN ROOF

PLANTS ROOM WITH VERTICAL
GREEN ENCLOSURE

VERTICAL GREEN WALL

OPENABLE TRANSPARENT
LOUVRES

20F

LANDSCAPE BED

VIEW BALCONY

VOID SPACE

15F

PLANTS ROOM WITH VERTICAL
GREEN ENCLOSURE

VERTICAL GREEN WALL

SKYLIGHT WITH OPENABLE
TRANSPARENT LOUVRES

10F

VOID SPACE

5F

Figure 1.3.11 Bio-integration of biotic (living) and abiotic (non-living) elements involved in examination of the areas of the building that could be used for such purposes

Source: Luella Yu.

could be replaced by more ecological systems. The transportation system is an opportunity in this scheme with a large proportion of the building devoted to that aspect. Functional areas are redistributed to the car park space, which is redesigned, and enhanced public transport and bicycles replace the displaced car-based transportation. Figure 1.3.12 shows some of the replanning that was carried out to the typical floor plate and to the building's service cores.

Finally, to make this possible, a new infrastructure is recommended. As seen in the DGI building, an eco-cell is provided, which is a secondary core to provide water reticulation through the façade. This system is employed in the original concept and has been extended in the VADS Plaza project (Figure 1.3.13).

The principles of supplying an eco-infrastructure were provided in terms of green natural systems, blue water systems, red human systems and grey engineering physical systems (see Chapter 1.2) and were supported by eco-design as a self-monitoring life-cycle measurement. This involved five steps to improve the VADS Plaza (Figure 1.3.14):

1   a new HVAC system to service three floors, floor-based;
2   an HVAC service space rezoned as a green space;
3   eco-cell technology with roof and floor-based water tanks;
4   new high efficiency chillers located in the basement to free up space in the tower;
5   displacement of the air-conditioning system integrated with the existing operable windows in the façade.

Overall, the challenges provided by eco-design for retrofitting are numerous. The concept of eco-mimesis provided a conceptual way of using nature as a metaphor for design thinking and providing an overarching framework. It was

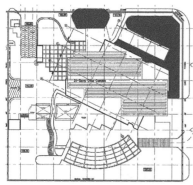

The existing green area is located mainly on site, partially on elevation and deck.

Existing green area: 505 sqm
Green ratio: 8%

Retrofit design, increase green area in four ways:
Green roof: 664 sqm
Green terrace: 664 sqm
Green deck: 498 sqm
Green wall: 2000+ sqm
New green ratio to reach 40%

Figure 1.3.12 Retrofitting the bio-integration of biotic (living) and abiotic (non-living) elements in the VADS Plaza, increasing the ratio of green areas

Source: Luella Yu.

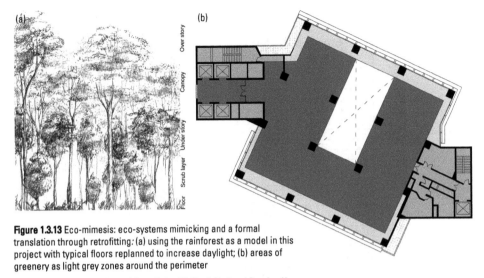

Figure 1.3.13 Eco-mimesis: eco-systems mimicking and a formal translation through retrofitting: (a) using the rainforest as a model in this project with typical floors replanned to increase daylight; (b) areas of greenery as light grey zones around the perimeter

Sources: (a) Christina Kim and Sara Najm; (b) Sunil Choi and Damian Kwang.

hard to see how that could be translated into all aspects of the building design. It did seem to provoke a far more ambitious approach, which meant reshaping of the building in terms of overall form, creating void decks and atria to improve ventilation and day lighting.

Urban infrastructure inertia, refitting and historical conservation

The scope and implications of new green technologies are still to be realised. This project focused more on the replanning aspects of green design to increase the integration of ecological systems. Emerging technologies are an interesting addition to the possibilities of retrofitting. In summary, this project demonstrates an approach to retrofitting using eco-design (Figure 1.3.15). It is an initial step but the breadth of the approach provides a new way of looking at obsolescence not in terms of a building design but also city planning.

However, a number of issues arose in the process of this project. First, the current inertia in both the building and city systems makes it difficult to propose retrofitting in this way. How can this inertia be addressed and a more ecological view taken of the city and its eco-systems?

Second, the VADS Plaza is a meritorious piece of architecture in its own right and represents important architectural ideas at the time. The question this raises is, to what extent should retrofitting be carried out, if at all? Would this compromise the historical significance of the building or indeed the environment it creates?

Third, the process of obsolescence of buildings and areas of the city remain present. The ebb and flow of economic and social forces within the city context may counter this and provide a basis for further change.

(a)

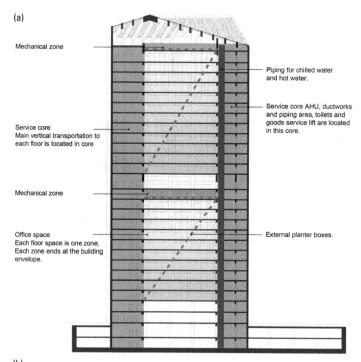

Mechanical zone

Service core
Main vertical transportation to
each floor is located in core

Mechanical zone

Office space
Each floor space is one zone.
Each zone ends at the building
envelope.

Piping for chilled water
and hot water.

Service core AHU, ductworks
and piping area, toilets and
goods service lift are located
in this core.

External planter boxes.

Figure 1.3.14
Providing an eco-
infrastructure:
(a) the existing system;
(b) the proposed
retrofitted system using
the five main steps to
using more efficient HVAC,
increasing landscaping
and new water storages
and services

Source: Sunil Choi
and Damian Kwang.

(b)

5 steps to retrofit the service core:

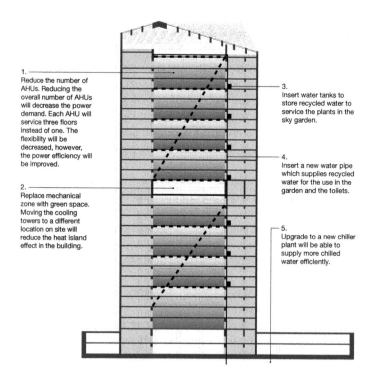

1.
Reduce the number of
AHUs. Reducing the
overall number of AHUs
will decrease the power
demand. Each AHU will
service three floors
instead of one. The
flexibility will be
decreased, however,
the power efficiency will
be improved.

2.
Replace mechanical
zone with green space.
Moving the cooling
towers to a different
location on site will
reduce the heat island
effect in the building.

3.
Insert water tanks to
store recycled water to
service the plants in the
sky garden.

4.
Insert a new water pipe
which supplies recycled
water for the use in the
garden and the toilets.

5.
Upgrade to a new chiller
plant will be able to
supply more chilled
water efficiently.

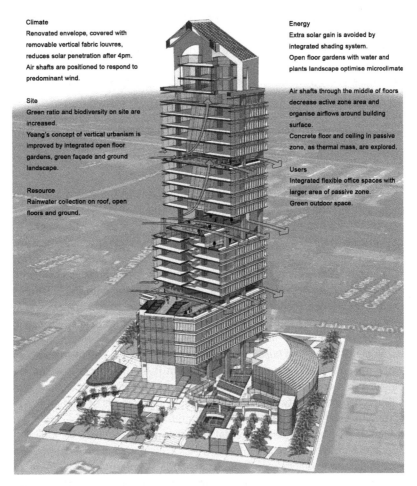

Climate
Renovated envelope, covered with removable vertical fabric louvres, reduces solar penetration after 4pm. Air shafts are positioned to respond to predominant wind.

Site
Green ratio and biodiversity on site are increased.
Yeang's concept of vertical urbanism is improved by integrated open floor gardens, green façade and ground landscape.

Resource
Rainwater collection on roof, open floors and ground.

Energy
Extra solar gain is avoided by integrated shading system.
Open floor gardens with water and plants landscape optimise microclimate

Air shafts through the middle of floors decrease active zone area and organise airflows around building surface.
Concrete floor and ceiling in passive zone, as thermal mass, are explored.

Users
Integrated flexible office spaces with larger area of passive zone.
Green outdoor space.

Figure 1.3.15 Holistic retrofitting proposal for VADS Plaza

Source: Castro Arturo, Gouniai Martha, Mansilla Roberto, Schukkert Tatiana, and Zhang Pengcheng (Aaron).

Understanding this process is necessary to better realise a view of eco-design and its application to retrofitting.

BUILDING OBSOLESCENCE

New definitions of building obsolescence

In a study called 'Measuring the gap: a user-based study of building obsolescence in office property', James Pinder and Sara Wilkinson have developed particular insight into the definitions of building obsolescence: 'A common definition of building obsolescence is that it is caused by physical deterioration of its fabric over time requiring corrective maintenance, that is, maintaining the functionality of the building' (Pinder and Wilkinson, 2000). However, Pinder and Wilson argue this definition is confusing because it is focused on physical decay and fails to distinguish between the *process* of obsolescence and the *causes* or circumstances linked to this process:

> To clarify, obsolescence is the *process* of becoming antiquated, old-fashioned, outmoded, or out-of-date. It describes a decline in utility that does not result directly from physical usage, the action of the elements or the passage of time. (2000)

They define obsolescence as '*caused* by changes in people's needs and expectations regarding the use of a particular object or idea'.

The concept of *utility* derived from this approach 'is the sense of usefulness, desirability or satisfaction. There is no single measure of utility, it is difficult to produce a rational, consistent and objective measure of obsolescence.' Relating this to the dimensions of sustainability and the environmental, social and economic factors, it is clear that different measures will prevail. Hence Pinder and Wilson suggest *economic obsolescence* in buildings is normally measured in terms of the real or nominal decrease in value of the building in the investment property market. They differentiate between *locational* and *building economic obsolescence*. The former occurs when an area – and the property located in it – suffer from devaluation because they are considered less fashionable or attractive by occupiers, and suffer a decline in utility resulting from changing standards of infrastructure, communications and environmental conditions. The latter is due to a disconnect between market rental and the rental required for building operations due to a building's intrinsic physical characteristics as determined by design and specification. This distinction is important since 'it is rarely possible for an individual building owner or occupier to remedy the causes of locational (economic) obsolescence, whereas building (economic) obsolescence can often be remedied by refurbishment'. The impact of this type of obsolescence is also social.

Social obsolescence in buildings

Social obsolescence in buildings, according to Pinder and Wilson, is associated with the needs of the occupants. This type of obsolescence is found through measuring the utility of office accommodation. There is clearly a need to distinguish between different types of office worker, who have different expectations of what constitutes a useful or satisfactory working environment. The impact of this type of measurement can foreshadow and direct the environmental management of the building utility. Hence it is clear that the involvement of the users in the process controlling the internal utility of the building, i.e. its indoor environmental quality, while also addressing the needs of the external impact on the environment, is becoming increasingly important to offset social obsolescence in buildings. However, through the process of retrofitting, the physical building provides a higher standard of technical performance, and the research looks at what system innovation is available and how these systems can be combined.

However, the indoor environment experienced by the occupants is also related to a range of psychological and behavioural factors. The pro-environmental attitudes of the occupants can affect the motivation for offset obsolescence, provided they have the ability to effect change. Hence the term 'passive buildings require active people' is a truth that is often said about buildings that are less automated in terms of providing comfort though air conditioning. This identifies some of the conventional boundaries and expectations associated with this type of obsolescence. Environmental obsolescence is another boundary that this is changing, when environmental forces start to impact on the function and operation of buildings. As the magnitude and rate change due to the costs of energy and climate events, this form of obsolescence will become more and more apparent and a driver for renovation and retrofitting.

Environmental building obsolescence

During the period of this research we have seen the emergence of *environmental building obsolescence*. This is the impact of new environmental performance standards on the investment property market. An example in this area is found in New South Wales, Australia. The New South Wales State Government reports that about 70 per cent of its agencies predominantly occupy office buildings. These offices account for 10 per cent of its building energy use. Space cooling, ventilation and lighting are the three most significant sources of greenhouse gas emissions, with lighting having the greatest potential for improvement.

However, it is rarely recognised that there are no standards for environmental retrofitting. This is beginning to change. In Europe,

> Seppänen and Goeders (2010) revealed there was [a] large variation in the energy performance regulations of the different countries. Not only the performance levels are different, but even the units in which the performance is measured are different. Primary energy, delivered energy, various energy frames and even CO_2 emissions are used.
>
> (Jagemar *et al.*, 2011: 14)

However, in recent years there has developed

> a common methodology, as the majority of the countries already use or are moving to use primary energy in the definition of energy performance in [kWh/m^2, a]. Many countries have prepared long-term roadmaps with detailed targets. Such roadmaps help the

industry to be prepared and committed to the targets. For example, in Norway, zero energy buildings are expected in 2027, but in the UK carbon neutral buildings already are planned in 2017.

(ibid.)

Furthermore, it appears through its Energy Performance Buildings Directive (EPBD) that in the EU methods are emerging to provide data on energy consumption, thorough steady state calculation to underpin the roadmap approach (ibid.).

The International Energy Agency Solar Heating Cooling Task Force 47 has been established to examine the renovation of non-domestic buildings. Its aims are to do the following:

1   develop a solid knowledge base on how to renovate non-residential buildings towards the Nearly Zero Energy Buildings (nZEB) standards in a sustainable and cost efficient way;
2   identify the most important market and policy issues as well as marketing strategies for such renovations (IEA, 2011).

The approach described here is largely a performance-driven approach and it remains to be seen whether these targets are achievable or indeed realistic. The plan is:

EU commitment to maintain the global temperature rise below 2°C, and its commitment to reduce, by 2020, overall greenhouse gas emissions by at least 20% below 1990 levels, and by 30% in the event of an international agreement being reached.

(ibid.)

However, there are no standards for renovation and retrofitting. Yet if there were, this could be a game changer in this process of meeting these policies. It is argued that by harnessing a retrofitting approach which is based on sound environmental principles and enabled through broader ecological design process, it would be possible to make significant gains in meeting these targets.

CONCLUSION

The main issues arising from the research in this section are, first, that in warm climates the effects of climate change are likely to cause localised heating. This exaggerates cooling problems associated with the activities of the high heat loads in buildings in these regions. A bioclimatic approach found in new buildings can be applied to retrofitting when renovation is needed due to physical obsolescence.

Standards and legislation are needed to ensure at this point that the building systems are retrofitted, not simply replicated. Replication of existing systems will not address other forms of obsolescence such as environmental or social obsolescence.

A more radical view is to apply the principles of eco-design to the retrofitting of buildings and to the area of the city in which the building exists. This creates eco-retrofitting as an urban problem impacting on planning and ecological zoning of the city. The finding from this work suggests that initial steps in this type of work are largely spatial and that the existing function zoning of the biotic and abiotic elements should be rebalanced. Furthermore the occupation and machine densities should be reduced to mitigate the high, localised heat loads that such concentrations create.

Hence, in this chapter, the four factors of bioclimatic context, climate change, building obsolescence and eco-design, are seen as significant forces shaping the process of retrofitting. They are precipitating the need for reevaluating our own approach to retrofitting while providing a direction for new sustainable policy, standards and tools to succeed in this task.

ACKNOWLEDGEMENTS

I would like to acknowledge the input from Master's students in the March programme at Sydney University and other universities; Siu Pong, Danny Sit, Christina Kim, Damien Kwan, Luella Yu, Sunil Choi, Sarah Najm, Mary Najm, Yibei Liu and also local students who attended the workshop at the International Islamic University of Malaysia (IIUM), for their input into the project on the VADS Plaza. In addition, students in the Sustainable Design Program provided work on the project, namely those in the Sustainable Design Program: Castro Arturo, Gouniai Martha, Mansilla Roberto, Schukkert Tatiana, Zhang Pengcheng (Aaron).

Also thanks should be given to Dr Nathan Groenhout who supported the workshop and to Dr David Leifer and Karen Sharpe who helped lead the workshop from AECOM Australia.

REFERENCES

Barry, R. and Chorley R. (2003) *Atmosphere, Weather and Climate*, London: Routledge.
Hamzah, T. and Yeang, K. (1994) *Bioclimatic Skyscrapers*, London: Ellipsis.
Hyde, R.A. (2000) *Climate Responsive Design: A Study of Buildings in Moderate and Hot Humid Climates*, New York: E & FN Spon.
Hyde, R.A. (ed.) (2008) *Bioclimatic Housing: Innovative Designs for Warm Climates*, London: Earthscan.
International Energy Agency SHC (2011) *Task 47*. Available at: http://www.iea-shc.org/task47/subtask/ (accessed 12 June 2011).

IPCC (2007) *Contribution of Working Group III to the Fourth Assessment Report of the Intergovernmental Panel on Climate Change*, edited by B. Metz, O.R. Davidson, P.R. Bosch, R. Dave and L.A. Meyer, Cambridge: IPCC.

Jagemar, L., Schmidt, M., Allard, F. Heiselberg, P. and Kuvnitski, J. (2011) 'Towards nZEB – some examples of material requirements and roadmaps', *REHVA European HVAC Journal*, 48(3): 14–17.

Jones, D.L. (1998) *Architecture and the Environment: Bioclimatic Building Design*, London: Laurence King.

Law, J.H.Y. (2001) 'The bioclimatic approach to high-rise building design: an evaluation of Ken Yeang's bioclimatic principles and responses in practice to energy saving and human well-being', BArch thesis, St. Lucia, Australia.

Olgyay. V. (1963) *Design with Climate: Bioclimatic Approach to Architectural Regionalism*, Princeton, NJ: Princeton University Press.

Parlour, R.P. (1997) *Building Services: A Guide to Integrated Design: Engineering for Architects*, Pymble, NSW: Integral Publishing.

Pedrini, A. (2003) 'Integration of low energy strategies to the early stages of design process of office buildings in warm climate', PhD thesis, St. Lucia, Australia.

Pinder, J. and Wilkinson, S. (2000) 'Measuring the gap: a user-based study of building obsolescence in office property', Sheffield Hallam University. Available at: www.rics.org/site/download_feed.aspx?fileID=1921&fileExtension=PDF (accessed 9 May 2012).

Powell, R. and Kurokawa, K. (1989) *Ken Yeang: Rethinking the Environmental Filter*, Singapore: Landmark Books.

Rajapaksha, U. and Hyde, R.A. (2002) 'Passive modification of air temperature for thermal comfort in a courtyard building for Queensland', in *Proceedings of the International Conference, Indoor Air*, Monterey, CA, USA..

Seppänen, O. and Goeders, G. (2010) *Benchmarking Regulations on Energy Efficiency of Buildings: Executive Summary*. Federation of European Heating, Ventilation and Air-conditioning Associations – REHVA, May 5, 2010. Available at: http://www.rehva.eu/en/375.towards-nzeb-some-examples-of-national-requirements-and-roadmaps (accessed 11 May 2012).

Szokolay, S.V. (2008) *Introduction to Architectural Science: The Basis of Sustainable Design*, Oxford: Elsevier.

Yeang, K. (1999) *The Green Skyscraper: The Basis for Designing Sustainable Intensive Buildings*, Munich: Prestel.

US Environmental Protection Agency (2008) *National Estuary Program, 2004–2006 Implementation Review Report*, June 19. Available at: http://water.epa.gov/type/oceb/nep/upload/2008_07_09_estuaries_pdf_2004-2005_irreportfinal_6_19_08.pdf (accessed 8 August 2011).

US Environmental Protection Agency (2011) *Determining the Causes of Ecological Impairment*. Available at: http://www.epa.gov/eerd/caddis.htm (accessed 23 Dec. 2011).

Wilkinson, S. and Reed, R. (2008) 'The business case for incorporating sustainability in office buildings: the adaptive reuse of existing buildings', paper presented at the 14th Annual Pacific Rim Real Estate Conference, Kuala Lumpur, Malaysia. Available at: http://www.prres.net/papers/ (accessed 8 August 2011).

Yeang, K. (2002) *Reinventing the Skyscraper: A Vertical Theory of Urban Design*, Chichester: John Wiley & Sons Ltd

Zarsky, L. (1990) 'Sustainable development: challenges for Australia', in Australian Government, *Service for Commission for the Future*, Canberra: Australian Government Printer.

1.4
SUMMARY
Richard Hyde

The challenges of addressing climate change are currently with us and dominate much of the thinking about design and planning. There are many suggested solutions; however, the proposition for ecological retrofitting is still to come into its own.

Definitions of green design vary from focusing mainly on reducing the environmental impact of buildings through a range of resources using efficiency strategies requiring change to the materials, energy and water efficiency measures. Many argue that this is not enough to really change the status quo; at best, these are the starting point, but in order to implement significant change, more comprehensive and far-reaching measures are needed. Hence, green design in recent years has had to move to a new level and find a new *modus operandi*. As Lewis Mumford reportedly said, all thinking worthy of the human condition is now ecological in nature.

This new interpretation of green design opens up the boundary conditions for design and seeks to address more complex and far-reaching aspects of climate change and its impact on ecology. Ecology, by definition, is not so much about the biological condition of flora and fauna that inhabit the planet but relationships between them, hence this sparks an investigation of the relations between the various types of flora, between flora and fauna, and so on, that form the biotic elements. In addition, it further encourages the relations between biotic and abiotic elements.

However, building design professionals are not by trained ecologists, so integrating this approach is not an easy task. Hence, Chapter 1.2 set out this new approach with some well-developed eco-design principles, which can be applied to new buildings and also to renovation and retrofitting of existing buildings.

However, we argue that eco-design is bioclimatic in nature and that as yet has not been applied to the retrofitting of existing buildings. This new *modus operandi* is the process of renovation and retrofitting of ecological systems to address climate change and other issues concerning the environment.

Chapter 1.3 examined the definitions of renovation and retrofitting in the context of climate change. With reference to a case study located in a tropical climate in Kuala Lumpur, Malaysia, it examined how a 1970s bioclimatic building can be retrofitted with ecological aspects to enhance its performance. The implications are far-reaching in terms of bioclimatic design. This project was the result of a design workshop carried out in 2010 to complete research work on bioclimatic retrofitting.

In summary, ecological design is still very much in its infancy. The totally green building or green eco-city does not yet exist. There is still much more theoretical work, technical research and invention, environmental studies and design interpretation that needs to be done and tested before we can have a truly green built environment. The conclusions reached are that we need to reevaluate the use of retrofitting as part of the planning and design process to improve the sustainability of our cities and their buildings. However, in order to significantly increase our ability to realise this potential a more holist eco-design approach is needed supported by new policy, standards and tools.

PART II
BIOCLIMATIC RETROFITTING

2.1
INTRODUCTION
Nathan Groenhout and Richard Hyde

Part II of this book examines the question of sustainable retrofitting at the building level and looks at the application of bioclimatic retrofitting to assist in addressing climate change. *Renovation* in this context is seen as replacing existing systems on a like-for-like basis (accounting for performance improvements that might be gained through model upgrades, improved manufacturing techniques and the like) while *systems retrofitting* normally involves replacement of these systems with new or alternative equipment or technology to meet higher performance expectations or standards. A series of questions arise over the selection of these strategies and solution sets which the following chapters seek to address.

Chapter 2.2 by Upendra Rajapaksha, Richard Hyde and Nathan Groenhout examines the strategies and solution sets which can be used for retrofitting. Limitations to previous studies have been advanced in terms of the prior lack of concern about the energy performance behaviour of interventions combined with their building characteristics, occupancy profile and climate change effects, i.e. what impact does an increase in extreme temperatures have on energy performance, and hence what is the most appropriate intervention to implement, such as a change in the air-conditioning system, improved shading on the façade, or changes to the microclimate surrounding the building to reduce insolation on the fabric and improve ventilation in the building.

An examination of existing buildings in relation to a bioclimatic methodology that can take account of these factors is presented as useful in quantifying their actual behaviour and to suggest appropriate retrofit solutions and explore their potential benefits. In warm climates the selection of strategies and solution sets to control the environmental heat loads is a major imperative if sustainable energy efficiency is to be achieved.

Chapter 2.3 by Richard Hyde and Indrika Rajapaksha looks at what sources of evidence can be used to select retrofitting strategies. Increasingly there is a demand from building owners, building managers and other building design professionals to provide evidence of the performance improvements provided by both the technical and non-technical solution sets used in the retrofitting process. This is called the application of an 'evidence-based' design approach to retrofitting. This chapter provides the basic framework for this approach to retrofitting. The chapter integrates and expands on this approach. It draws on the previous chapter which presents a process for developing the solution sets, and in the first section of this chapter more discussion is provided on the sources of evidence for retrofitting, while the second section discusses the integration of reference projects, and the final section describes a process for implementation. In the retrofitting process, this starts from existing building.

Chapter 2.4 by Indrika Rajapaksha examines the performance improvements of the solution sets using computer simulation to predict actual performance improvements. With the advances in building simulation tools, it is now possible to model the energy performance of possible solution sets for retrofit buildings. In this chapter the approach is applied to a particular existing building project. The selection of the building project is a 'Critical Case' building, as the evidence from it can be used not only to develop an operational plan for the building under investigation, but also can be generalised for application to a wider number of buildings with similar characteristics and levels of obsolescence.

Using the existing building as a base case, a calibrated building energy simulation model was developed, using actual operational data obtained from the building managers. This calibrated model was then used to determine the predictive energy performance improvements of the retrofitting solutions proposed in Chapter 2.3. This chapter explains the methodology and how it was tested and applied to the retrofitting of a 23-storey, ageing office building in the central business district of Brisbane, Australia. It characterises a 'critical case for retrofitting'. Two stages are used: first, defining and establishing the baseline for technological obsolescence of the critical case, and, second, applying the modelling methodology to test the best case retrofitting strategies.

Chapter 2.5 by Francis Barram presents a discussion of the relative merits of the different methods of economic analysis available to assess the benefits in designing retrofits of existing buildings under economic constraints. This case study demonstrates that further research is needed on the development of a standard economic analysis test that could used in developing an energy efficiency retrofit policy for government or asset owners. The existing lack of policy in this area means that significant energy reduction opportunities are currently being lost. The chance to capture energy efficiency savings during major renovations in the Australian non-residential building stock happens only very irregularly, so when the opportunity is overlooked, it is lost until the next major renovation.

2.2
DESIGN SOLUTION SETS FOR BIOCLIMATIC RETROFIT

Upendra Rajapaksha, Richard Hyde and Nathan Groenhout

INTRODUCTION

Buildings and building-related activities are responsible for more than 40 per cent of global energy demand in the form of electricity and therefore carbon dioxide emissions (IPCC, 2007). Further, carbon dioxide emissions associated with existing buildings are estimated to be increasing, and are now an important policy and design issue for all governments around the world.

With the projected climate change effects associated with increasing carbon dioxide in the atmosphere, combined with the demand for more comfortable indoor environments in offices, there is a growing concern about increasing energy consumption in existing commercial buildings and the likely adverse impacts on the environment as a consequence.

With a large proportion of existing commercial buildings in major cities in the world being over 20 years old, much of the high-rise building stock presents a high level of unrealised opportunity to improve energy efficiency and relieve pressure on energy supply markets. Exploring the unrealised potential for energy efficiency in existing commercial buildings is arguably the most significant action the commercial building sector can take to reduce dependency on non-renewable energy resources, and on reducing greenhouse gas emissions.

Design solutions for energy-efficient retrofitting which are based on a 'bioclimatic approach', that is, they address the effects, and mitigate the causes, of climate change, are currently needed if significant environmental impacts in the built environment are to be reduced and avoided in the future.

What is currently missing is an integrated design approach, which uses both non-technical (human-based) and technical (technology or physically based) systems to achieve environmental improvements in existing buildings. Many combinations of systems are possible, so part of the design process is usually to research and select particular combinations of systems (known as solution sets) to evaluate their potential to create improvements.

We can define a 'solution set' in a mathematical context, where it is the set of values that satisfy a given set of equations or inequalities. Hence, when applied to buildings, it refers to the relationships between building systems that can be optimised for particular goals. We argue that the current inequality lies in society's set of values concerning climate and the biosphere, and suggest a recalibration of the given set of equations or inequalities in the context of buildings through the adoption of both technical and non-technical systems (Hyde, 2010).

As seen elsewhere in this book, there is a diverse range of building types, so a design approach that provides a process and some basic solution sets that can be used in retrofitting will have value to the designer who is tasked with developing a sustainable retrofit for an existing building. Hence, this chapter is focused on how design solution sets based on a 'bioclimatic approach' can improve the environmental performance of existing commercial buildings and thereby improve the energy performance and achieve emission reduction targets.

The chapter establishes a development process for bioclimatic design solution sets for an energy-efficient retrofit and environmental recalibration of existing commercial buildings in warm climates. This process involves consideration of an environmentally sensitive interaction between climate, occupants and building design. The first section of this chapter discusses the policy context, which is the primary driver for using bioclimatic design in the context of retrofitting. Second, a number of types of solution sets are identified and the indicators for selecting these solution sets are given. The final section of the chapter provides examples of how these solution sets have been applied.

POLICY CONTEXT

The design approach to retrofitting has been grounded in the current changes within the existing policy context. While there is a general move to establish energy efficiency targets, the extent to which this is linked to retrofitting and bioclimatic issues is unclear, and this section examines this question.

Meeting energy efficiency, emission reduction targets and carbon neutrality
According to the IPCC (2007), buildings and building-related activities are responsible for more than 40 per cent of the energy demand in the form of

electricity and therefore carbon dioxide emissions. Furthermore, CO_2 emissions associated with existing buildings are estimated to be increasing rather than decreasing and hence emissions are becoming an important policy and design issue in order to achieve the often ambitious reduction targets in greenhouse gas emissions set internationally (de Wilde and Coley, 2012).

With the projected climate change effects due to increasing carbon dioxide levels in the atmosphere and the demand for more comfortable indoor environments in offices, there is a growing concern about increasing energy consumption in existing commercial buildings and its likely adverse impacts on the environment.

With a large proportion of the existing commercial buildings in major cities in the world being between 20 and 30 years old, a major issue is that the existing building stock is becoming less energy-efficient due to increased building obsolescence. Without retrofitting, these buildings offer a high level of unrealised energy efficiency opportunities, placing more pressure on the existing energy supply market. Exploring the unrealised energy efficiency of existing commercial buildings is arguably the most significant action the present commercial building sector can take in its aim to reduce dependency on non-renewable energy resources, which produce GHG emissions.

In order to achieve a reduction in both energy use and its associated carbon emissions due to buildings, an integrated energy-efficient sustainable retrofitting strategy is required. This approach should focus on the various components of energy use in buildings. Hence, application of design interventions, that are based on standards in the context of bioclimatic design, need to be considered in the sustainability of the construction and operation of buildings. Objective standards for retrofitting are needed to reduce energy use in buildings and subsequently help realise national and global obligations in emission reduction and environmental protection.

Achieving efficiency in energy use will lead to the emergence of building practices that result in an improved state of energy consumption, and this has been a key area of concern in the past few years. The expected global environmental problems have become more severe and have encouraged building designers and owners to go beyond this effort and produce new building practices for zero carbon emission potential. Based on the current high energy use patterns in the majority of existing commercial buildings in most developed economies, identifying methodologies and design solutions for improving energy efficiency has already become more imperative. The use of bioclimatic design is one way of working towards this target.

BIOCLIMATIC DESIGN FOR RETROFIT

Bioclimatic design

Bioclimatic design is focused on the interaction between climate, occupants and building design. The approach developed by Olgyay, which considers the way buildings filter and modify the external climate for occupants' comfort, is seen as an appropriate way to deal with energy efficiency opportunities in retrofitting existing commercial buildings (Olgyay, 1963). Part of the problem comes from the process of building design over the past 50 years, which has moved away from this approach and hence led to buildings that rely on high energy use systems to provide occupant comfort.

The high energy use approach comes about in an attempt to respond to two factors. The first is the response to impacts from outside the building to maintain thermally comfortable indoor temperatures. The second is the response to impacts from the components and systems inside a building for the benefit of occupants' comfort. In both responses, bioclimatic design involves four equally important interlocking variables, i.e. climate, biology, technology and architecture. In this process, the building envelope, section and form are the main bioclimatic modifiers that can promote the following:

- reduction in negative impacts from outdoor radiant heat and elevated air temperatures in the form of summer heat gain;
- optimisation of the cooling effects of ventilation in summer;
- increases in heat loss from building interiors in summer;
- reduction in heat loss from building interiors and optimisation of solar gain in winter;
- optimisation of daylight efficacy in building interiors in both summer and winter to reduce internal loads from lighting.

These effects can be seen as the essential objectives of a bioclimatic retrofit approach. Conventionally, bioclimatic design is often considered to be about aspects of the building such as the microclimate, the form and the fabric, often in isolation. With bioclimatic retrofitting, the approach is concerned not only with looking at the elements of the building but with controlling different sources of thermal loads, for example, the building interior's equipment usage. Therefore, bioclimatic retrofitting requires an understanding of the thermal performance of an existing building in the context of thermal load characteristics, and, thus, this becomes the basis for selecting new solution sets (Figure 2.2.1). For example, surface temperatures on the exterior of buildings in moderate–hot climates can reach 50–60°C due to insolation (Hyde, 2000), hence addressing this surface temperature is a major impera-tive with the passive low energy design of these buildings, through the inclusion of façade shading by building or landscape elements, addition of insulation, changes in the building and the like. However, in the past it has been common practice to reduce the effects of these thermal loads by the use of active or engineered mechanical HVAC systems.

Figure 2.2.1 Bioclimatic design to mediate between climate and building interior for controlling thermal loads; defensive outward heat transfer in summers and inward-promoting heat transfer in winter climates

New buildings and existing buildings: dissolving the duality

In new buildings the practice of using active systems as the mechanism for dealing with high environmental and internal loads is now less common. The thermal performance of new buildings can minimise energy use through the integration of 'architectural design' and 'engineering systems' into the building design process. In this way, bioclimatic design can minimise the demand for energy for cooling by reducing heat from the environmental loads and minimising the magnitude of the internal loads. The efficiency obtained from the reduction in active systems provides supply side efficiency. This approach further encourages the use of energy produced by renewable energy resources since the energy needs of the building are reduced and the fraction of energy, which can be provided by renewable technology, becomes more meaningful (Rajapaksha, 2004).

An appropriately designed, climate-responsive building can consume less energy than a conventional building, while still providing comparable comfort levels. In new climate-responsive building projects, key interventions that affect the sustainable use of energy in construction and operation are made in the early design stages. These typically involve the manipulation of architecture, masterplanning, microclimate, building plan and sectional forms, building envelope and the choice of efficient engineering systems.

Furthermore, the influence of bioclimatic design can become more effective if the ratio of surface area to volume is high because of the larger potential interaction between climate, building and occupants. However, retrofitting this kind of approach into existing buildings, which have the reverse situation, that is, low surface to volume ratio, may still be an option, as seen in Table 2.2.1.

A comparison of design and building operation in new buildings and existing buildings shows that the order may be reversed with bioclimatic retrofitting. This is due to the fact that in new buildings the cost of implementing the most effective strategies is lower if they are included from the start, while the cost of making changes to the built form are normally prohibitively high in the renovation or retrofitting of existing buildings and hence in retrofitting the strategies focus on the occupants and the smaller, less costly changes such as engineered systems.

Therefore, the integration of bioclimatic design in existing buildings has very different implications to that of new buildings because the inherent

Table 2.2.1 Ordering of strategies for bioclimatic design interventions to control thermal loads and improve thermal performance efficiency

New buildings	Existing buildings
1. Building microclimate	6. Building microclimate
2. Plan form	5. Plan form
3. Building section	4. Building section
4. Building envelope	3. Building envelope
5. Engineering systems and appliances	2. Engineering systems and appliances
6. Occupants	1. Occupants

characteristics of the existing building generally preclude major changes. The utility and potential effectiveness of these strategies may vary depending on the building type and architectural design of the building, but generally include the upgrade or introduction of interventions to the envelope, and efficient and green engineering systems with the participation of occupants. The design can have a greater impact on a building's energy demand but this is not generally an easy task as the plan form, orientation and structural system of such buildings are already in existence and the energy needed to change many systems can be higher than the energy savings achieved through the changes. How the energy loads are accommodated within the building is a common denominator in discriminating between these factors.

Given the number of factors at interplay here, a potential framework for selecting strategies and providing design solution sets for effective thermal performance in existing buildings is needed.

Recent research on retrofitting of non-domestic buildings in moderate climates suggests improvements to building sectional details can contribute to energy conservation, lower environmental impact and improved indoor climate (Pfafferott et al., 2004). Research by Hyde et al. (2009) presents methodologies for application of bioclimatic and green technologies in existing large-scale office and non-residential buildings with the aim of developing design guidance which will demonstrate opportunities for synergies in technologies – both passive and active – to reduce energy demand and negative environmental impacts. The outcome from this research focused attention on using the energy loads as determining criteria.

BUILDING CHARACTERISTICS: THE GENERATORS FOR RETROFIT

Building characteristics, both physical and non-physical, are important parameters affecting thermal performance and therefore the energy use of a particular building. These can be considered as thermal load-contributing characteristics, since, as the name implies, they add to the thermal load of the building. The attempt to understand thermal load characteristics and their impact on existing buildings can lead to a basis for an environmental retrofit brief.

Internal and external loads

With energy performance improvements as the main target of retrofit interventions, thermal load characteristics can be grouped into the following two broader categories in order to determine appropriate retrofit options:

- *internalities*: elements of the building and occupancy that contribute to internal loads;
- *externalities*: elements external to the building that contribute to environmental loads within the building.

In high-rise non-domestic buildings with tall façades and small roof area in relation to the external façade area, the effects of externalities between indoor environment and outdoor climate is significant through fenestrations. Moreover, non-domestic buildings are usually more densely occupied than residential buildings. This leads to greater negative impacts from internalities such as process and equipment loads and occupant loads. These internal loads, however, can also be manipulated to optimise indoor environments, depending on the climatic characteristics of the building location.

Taking the building–climate relationship into consideration, there are three ways that the environmental and internal load transfer can be considered in terms of externalities: (1) conduction and absorption through the opaque surfaces; (2) conduction and radiation through the glazed areas; and (3) convection through openings in the building section or envelope.

Building characteristics can be identified in terms of the 'contributing' and 'controlling' aspects of their thermal load. Components that can control thermal loads provide significant benefit in developing retrofitting solutions as they assist in the realisation of more sustainable building operation by reducing the need for active climate control solutions. Table 2.2.2 summarises typical building characteristics that have an impact on the building's energy use. A division is made between heat load contributing characteristics and heat load controlling characteristics. Because energy performance and potential retrofitting options are expected to depend on these characteristics, they are identified as important factors to be investigated in any retrofit process.

These thermal load characteristics are inter-related to each other. Therefore, retrofits need to span across all areas of concern but take into account the architectural significance (see below), building usage and climate.

Architectural effects and typology

The physical characteristics of a building's design can also be categorised with respect to the architectural significance of the building's dependency

Table 2.2.2 Some examples of heat load (HL) contributing and controlling characteristics of a non-domestic building

Thermal load contributing characteristic	Contributing component	Impact on indoor environment and energy use
Internal loads and occupancy (internalities)	Lighting	Heat emission to occupied spaces
	Equipment	Heat and moisture emission to occupied spaces
	Computers	Heat emission to occupied spaces
	Occupancy	Affect the usage of lighting and equipment and contribute heat and moisture to occupied spaces
Environmental loads (externalities)	External envelope	Transfers heat from outside to inside (or inside to outside in winter) and increases energy use through requirement for space cooling (or heating)
	Internal mass	Retains and stores heat which can contribute to high energy demand for space cooling
	Building microclimate	Non-uniformity of climate around a single building can create heat pockets, wind shadows and limit shading etc.
	Building section	Geometry with unshaded openings promotes heat transfer and without proper aerodynamics traps heat inside through stagnation
	Infiltration (building leakage)	Increases heat and moisture transfer between outside and inside increasing the need for space cooling
Heat load controlling characteristic	Contributing component	Benefit to energy savings
Active mode	Efficient HVAC	Optimises energy use and reduces waste heat.
	Air handling units and chillers	Co-generation system may use heat to drive air-conditioning process (absorption chillers) or to generate energy
	Outside air cycle	Uses ambient outside air when external conditions are appropriate to provide comfort conditions inside.
	Adaptive (floating) set point temperature	Reduces energy use of air-conditioning plant by setting the supply air temperature closer to the ambient temperature
Passive mode with Bioclimatic influence	Building design including plan form, section and envelope	Minimises heat gain (loss) from externalities and optimises the indoor environment while increasing the energy efficiency

on the environment/context, planning and internal structuring. Therefore, office buildings can be described as:

- *heavy or light mass*, depending on the structure and material of construction;
- *deep plan or narrow plan*, depending on the size of the floor plate and its proportions, and the structure of the building with respect to cores, amenities and circulation;

- *open plan*, consisting of free spaces and minimum partitioning, or compact and cellular, consisting of smaller enclosed spaces (Dascalaki and Santamouris, 2002).

Selection criteria for retrofit solution sets

The application of a bioclimatic approach to retrofit design solution sets depends on the existing building and determination of criteria and changes which offer the best solution or solutions to minimise environmental impacts and reduce energy use. To select the appropriate solution sets it is necessary to look at the sources and types of energy loads. Office buildings range from being *internal load dominant* to *environmental load dominant*. The magnitude of these loads and their proportionate contributions to thermal load profile vary significantly due to the building characteristics. Therefore, interventions for rehabilitation that address thermal performance can take the solutions in diversified directions. The selection of retrofit criteria for re-calibration of thermal performance should therefore address the load dominancy pattern, as shown in Figure 2.2.2.

Performance indicators for internal loads

Total energy use patterns including cooling and heating loads from internalities must be taken into account as performance indicators. Solutions are required to reduce energy footprints in both non-technical and technical

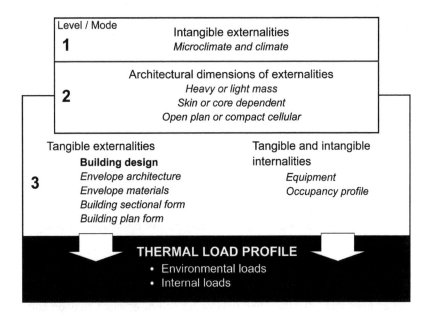

Figure 2.2.2 Availability of diversified retrofit options through three different modes of influences on buildings' environmental loads and internal loads

areas. Design solutions that address internal loads should consider the following aspects:

- non-technical aspects:
  - occupancy: the number and distribution of occupants as well the diversity of the occupancy profile over time;
  - operational profiles, including hours of operation of plant and equipment compared to occupancy, i.e. is the HVAC system operating after hours when there are no occupants in the building?
- technical systems:
  - HVAC systems, including the types of installed systems and the modes of operation
  - lighting systems;
  - other equipment and plant, including items such as lifts and pumps

The harmony between non-technical and technical components can create an optimised environment inside buildings and lead to greater improvements in energy savings.

Performance indicators for external loads

Total energy use and the effects on environmental loads on the building cooling and heating loads from externalities must also be taken into account when working on solutions to reduce energy footprints. Design solutions for dealing with both environmental and technical factors as follows:

- environmental factors:
  - microclimatic effects due to climatic influences, landscape, adjacent built forms and climate change
  - daylight
  - wind
  - humidity
  - solar radiation
- technical factors:
  - envelope architecture
  - material component
  - building section, including internal planning and zoning
  - building plan form.

Bioclimatic influence on design solutions has come to be understood as an interplay of all these factors. Bioclimatic building renovations should consider and address this interplay when developing energy-efficient solution sets. Energy performance improvement is the main target of retrofitting, and these building characteristics are found to result in various levels of intervention,

depending on the building–climate relationship, to control building indoor environments. Retrofit interventions can be effective in the following areas:

- exploiting the bioclimatic influence to manipulate environmental loads;
- implementing active system efficiency with bioclimatic influence to manipulate internal loads;
- modifying occupant behaviour and operational profiles in building operation to reduce energy use.

From this, a number of types of retrofitting solution sets have been developed (Hyde *et al.*, 2009).

RETROFITTING DESIGN SOLUTION SETS

Bioclimatic solution set types

There are five major types of solution sets that have been identified as providing viable design options for sustainable renovation and retrofitting.

1. *Microclimate enhancement and external passive systems* – use of the building external envelope as a third skin between humans and climate in order to reduce solar gain and promote ventilation in summer, capture or retain heat gain in winter and promote daylight in both seasons.
2. *Internal passive systems* – use of building form/sectional characteristics and planning of spaces to support Type 1, capturing optimum ventilation, daylighting and manipulating/controlling heat capture from environmental and internal loads.
3. *Active systems* – efficiency of plant and equipment, reduction of air-conditioned space.
4. *Synergies between active and passive systems* – use of mixed mode where the envelope/façade and air-conditioning system can be linked with human interaction for efficient control of systems.
5. *Behavioural change* – modifying building occupants' behaviour to reduce energy use through active participation in their space (e.g. switching lights off when leaving or when daylighting is sufficient) and consideration of non-technological aspects of building management and operation.

APPLICATION OF BIOCLIMATIC SOLUTION SETS

Microclimate enhancement

Enhancing the effects of the local microclimate can optimise the use of passive cooling and heating in building design. The microclimate around a single building is often diverse, and research has indicated that a wide variety of different microclimatic effects with regards to air temperature, solar radiation, shading and wind effects can occur around a single building (Gokhale, 1997). This non-uniformity of the microclimate can be attributed to factors such as the building's plan form, its sectional form, its orientation, and the design components of the microclimate.

For summer cooling, ventilation through shading is encouraged as it can reduce heat transfer from solar-irradiated surfaces, reduce the radiant temperatures of surrounding surfaces and subsequently reduce the dry air temperature of the incoming air (Givoni, 1994). Combining the use of thermal mass and shading in microclimates can reduce incident solar radiation and surface absorption. Modifying microclimates around existing buildings with shading and thermal mass can promote the exchange of heat by convection with ambient air. The consequence of this is that the air and surface temperatures in these microclimate zones tend to be lower than the corresponding ambient levels in adjacent areas. This retrofit effort outside the building helps to reduce outdoor–indoor heat flow through both ventilation and skin conduction and thus minimises indoor overheating in naturally ventilated buildings, and can even reduce the thermal loads in air-conditioned spaces. Therefore, linking the enhanced favourable (or reduced unfavourable) effects of microclimates to the building interior and the planning of internal spaces through architectural design becomes significant (Figure 2.2.3).

For winter heating, microclimates can increase solar exposure to building surfaces and interiors. A retrofit effort aimed at reducing the density

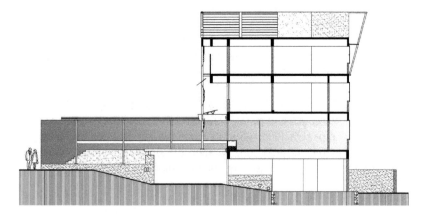

Figure 2.2.3 Nikini Automation, Colombo, linking effects of microclimates to building design, section and internal planning; buffer zone outside the left envelope reduces unfavourable effects of microclimate such as heat transfer

of the built mass and increasing sky view factors on the south and north directions can ensure more solar collection and increases in the air temperatures into the buildings in winters. In warm climates, however, caution needs to be applied to ensure that summer overheating does not occur as a consequence.

Mapping microclimatic effects

Mapping microclimatic enhancement can be used to identify the applicability and effectiveness of this input measure as a bioclimatic intervention in retrofitting. For example, the impact of microclimate effects on internal planning can be more effective and easily visible in low-rise and single-storey buildings whereas high-rise buildings, due to their height, offer less potential for the influence from enhanced microclimatic effects unless the façade is designed to accommodate the effects of landscape (see Chapter 1.3) (Figure 2.2.4).

Reshaping and re-calibrating the form and fabric

Reshaping the form and envelope can be beneficial in improving the thermal load profiles inside buildings and thus increasing energy efficiency. To achieve this, passive strategies can be used to regulate the visual and thermal comfort expectations of occupants. This is possible by manipulating the building form (both plan and sectional) and the thermal properties of the envelope (Figure 2.2.5). Good renovation efforts of both the building form and envelope are considered to be:

- promotion of comfort ventilation due to air flow effects;
- prevention of indoor heat gain, removal or transfer of indoor heat gain into heat sinks combined with night cooling or the like (thermal mass effects) during summer;
- promotion of optimum daylighting to interiors without heat gain in summer and without heat loss in winter.

The application of reshaping the built form and envelope in renovations depends on the acceptability of passive systems such as the cooling effect of ventilation (*day and night*), evaporative cooling (*direct and indirect*) and

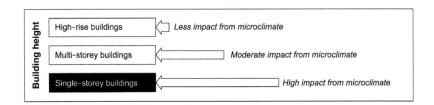

Figure 2.2.4 Mapping the effectiveness of microclimate enhancements on buildings

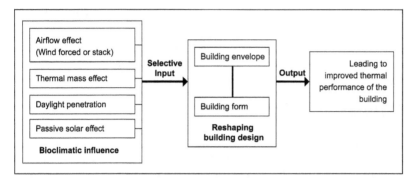

Figure 2.2.5 Mapping interventions to reshape form and envelope to enhance bioclimatic influence: effects of airflow, thermal mass, daylight penetration and passive solar

passive solar (*direct and indirect*) with the climate type and building usage. The potential of modifying building envelopes as a renovation measure has begun to be demonstrated with medium-rise office buildings. Design priorities to restrict the dynamics of external heat loads with an average ratio of glazing to façade area of around 43 per cent, external sun shading devices, insulation, solar control glass, and removal of internal heat loads with night ventilation and slab cooling have shown that the primary energy use of office buildings can be reduced to about one-third of the average building stock and kept within a limit of 100 kWh per net floor area per year (Voss *et al.*, 2007). This includes energy for heating, ventilation, cooling and lighting as well as auxiliary energy for other services and energy dissipation by conversion from primary energy to end energy. These results justify the generalisation of these interventions for office retrofits.

Response to airflow effects in warm climates

Airflow effects within building envelopes can be an effective intervention to reduce the impacts of heat loads on buildings (Givoni, 2011) by moving air through buildings either by wind-driven ventilation or by stack-driven ventilation (known as the 'stack effect'). Retrofitting interventions for airflow effects can be integrated into existing buildings to improve linking conditions between inside and outside, and to regulate both environmental loads and internal loads through the following means:

- modifying the wind permeability of the building's plan and sectional form to promote comfort ventilation and remove heat from inside in summer (wind-driven ventilation);
- altering the sectional form of the building to promote the buoyancy action of warm air from low level openings to high level opens to promote night cooling and remove heat in summer, as well as provide air movement (stack-driven ventilation).

Renovation options for building forms can also regulate the wind climate around a building. Reshaping the building envelope to create positive pressure fields, combined with the effect of thermal mass in good shading on the windward sides and improvements to internal planning for wind permeability can produce favourable summer comfort conditions (Figure 2.2.6).

Summer thermal comfort can be improved by airflow effects generated by building sectional forms involving internal passive areas such as atriums/courtyards and high thermal mass ceilings inducing stack ventilation. This is more applicable in retrofit efforts to medium-rise buildings. Even under moderately extreme outdoor temperatures, comfortable conditions can be reached through the passive cooling potential of the building envelope in place of technical installations of the building services system. Research suggests that comfort standards need to be revised because occupants show an adaptation to their thermal environment and acceptance of higher indoor temperatures in non-conditioned buildings when they can control the climate to some extent. To this end, the adaptive model of thermal comfort (de Dear and Brager, 2002) has great potential for conserving energy in buildings and allowing buildings to begin the process of adapting to climate change (Kwok and Rajkovich, 2010). Currently the adaptive model of thermal comfort could be applied to air-conditioned buildings with the view of reducing energy consumption in retrofitting (de Dear, 2009).

Response to thermal mass effect

Thermal mass in the building form and envelope affects the process of heat flow from both environmental and internal loads and thus both indoor air temperature and internal surface temperatures. It is an important design component in bioclimatic retrofitting that dictates the thermal response of the building and assists with the means of storing heat. Retrofitting interventions to reshape the thermal mass component of the building form and envelope can produce two major thermal modifications to improve energy efficiency in summer:

- The thermal mass in the façade acts like insulation and reduces the amount of heat penetration through the envelope, thus reducing the peaks of the indoor air temperature. *This modification reduces the heat transfer, improves indoor thermal environment and reduces the need for energy-intensive space cooling during a typical day.*
- The role of the thermal mass can be extended into the night period. When the thermal mass is coupled with night ventilation and effective daytime shading, it can act as a heat sink during the following day. *This modification prevents an increase in indoor air temperatures due to both environmental loads and indoor loads and reduces the need for energy-intensive space cooling.*

Figure 2.2.6 (a) A retrofit effort made to College House, University of Colombo, to improve wind permeability and shading without heat gain through (a) existing building envelope; (b) and (c) building with improved envelope and courtyard at the centre of the form

Source: Jayaweera (2009).

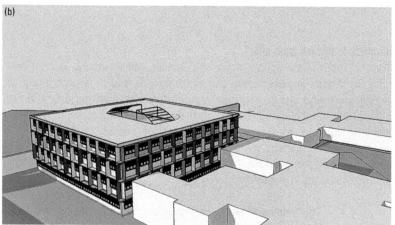

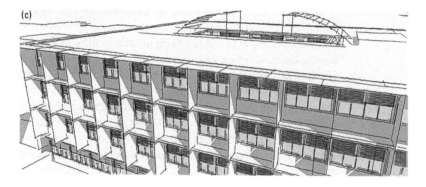

In winter, thermal mass is beneficial as a heat storage material. The stored heat is released to the interior of the building when the internal temperature drops below the surface temperatures in the evening. Moreover, quantitative variables such as the thickness and thermo-physical properties of thermal mass can regulate the heat storage capacity, heat conductivity and emissivity and therefore the indoor environmental conditions. Again, however, caution must be applied to ensure that thermal mass does not retain too much heat, preventing the building from cooling adequately overnight, resulting in elevated temperatures the following day.

In warm climates, envelope elements exposed to direct sunlight should be lightweight and insulated to prevent heat storage, while thermal mass should be well shaded or located internally to enable it to work properly as a heat sink. It is important to have a mechanism for removing the heat from the thermal mass during periods of no occupancy or when passive heating is required.

Daylight penetration

Daylighting design in the retrofitting of commercial office buildings cannot be looked at in isolation but needs to also consider the characteristics of the building form and envelope. It has been shown that floor plate depth, floor plan layout, window area and orientation with respect to glare and reflectivity can have a significant effect on lighting levels at the desk level (Viljoen et al., 1997). Building sections and envelopes for daylight penetration in commercial office buildings can be used to provide more comfortable indoor working environments while optimising daylighting with acceptable indoor luminance levels (Figure 2.2.7), thereby reducing energy consumption by reducing the need for artificial lighting.

Façade interventions such as light shelves, prismatic glass and holo-graphic coatings with shading are effective in improving visual comfort in office interiors. These interventions can be integrated even with artificial lighting in retrofitting existing buildings while improving the daylight conditions and providing substantial energy savings (Steemers,1994). With extremely high levels of ambient luminance (> 120,000 lux) in warm climates such as sub-tropical or tropical, building façades with severe shading characteristics com-bining glare cutting and light redirection can prove successful in promoting daylight, and reducing radiant heat (Edmonds et al., 2002). The relatively higher luminous efficiency of daylight generates less heat than most electric lighting systems and therefore reduces energy consumption (Lam, 2000).

A study carried out in Hong Kong on 35 high-rise commercial office buildings in a densely built business district shows the benefit of daylight performance with respect to daylight factor, glare index and room depth and their energy implications. It has been found that façade designs incorporating innovative daylighting systems such as light redirecting panels and light pipes, even in deep plan offices, can contribute to saving over 25 per cent of the total electric lighting use (Li and Tsang, 2008).

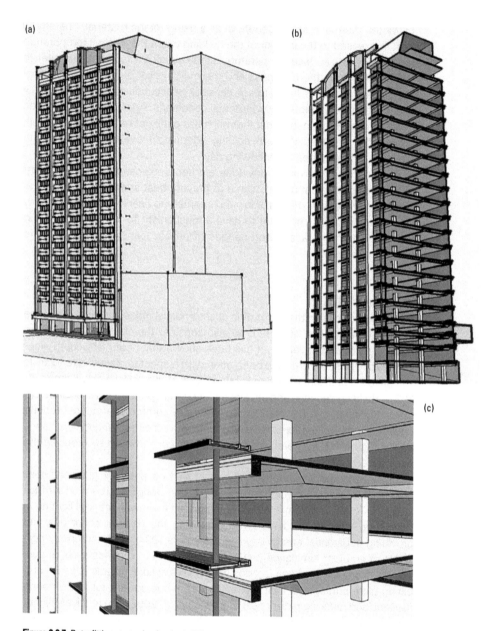

Figure 2.2.7 Retrofitting strategies for the building envelope of a high-rise office building (160, Ann Street, Brisbane, Australia) with more daylight efficiency and effects in reducing environmental and internal loads, (a) existing building, (b) improved envelope with more daylight penetration, (c) detailed window bays

Interventions that harness daylight in situations where there are architectural limitations in the existing windows in the façades and deep floor plate forms include the use of light pipes and mirror systems. Despite the high capital costs associated with this type of external intervention, optical daylight systems, both light pipes and sun lighting mirrors, can deliver natural light while contributing to reducing operational energy use in office buildings (Kim and Kim, 2010). Other solutions have shown some promise in the potential use of active façades for daylight penetration in retrofitting high-rise office buildings.

Passive solar effect

Improvements to existing building forms and envelopes to harness passive solar effects to meet heating requirements in winter can have some positive influence on the indoor thermal conditions. These are:

- direct solar gains through equator-facing glazed areas in the building façade;
- solar gain reflected by low-mass surfaces in the envelope to internal spaces;
- heat conducted through the building's non-opaque surfaces;
- heat released from thermal mass contained in the floors, walls and ceilings into the indoor environments.

These effects are obviously undesirable during summer months, and retrofit interventions that are intended for passive solar heating may not work at their optimum in buildings with retrofitting interventions that are primarily designed for summer climatic conditions. Therefore, optimising passive solar heating and cooling in conjunction with the effects of thermal mass and airflow for summer in a holistic manner becomes a more challenging task.

Active system efficiency with bioclimatic influence to manipulate internal loads

There are energy-saving benefits in renovating building services and active systems and this has been the most common approach used in retrofitting multi-storey and high-rise commercial buildings. With the involvement of engineering professionals, active system and service upgrading arguably has become the most common retrofit practice.

The large amount of internal heat gain from occupancy, lighting and environmental loads through the envelope in existing commercial buildings raises concern over the ability to reduce energy consumption. Buildings with environmental load dominancy and core-dependent buildings with internal load dominancy are likely to depend on active systems for indoor climate control. Focusing on the operational stage of buildings, improving the efficiency of energy-driven active systems could result in significant energy savings. The study by Lam (2000) demonstrates that energy use due to internal loads from lighting, equipment and occupancy accounts for more than 50 per cent of the total energy use in typical air-conditioned high-rise office buildings.

Recent research (Hyde *et al.*, 2009) substantiates a methodology for retrofitting internal load-dominant high-rise office buildings in warm climates. The work shows a trend towards a focus on issues such as improving the efficiency of active systems, their operation and service in buildings. The critical case building, a 23-storey office building in Brisbane, Australia, used in this project has demonstrated a criticality of high levels of thermal loads, cooling loads and therefore energy use. The major contributors for this behaviour are:

- heat loads from outdated installed appliances and lighting equipment;
- occupancy and operational profiles that are outside the recommended standards;
- degrading of HVAC system over time through poor maintenance or lack of preventative maintenance.

These behaviours have resulted in very high percentages of total heat gains representing high levels of energy consumption and putting the energy use function of the building at risk of becoming obsolete long before its physical life has come to an end. It is evident that much of the existing commercial building stock in commercial building districts in most developed and developing economies performs very inefficiently in terms of energy and this critical case study is an ideal example to show this trend.

Interventions that were introduced to the critical case study of 60 Ann Street, Brisbane (Figure 2.2.7) to explore unrealised energy efficiency in equipment, artificial lighting systems and HVAC are measures recommended in the Building Code Australia (BCA). It is found that total energy savings due to these interventions are as follows:

- 9.4 per cent due to efficiency improvements of installed appliances (plug loads);
- 3.1 per cent due to efficient lighting systems with photo electric dimming control systems;
- 3.8 per cent due to daylight sensors for artificial lighting that switch lights off when ambient levels are sufficient;
- 18.4 per cent due to improvements in HVAC systems such as increasing the co-efficient of performance of the chiller.

More results of this project are presented in Chapters 2.4 and 2.5.

Heat sink effect with mechanical systems

The heat sink effect of thermal mass can be coupled with a mechanical system to form a retrofit design solution to absorb internal heat gains from equipment. Convective heat transfer between the thermal mass and air, absorbing internal heat gains, can be effective when the thermal mass is properly integrated to the building interior through the retrofitting process. For example, the surface convective coefficient of thermal mass can be effective when high mass ceilings are placed in areas where there is a potential for internal heat gain from equipment. Pumping chilled water through pipes embedded in the slab floor or the use of night purge ventilation can be utilised to remove the heat from the thermal mass and 'reset' the thermal mass.

Occupancy behaviour and operational profiles in building operation

Encouraging the building industry to take greater consideration of occupant behaviour in building operation is also one approach that is particularly effective. Greater efficiency could be achieved if the retrofit interventions were to follow occupancy behaviour. Energy audit information of office building characteristics (McGuire *et al.*, 2009) have highlighted that the energy demand during a typical day in an office directly varies with the occupancy.

The influence of building occupants who can control their own lighting, heating and cooling on the energy consumption of buildings is also visible (Yun and Steemers, 2009). There is evidence that occupants in these indoor environments are less prone to take time off work, thereby reducing absenteeism (Santamouris, 2002). These controlling systems/occupant behaviours may vary from individual actions affecting a specific operational practice, to combinations of actions on areas of energy consumption.

Possible scenarios that control internal loads and thus reduce energy use may be as follows. Occupant decisions on the choices made about required comfort standards: changes in occupant behaviour in the context of retrofitting office buildings can become a significant concern in the design and implementation of retrofit strategies. Controlling internal loads and conserving energy in this way are observed due to the following:

- increased indoor air (set-point) temperatures;
- change of operational schedules;
- control mechanisms involved with heating, cooling and lighting systems;
- selection of energy-efficient (star-rated) appliances: small plug loads such as computers, instantaneous water heaters, and desk equipment can be a major contributor to both internal heat loads and energy consumption;
- sub-metering: to enable measurement and hence optimisation of energy use.

PREDICTION AND QUANTIFICATION OF EFFICIENCY OF E-DESIGN SOLUTION SETS

Computer-based simulation tools can be used to predict and quantify potential energy efficiency benefits of proposed retrofitting solutions. The LTV tool (Hyde and Pedrini, 2002) and Design Builder (http://www.design builder.co.uk) are examples of such simulation tools. The advantage of using LTV is that it can be used to develop results quickly, although it only examines a limited number of variables. LTV is useful for examining variables which are organised in the planning of the buildings. These parameters include: plan depth (*shallow or deep plan*), orientation (*north, south, east, or west*), window-to-wall ratio (*opaque versus transparent elements*), shading (*full shading to no shading*) and passive zoning (*utilising natural ventilation and daylight versus active zones*). A more detailed approach to simulation is discussed in subsequent chapters of this book.

CONSIDERATION OF OTHER FACTORS IN DEVELOPING SCENARIOS FOR RETROFIT

Selecting design interventions for retrofit cannot be looked at in isolation. Although there may be several options available to improve energy consumption, certain parameters with varied objectives play significant roles in a single retrofit project. Prioritising these multiple factors which deliver a variety of benefits is the next step to enable comparison and to assist in identifying the most suitable retrofit interventions. A collection of options can then form a solution set for testing and analysis (Figure 2.2.8). A triple bottom line approach to technologically appropriate retrofit efforts as identified by Rey (2004) includes:

- providing environmental benefits in terms of energy savings and meeting emission reduction targets;
- linking social relevance with user requirements;
- achieving financial viability with cost considerations.

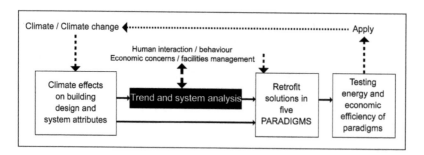

Figure 2.2.8 A map showing all the stages of a total retrofit process

Each of these has direct or indirect impacts in terms of energy benefits in retrofit design solutions. Therefore, it is necessary to give greater consideration to social and financial outcomes, not just environmental outcomes when assessing energy benefits from different possible interventions (Figure 2.2.9). A method to include the cost effectiveness of energy-saving measures in the re-calibration of existing buildings over the simple payback period of retrofit measures and including the economic efficiency in respect to conserved energy due to retrofitting has become a priority (Galvin, 2010). This is important as the tightening of standards applying to retrofitting have a strong effect on escalating the cost threshold, that is, the minimum cost required for a retrofitting.

Three possible application levels have been identified when considering potential architectural and engineering strategies for retrofitting:

1   *Stabilisation strategy or minor retrofit*: These are interventions that do not fundamentally modify the architectural character, substance or appearance of the building (Ray, 2004). Solutions in this set allow quick and easy formulation of cost-effective interventions which are straightforward in terms of changes to the building's architecture and engineering technology. The Property Council of Australia (PCA, 2009) outlines several quick 'wins' which can produce immediate benefits.

2   *Substitution strategy or intermediate retrofit*: A substitution strategy consists of interventions that provide an efficiency improvement to a particular energy-related element or system and may simultaneously transform the substance and the appearance of the building (Rey, 2004). Substitution strategies are distinguished by a large degree of transformation to components of the building. In this case the interventions imply a ratio between cost and energy performance. A PCA study in 2009 presents a large number of interventions with differing costs, benefits and sustainability outcomes.

3   *Restitution strategy or major retrofit*: This major refurbishment strategy involves a complete transformation of the building's energy-related characteristics or appearance but maintains a large portion of the original substance. The degree of intervention may vary with the building typology and the age of the

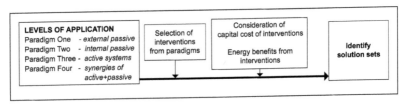

Figure 2.2.9 Conceptual framework for identifying cost- and energy-efficient design solution sets for retrofit

building. A comprehensive manual of energy upgrades to services systems is offered in the *CIBSE Guide F* (CIBSE, 2004).

The financial feasibility of design solutions may be amortised more quickly where there is also an intention to improve the building's image. Using a multi-criteria analysis approach to selecting solution sets implies that 'win–win' situations will arise when all related stakeholders contribute to retrofit efforts. This addresses the framework for retrofitting existing commercial buildings in the energy sustainability context and also provides new methodologies in understanding the process and then prioritising retrofit interventions.

FROM ENERGY EFFICIENCY TO CARBON NEUTRALITY

The integration of energy-efficient design solution sets to the retrofitting of existing buildings can be seen to improve energy sustainability in stages. Depending on the availability of funds, a retrofit project may use a mix of design solutions to improve the demand efficiency in the first stage. A total scenario can then be developed with the commensurate impact and energy-saving benefits determined. The decrease in required cooling or heating loads (the energy-efficient state) for the building with different retrofit solution sets can then be determined for both the existing and potential future climatic conditions. An additional step may then be considered to improve the energy performance of the building from an 'energy-efficient state' to a 'carbon neutral state' with the provision of a moderate amount of renewable or 'green' power (Figure 2.2.10).

The concept of carbon neutrality can be implemented in the retrofitting of existing buildings based on a methodology accounting for the following:

- embodied energy in construction material and components used during the retrofit;
- energy use in building operation as a result of implementing the design solution set interventions;
- energy produced from renewable energy resources and systems;
- environmental impacts due to construction and operation of retrofit interventions.

IMPLEMENTATION

It is essential that any existing office building being considered for upgrading to a higher environmental performance rating and then further to a zero carbon emission state should have a proper diagnosis of thermal load characteristics undertaken to understand its thermal and energy behaviour. From an environmental sustainability perspective, once thermal load profiles and their contributors are determined, design solutions for retrofit, including their potential contribution in manipulating thermal loads associated with energy savings and compliance with minimum performance standards, can be identified.

An outline of the process of implementation of interventions in both the building design and the non-technological aspects aimed at cutting the demand for energy is shown in Figure 2.2.10:

- Phase 1: Demand cut (aiming to reduce up to 60–70 per cent of total energy consumption):
  - non-technological interventions to control internal thermal loads in summer including behavioural change;
  - use of internal loads in winter to reduce the need for active heating;
  - technological and design interventions to manipulate environmental loads in summer or winter periods;
  - upgrade of HVAC systems to improve efficiency.
- Phase 2: Use of renewable energy sources (use of green power to meet the balance of the energy demand):
  - use of renewable energy technologies to achieve carbon neutrality.

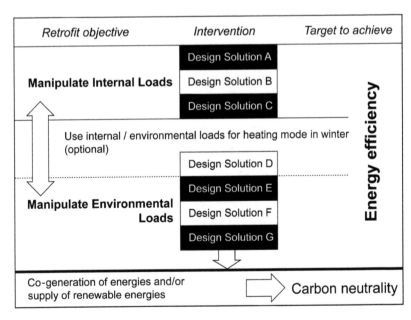

Figure 2.2.10 Implementation: manipulating thermal loads with design interventions and integration of renewable energies for carbon neutrality in stages (order of interventions may change between internal and environmental loads)

CONCLUSION

The framework presented in this chapter details the areas in which a mix of design solution sets can be integrated to control the environmental links between building, climate and occupants for energy sustainability. The areas include consideration of both internalities and externalities that affect the thermal load profile, i.e. *internal loads and environmental loads*. Improvements to the energy performance can be targeted to transform buildings from highly air-conditioned, energy-intensive structures to low-energy, passive or mixed-mode buildings, and then finally moving to a zero carbon emission state with the inclusion of a moderate amount of renewable or 'green' power over a period of time.

It is suggested developing the retrofit approach by understanding and prioritising the sources of thermal loads, considering the economic benefits and potential to address climate change effects as well as the ability to utilise green power to meet the ultimate energy demand.

These conclusions have some generic implications to any high-rise commercial building in warm climates. Using this framework, existing commercial buildings in a CBD can be assessed and ranked with regard to the potential they offer for retrofit.

ACKNOWLEDGEMENTS

The research work that underpins this chapter was supported by an Australia Research Council (ARC) Linkage grant LP0669628, 'Exploring Synergies with Innovative Green Technologies: Redefining Bioclimatic Principles for Multi Residential Buildings and Offices in Hot and Moderate Climates', initially from 2006 at the University of Queensland and subsequently at the University of Sydney from 2008 to the conclusion of this project in 2012. During this time the research work has been actively carried out in collaboration with international partners Llewelyn Davies Yeang and T.R. Hamzath & Yeang, AECOM and ENSIGHT. The industry partners were joined in 2009 by INVESTA.

REFERENCES

Building Codes of Australia Board (BCA) (2004) *Building Code of Australia*, Canberra: Australian Government. Available at: http://www.bcillustrated.com.au, accessed 13 May 2012.

CIBSE (2004) *Energy Efficiency in Buildings, Guide F*, London: Chartered Institution of Building Services Engineers. Available at: http://www.bom.gov.au, accessed 13 May 2012.

Dascalaki, E. and Santamouris, M. (2002) 'On the potential of retrofitting scenarios for offices', *Building and Environment*, 37: 557–67.

de Dear, R. (2009) 'Retrofitting comfort', presentation at the Mini Conference on bioclimatic strategies for retrofitting commercial buildings in warm climates, The University of Sydney, 3 August.

de Dear, R. and Brager, G.S. (2002) 'Thermal comfort in naturally ventilated buildings: revisions to ASHRAE Standard 55', *Energy and Buildings*, 34: 549–61.

de Wilde, P. and Coley, D. (2012) 'The implications of a changing climate for buildings', *Buildings and Environment*, 45(9): 1–7.

Edmonds, I.R. and Greenup, P.J. (2002) 'Day lighting in the tropics', *Solar Energy*, 73(2): 111–21.

Galvin, R. (2010) 'Thermal upgrades of existing homes in Germany: the building code, subsidies and economic efficiency', *Energy and Building*, 42: 834–44.

Givoni, B. (1994) *Passive and Low Energy Cooling of Buildings*, New York: Van Nostrand Reinhold.

Givoni, B. (2011) 'Indoor temperature reduction by passive cooling systems', *Solar Energy*, 85(8): 1692–726.

Gokhale, M. (1997) 'Between the building and the microclimate', in *Proceedings of ANZAScA Conference*, pp. 27–39.

Hawkes, D. (1982) 'The theoretical basis of comfort in the selective control of environment', *Energy and Environment*, 5: 127–34.

Hyde, R.A. (2000) *Climate Responsive Design: A Study of Buildings in Moderate and Hot Humid Climates*, New York: E & FN Spon.

Hyde. R. (2008) *Bioclimatic Housing: Innovative Designs for Warm Climates*, London: Earthscan.

Hyde, R.A. and Pedrini, A. (2002) 'LTV Design Tool', in *Proceedings of the Experts Meeting*, Singapore: Department of Building, NUS Singapore.

Hyde, R., Rajapaksha, I., Groenhout, N., Barram, F., Rajapaksha, U., Abu Nur Mohammad Shahriar and Candido, C. (2009) 'Towards a methodology for retrofitting commercial buildings using bioclimatic principles', paper presented at the 43rd ANZAScA Conference, The University of Tasmania, November.

IPCC (2007) *Climate Change Synthesis Report: An Assessment of the Intergovernmental Panel on Climate Change*, ed. Allali Abdelkader *et al.*, Valencia, Spain, 12–17 November.

Ismail, L.H. and Sibley, M. (2006) 'Bioclimatic performance of high rise office buildings: a case study in Penang Island', in *Proceedings of the 23rd Conference on Passive and Low Energy Architecture, Geneva, Switzerland*, 6–8 September.

Jayaweera, N. (2009) 'Retrofitting non-domestic buildings in tropical climates for environmental sustainability', unpublished MSc (Architecture) dissertation, University of Moratuwa.

Kim, J.T. and Kim, G. (2010) 'Overview and new developments in optical day lighting systems for building a healthy indoor environment', *Energy and Environment*, 45: 256–69.

Kwok, A.G. and Rajkovich, N.H. (2010) 'Addressing climate change in comfort standards', *Building and Environment*, 45(1): 18–22.

Lam, J. (2000) 'Energy analysis of commercial buildings in subtropical climates', *Building and Environment*, 35: 19–26.

Li, D.H.W. and Tsang, E.K.W. (2008) 'An analysis of day lighting performance for office buildings in Hong Kong', *Energy and Environment*, 43:1446–58.

McGuire, T., Vale, B. and Vale, R.J.D. (2008) 'Meeting climate change in today's office buildings', in N. Gu, L.F. Gul. M.J. Ostwald and A. Williams (eds) *ANZAScA 08 Innovation, Inspiration and Instruction: New Knowledge in the Architectural Sciences*, Newcastle, Australia: ANZAScA, pp. 203–10.

Olgyay, V. (1963) *Design with Climate: Bioclimatic Approach to Architectural Regionalism*, Princeton, NJ: Princeton University Press.

Pfafferott, J., Herkel, S. and Wambsgan, B. (2004) 'Design, monitoring and evaluation of a low energy office building with passive cooling by night ventilation', *Energy and Buildings*, 36: 455–65.

Property Council Australia (PCA) and Arup (2009) *Existing Buildings, Survival Strategies: A Toolbox for Re-Energising Tired Assets*, Thousand Oaks, CA: Sage.

Rajapaksha. U. (2004) 'An exploration of courtyards for passive climate control in non-domestic buildings in moderate climates', unpublished PhD thesis, the University of Queensland, Australia.

Rey, E. (2004) 'Office building retrofitting strategies: multi-criteria approach of an architectural and technical issue', *Energy and Buildings*, 36: 367–72.

Roaf, S. (2003) *Ecohouse 2: A Design Guide*, Oxford: Architectural Press.

Santamouris, M. (2002) 'Office: Passive retrofitting of office buildings to improve their energy performance and indoor working conditions', Editorial, *Building and Environment*, 37: 555–6.

Steemers, K. (1994) 'Day lighting design; enhancing energy efficiency and visual quality', *Renewable Energy*, 5(11): 950–8.

Steemers, K. and Yun, G.Y. (2009) 'Household energy consumption: a study of the role of occupants', *Building Research and Information*, 37(5–6): 625–37.

Triantis, E., Morck, O., Erhorn, H. and Kluttig, H. (2004) 'Environmental retrofitting of educational buildings—architectural aspects of an international research project', paper presented at 21st PLEA Conference: Sustainable Architecture, 19–21 September 2004, Eindhoven, The Netherlands.

Viljoen, A., Dubiel, J., Wilson, M. and Fontoynont, M. (1997) 'Investigations for improving the day lighting potential of double-skinned office buildings', *Solar Energy*, 59(4–6): 179–94.

Voss, K., Herkel, S., Pfafferott, J., Löhnert, G. and Wagner, A. (20070 'Energy efficient office buildings with passive cooling: results and experiences from a research and demonstration programme', *Solar Energy*, 81: 424–34.

Yun, G.Y. and Steemers, K. (2008) 'Time-dependent occupant behaviour models of window control in summer', *Building and Environment*, 43(9): 1471–82.

2.3

AN EVIDENCE-BASED DESIGN (EBD) APPROACH TO SELECT RETROFITTING STRATEGIES

What sources of evidence can be used to select retrofitting strategies?

*Richard Hyde and
Indrika Rajapaksha*

INTRODUCTION

More than half of the central business district (CBD) building stock in Australian cities comprises ageing office blocks. The majority of this existing commercial high-rise building stock in Australia is found to be carrying a high level of unrealised, cost-effective, energy efficiency opportunity, posing more pressures on the existing energy supply market (Warren Centre, 2009). The Warren Centre study found that the average annual energy consumption in these buildings falls between 230–270 kWh/m². With a large proportion of existing commercial buildings in CBD office markets being over 20 years old (LaSalle, 2010), and with high energy consumption patterns, the majority of commercial properties are approaching both physical and environmental obsolescence, which poses a serious threat to global attempts to meet emission reduction targets. The LaSalle study found that the focus would need to be on at least 80 per cent of the older stock, where retrofits provide greater opportunity to achieve meaningful emission reductions.

Increasingly, there is a demand from building owners, building managers and other building design professionals to provide evidence of the performance improvements of both the technical and non-technical solution sets used in the retrofitting process. This is called the application of an 'evidence-based' design approach to retrofitting. This chapter provides the basic framework for this approach. Chapter 2.4 describes the application of this approach to a 'critical case study' building.

WHAT IS EVIDENCE-BASED DESIGN?

> Evidence-based design is an approach for the conscientious, explicit and judicious use of current best evidence from research and practice in making critical design decisions, together with research and practice in making critical decisions together with [an] informed client, about the design of each individual and unique project.
>
> (Hamilton and Watkins, 2009: 9)

The crucial features of EBD are:

1 Fact-finding and identification of the sources of evidence.
2 Identification of a reference project as a source of evidence.
3 Development of a process for design to use outcomes from the fact finding (ibid.).
4 Increased trust and credibility in the design decision-making process.

There are some concerns that this approach has its limitations. First, it is felt that it will reduce creativity through the uncertainty that comes about because of a lack of understanding of research in practice. Second, there is a level of unfamiliarity in practice with the integration of research into the design decision-making. However, in Chapter 2.2 a process for integrating research information into practice was discussed, which, in complex and intractable areas of design such as retrofitting, provides an important basis for moving forward (ibid.: 17). This chapter integrates and expands on the approach given in Chapter 2.2, which provides a process. However, in the first section of this chapter more discussion is provided on the sources of evidence for the retrofitting process. The second section discusses the integration of reference projects, while the final section describes a process for use.

SOURCES OF EVIDENCE

This work is based on a research project undertaken by the authors, examining the advanced renovation of commercial buildings in warm climates. They developed a methodology to address climate change effects on existing commercial buildings and to improve their energy efficiency. The work was based on an existing case study of a multi-storey building in Brisbane (Hyde et al., 2009).

The work has shown that global warming is likely to have a negative effect on existing commercial buildings because the majority of such buildings were not designed to address this kind of change. Further, there is increased recognition of the important role of system performance as well as highlighting the impact on energy demand of occupant/owner behaviour on system management and comfort. Since these relationships are considered in bioclimatic design investigations of existing buildings in the context of a methodology which can take account of these areas, it will be useful to quantify the actual thermal behaviour of buildings and then to suggest retrofit design solutions and explore their potential benefits. Two main sources of evidence are found: (1) monitoring building performance; and (2) simulation of building performance. Table 2.3.1 shows the application of these to the different stages of retrofitting.

This approach is examined by using case studies, which include studies of the *stages of diagnosis* in order to be able to focus more accurately on the *stages of prediction*. Case studies are used to explore a number of issues in respect to the four activities of the proposed methodology.

Building monitoring: diagnosis for trend analysis of energy use

Existing buildings can become obsolete due to many reasons, including inefficiency of the indoor climate control measures, the services and systems being disabled and unable to control environmental and internal loads. In particular, building envelopes that are not responsive to the climate can contribute to overheating in summer and underheating in winter, resulting in

Table 2.3.1 Sources of evidence for retrofitting

Stages of retrofitting	Sources of evidence
Diagnosis for retrofit	
1. Climate influences and climate change effects on existing buildings	Simulation
2. Trend and system analysis of actual energy consumption behaviour of existing buildings.	Monitoring
Prediction of retrofit	
3. Identify design solution sets for retrofit including energy, economic and social benefits	Simulation and Monitoring
4. Test efficiency and effectiveness of these solution sets and quantify the benefits	Simulation and Monitoring

more energy use required to deal with both environmental loads and internal loads. Diagnosing energy use patterns can be considered as problem iden- tification in the context of environmental retrofitting. Measuring actual energy consumption patterns is the means to identify the thermal load profile of the building and includes the following:

1   total energy consumption and energy footprint of the building in KWh/m^2/annum;
2   comparison of the above with standards in respect to energy footprints of the building type in order to determine the level of criticality and energy sustainability;
3   comparison of at least one full year of electricity bills and aggregated sub-metering data of the building concerned, to reveal annual average electricity end-use breakdown for:
    i   HVAC systems with cooling and heating energy loads;
    ii tenants' energy consumption (plug loads for lighting and equipment).
4   lifts and other lifting systems;
5   seasonal variations in the total energy consumption for summer and winter;
6   identification of any anomalies in the data due to faults or failures, low occupancy, or other mitigating events such as unusual climatic occurrences.

This approach is useful in that it provides an overall picture of the building performance on a yearly basis and can be compared to the performance of other similar buildings. However, it only really provides a snapshot of per- formance. It does not necessarily help identify areas of weakness in the systems and sub-systems within the building or their impact on performance over the year. Hence it is recommended that systems' breakdowns be provided as shown in Figure 2.3.1.

This approach was applied to the critical case building and the data collected from monitoring of the case study is shown in Figure 2.3.1. This provides a snapshot of the performance of the building for the year 2005– 2006 of 5226 MW h electricity and 332 GJ of gas at a cost of approximately A$400,000 per year (in 2006 dollars). The building was rated between 0 and 1.5 Stars according to the Australian Building Greenhouse Rating (ABGR, which has now been replaced by the NABERS scheme). The systems breakdown as shown in Figure 2.3.1 shows area of improvement such as identification of the 9 per cent of energy consumption unallocated to a specific item of plant or equipment. The Sustainable Energy and Greenhouse Action Plan (SEGAP, 2001) in Queensland established a reduction target of 50 per cent greenhouse gas emission between 2006 and 2026, the imple- mentation of a 4.5 Star ABGR star rating for energy-efficient commercial buildings from 2010, and by 2020 all Queensland's government office build- ings are to be carbon neutral.

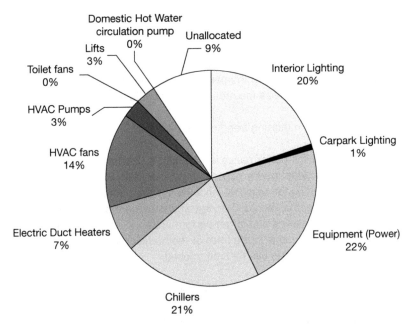

Figure 2.3.1 Total electricity consumption of case study building, April 2005 to April 2006. This is broken into sub-systems which assist in assessing the performance of these sub-systems, which can be compared to benchmarks to determine the need for retrofitting. In this case, the high use of pumps and fans would indicate a need to replace these sub-systems.

Source: Bannister and Duponchel (2006).

To help meet this target, more can be done with the data, collected over time, that comes from the building monitoring systems, such as energy and water meters. This has been found to provide a useful source of evidence and when arranged in a particular timeline can be used for analysis for two purposes:

1   To identify a major trend in the data, which will assist in discerning whether the building is achieving its performance goals. By looking at the data at the beginning and end of the timeline (usually longer than one year and accounting for seasonal variations), we can discern whether the performance factors are going up or down, relative to particular performance targets.
2   To identify on the timeline whether there are minor trends, which show the intervention of other factors and their effects on performance factors.

In Figure 2.3.1 the influence of a number of retrofitting interventions can be seen, plus the cost of these interventions and the initial energy reduction. The main sources of evidence are identifiable.

In some fields of study, the term 'trend analysis' has a more formal defined meaning; however, in the case of retrofitting, it can be harnessed to an 'action research approach' to examine the impact of retrofitting strategies on the energy performance of the building. This is not only a useful tool for diagnostic work in identifying the performance benchmarks for the project, but is also possible to utilise the monitoring data in fault identification during normal operation.

The steps in utilising trend analysis are:

1   Baseline performance is identified in the first year of operation.
2   The target performance, as a standard across all years, with a variation for seasonal influences is then determined.
3   The actual recorded energy and data are then mapped against the baseline and target performance.
4   Interventions in the form of the technical and human systems adjustments can then be tracked.

The result of this approach is to allow the tracking of initial energy savings and capital costs (Table 2.3.2). So for a retrofitting project the owner may invest a million dollars in the upgrade of selected systems and then rely on energy cost savings over a defined period to pay for the remainder. The trend analysis method allows the interventions, cost and performance over time to be tracked, in this case, a capital investment of A$730,000 in energy systems retrofitting yielded annual energy savings of A$370,000, an average monthly saving of $A30,800.

Hence, this demonstrates that coupling trend analysis with technological interventions can be mapped against sustainability retrofitting factors for use with associated economic considerations. Additional information is needed to relate causes and effects to the trends seen here, in particular, the effects of occupant behaviour (Roussac, 2009). To operate this kind of approach the use of a critical case methodology is proposed.

Table 2.3.2 Tracking of energy and water retrofitting initiatives against initial savings and costs as seen in Figure 2.3.1

Retrofitting interventions	Costs of initiative	Initial water and energy saving
1. Installed sub-meters	$50k	8 MJ/m²
2. Replace O/Q Dampers	$72k	10 MJ/m²
3. BMS field controllers upgrade	$93k	10 MJ/m²
4. Chiller replacement	$221k	20 MJ/m²
5. BMS head-end upgrade and fine-tuning controls	$22k	5 MJ/m²
6. Permafrost applied to chillers	$36k	20 MJ/m²
7. Individual floor dampers for skin AHU installed	$25k	10 MJ/m²
8. Base building lighting controls installed	$25k	10 MJ/m²
9. Waterless urinals	Not known	50 L/m²

Source: Roussac (2009).

The critical case

As seen above, there have been numerous studies to understand the energy efficiency issues associated with the existing building stock, primarily through surveys of large samples of buildings. These studies often centred on trying to identify a typology and then to compare buildings across this typology. However, it is often difficult to disaggregate the data on specific buildings or to test possible strategies for improvements using this approach. Hence, a case study methodology was developed to avoid the limitations, yet there are various types of case study approaches that can be used. Five common misconceptions about case study research are:

1   Theoretical knowledge is more valuable than practical knowledge.
2   One cannot generalise from a single case; therefore, the single case study cannot contribute to scientific development.
3   The case study is most useful for generating hypotheses, whereas other methods are more suitable for hypotheses testing and theory building.
4   The case study contains a bias toward verification.
5   It is often difficult to summarise specific case studies.

A scientific discipline without a large number of thoroughly executed case studies is a discipline without systematic production of exemplars, and a discipline without exemplars is an ineffective one. The term 'critical' here relates to a specific instance when problems in the field are contained in one case. This makes the case study more important, and perhaps critical, to an audience. Understanding how the issues are resolved in the critical case can play a significant role in the debate as to how the field might further address similar problems in the future.

Thus 'critical case' analysis is different from a standard case study in that it provides evidence relating to more general characteristics of the issues of concern. Hence the critical case in this research is an example of many of the problems facing our ageing urban building stock. As an example, it can contribute to the cumulative development of knowledge through the ability to summarise the findings into general propositions and theories (Flyvjberg, 2006). A study such as this contributes through its ability to delve into particular cause and effect scenarios which are central to the decision-making process. Up to this point the methodology has not been clearly defined though the issues of developing a clear evidence-based approach have been articulated.

In retrofitting, the effectiveness of strategies to improve the energy performance of an office building depends on specific characteristics related to its architectural structure, its operational features and relationship to the surrounding environment. Thus, the critical case demonstrated in this chapter embodies the specific obsolete characteristics of the built form and inefficiencies in energy performance suitable for retrofit and renovation (Hyde *et*

95

al., 2009). The modelling methodology used to analyse the critical case was developed to complement this direction.

Whole building integrated monitoring and simulation methodology

Research has repeatedly shown that the discrepancy between predicted and actual building energy use is a problem when using simulation to develop retrofitting, or even new design solutions for buildings. This discrepancy can then significantly impact on the potential to achieve carbon reduction targets. Therefore, while some studies suggest methodological improvements to environmental standards in energy use, they suggest that these studies need to be based on empirical data. The perspective from other studies suggests using theoretical data derived from simulation studies (Roberts, 2008) since the simulation can be used for parametric analysis of discrete variables across a range of different scenarios and therefore more readily demonstrates the impact of various interventions both individually and collectively. This illustrates the type of current controversies in the field surrounding the approaches to evidence-based design.

The conventional approach is to develop a 'whole building energy model', which represents all relevant operational systems, calibrated against operational data from the actual building. This is then simulated with regard to energy flows and an understanding of the predicted operational energy consumption patterns achieved. In order to assess performance, the results are compared to a reference building or larger sample of buildings.

Difficulties were experienced trying to apply this approach to a large building sample for a number of reasons:

1   A large and detailed amount of information regarding the specification of the buildings and their form and fabric is needed to be successful.
2   The building typology for retrofitting is highly diverse, hence trying to compare the energy performance across different buildings is problematic

It is useful to follow a 'critical case' building which can be used to provide an exemplar for wider practice. While it is difficult to generalise more broadly from the single case, it is argued that the nature and diversity of buildings suitable for retrofitting have defied previous attempts at classification. Hence it is proposed that the critical case be used to provide the evidence-based problem domain for retrofitting. A calibrated Forward Simulation Model (FSM) of the building is required to ascertain the predictive energy performance improvements of the retrofitting solutions.

Building simulation: predicting retrofitting improvements

The building information and physical parameters shown in Table 2.3.3 are examined with regard to energy and using a Whole Building Calibrated

Table 2.3.3 Evidence-based model

Inputs	Mechanisms	Outputs
Filters	*Fabric*	*Objective*
Climate	Structure	Water
Standards	Façade	Energy
Site	*Services*	Waste
Legislation	HVAC	*Subjective*
Function	Lighting	Comfort
Office	Controls	
Form	*Appliances*	
Deep plan	*Occupant*	
Narrow plan		

Simulation Methodology, the energy outputs were examined. This is used to create a credible building simulation model for testing retrofitting solution sets. The outline of the approach is as follows:

1   *Inputs*
    i   Identify the existing building as a critical case and collect weather data and building parameters including information from the monitored energy data.
    ii   Use the monitored energy data to model and analyse the energy loads of the critical case building.

2   *Process: calibration*
    i   Create a building simulation model of the critical case using a simulation tool. Simulate the operation of the building to create a second, simulated set of energy data.

3   *Outputs*
    i   Calibrate the building simulation model by looking at the discrepancies between the outputs from the second model against the actual data used in the first model. If discrepancies are found, the model is checked with reference to the source data.
    ii   A calibrated building simulation model is then created which provides the credible evidence source for the next part of the process.
    iii   This process creates the base case building model.

3   *Inputs*
    i   Feasible retrofitting solution sets for the critical case and changes to the weather data due to climate change are identified. Changes to the building and weather parameters are made in the input data.

4   *Process: testing retrofitting solution sets*
    i   The calibrated building simulation model is used to test performance improvements for the various interventions identified in the solution sets.

ii   Solution sets are added incrementally starting with the non-technical and then moving onto the technical. Non-technical solution sets are used first as they are often the least costly and easiest to implement.

5   *Outputs*

i   The cumulative performance improvements, resulting in reductions in the energy loads are developed as a sequence of scenarios. This creates the 'best case' retrofit for the critical case.

This approach has emerged as an appropriate method to better direct and manage the retrofitting process. Furthermore, it is argued that a computer-based simulation package is needed to assess the predicted quantitative outputs of aspects of the retrofitting process, such as the energy behaviour of specific retrofit solutions, without the need for costly prototyping or physical testing. This chapter outlines some of the underlying thinking behind this approach.

The FSM approach is preferred to predicted simulated energy performance data as it is thought to be more valid than simple simulation of a building, since it comprises both theoretical and empirical data from the existing building. It is argued that, in a retrofitting context, this gives a robust approach to evaluate the possible retrofit solutions in terms of energy performance optimisation, hence increasing confidence and credibility in the evidence base.

Retrofitting for performance improvement of a building requires a calibrated building energy simulation model to assess and prioritise retro-fitting solutions in relation to the energy savings achieved through retrofitting. Calibrated simulation requires a systematic approach that includes the development of the whole-building simulation model, collection of data from the building being retrofitted, and the coincident weather data. The calibration process then involves the comparison of selected simulation outputs against measured data from the systems being simulated and the adjustment of the simulation model to improve the comparison of the simulated output against the corresponding measured data. Most of the calibration methodologies require onsite measurements and monitoring of the building's energy use. Pan *et al.* (2007) explain the calibration methodology and the importance of the reassessment of internal loads and operational schedules in generating a perfect fit simulation model to the actual building profile and internal loads.

Figure 2.3.2 shows the modelling methodology developed to assess the performance improvements of the retrofit solution sets in this project. There are five main phases of the modelling methodology. The first four stages – (1) energy baseline; (2) energy billed; (3) water baseline; and (4) water billed – are used to create the base case, and the fifth stage creates the best case as follows.

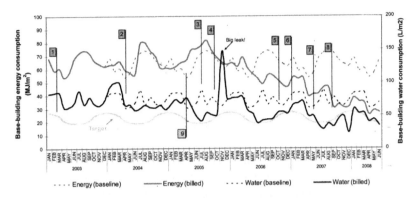

Figure 2.3.2 Building monitoring: diagnosis for trend analysis of energy use (Source INVESTA). The vertical lines represent the following interventions: 1. Installed sub-meters, 2. Replaced O/A Dampers, 3. BMS field controls upgraded, 4. Chiller replaced, 5. Upgraded and the fine-tuning of the controls systems on the Building Management systems server, 6. Permafrost applied to chillers, 7. Individual floor dampers for skin AHU installed, 8. Base building lighting controls installed, 9. Waterless urinals installed.

'Base case' and 'best case' development

We have developed five main activities in the development of the approach:

1   Input data collection and monitoring of energy use
    i   Baseline energy assessment against selected perform-
        ance standard
2   Model development
    i   Critical Case Input Model (CCIM)
    ii   Forward Building Input Model (FBIM)
3   Simulation
4   Calibration
    i   Forward Simulation Input Model (FSIM)
5   Best case development: prioritisation and testing of retrofit
    solutions
    i   Data-driven Simulation Models (DDSMs) of solution sets
    ii   Best practice energy assessment against selected per-
        formance standards.

The details of the stages are as follows.

Critical Case Input Model (CCIM)

The components of the physical characteristics of the critical case are used as the input data for development of the Critical Case Input Model (CCIM) of the building.

•   *Weather information*: This comprises the input data at different
    levels represented by building (site), blocks, zones and surfaces.
    The building level requires all site-specific information including
    the location, orientation, outside wind environment and weather

data. The weather data file from the International Weather for Energy Calculations (IWEC) in Energy Plus Weather (EPW) format was used for the simulations in this study. The original weather data file of the World Meteorological Organisation monitoring station for Brisbane Airport was used to generate ASHRAE (American Society of Heating, Refrigerating and Air-conditioning Engineers) and IWEC (International Weather for Energy Calculations) weather data files for use in the simulation tool (EnergyPlus 2010). See Chapter 2.4 for further details.

- *Form and fabric information*: The building physical data such as the building plans, construction details and elements of the building envelope, orientation, ground surfaces and its imme-diate surroundings were collected from an on-site survey, and review was taken of architectural drawings, building photo-graphs and audit reports on the building structure and façade. The lighting, appliances and HVAC system equipment and operational data were obtained from mechanical drawings, equipment inventories, and on-site walk through. Other data and information were gathered through on-site surveys, inter-views with occupants, engineers, maintenance personnel and the fulltime building supervisor.

- *Monitoring data*: Energy utility data was obtained from the energy management and audit consultant, who provided sub-metering data of tenants' light and power use, HVAC systems, common area lighting and lifts.

Model development

The *ASHRAE Handbook of Fundamentals* (1997) has classified different model types according to the method of building energy analysis. They are Forward, Inverse and Hybrid models. Forward models involve physical models and are used to model existing buildings in calibrated simulation software such as DOE-2 (LBL, 1980; EnergyPlus 2010), Trnsys (Klein, 2007) and EnergyPlus (US DOE, 2009). In forward modelling, a thermodynamic model of a building is created using fundamental engineering principles to predict the hourly energy use of a building for a year. The inverse models are created using measured data and rely on regression analysis (Fels, 1986; Reddy et al., 1999; Kissock et al., 2002). In Inverse modelling, an empirical analysis is conducted on the behaviour of the building as it relates to one or more driving forces or parameters. The two primary types of inverse models are steady state and dynamic inverse models (Doty and Turner, 2009).

The third category is Hybrid models. Hybrid models are models that contain forward and inverse properties. When a dynamic simulation program is used to simulate the energy use of an existing building, one has a forward analysis method that is being used in an inverse application, i.e. the forward simulation model is calibrated or fitted to the actual energy consumption data from a building. Thus the building energy simulation model of this study is a hybrid multi-zone model, which simultaneously simulates different thermal zones and multi-zone HVAC systems.

Simulation

Whole building calibrated simulation approaches normally require the hourly simulation of an entire building, including the thermal envelope, interior and occupant loads, and primary and secondary HVAC systems. This is usually accomplished by using a general-purpose simulation program such as BLAST, DOE-2, Energy plus or E-quest, or some similar proprietary program. Such programs require an hourly weather input file for the location where the building is being simulated. Calibrating the simulation refers to the process whereby selected outputs from the simulation are compared and eventually matched with measurements taken from an actual building. A number of papers in the literature have addressed techniques for accomplishing these calibrations and include results from case study buildings where calibrated simulations have been developed for various purposes (Norford *et al.*,1994; Pedrini *et al.*, 2002; Pan *et al.*, 2007; Iqbal and Al-Homoud, 2007; Raftery *et al.*, 2009).

Calibration

The percentage difference is the simplest numerical evaluation used in the practice. The computer model is usually considered calibrated when the model prediction error is within 10 per cent of actual data during the calibration process (Diamond and Hunn, 1981; Pratt, 1990; Kaplan, 1992b; McLain *et al.*, 1993; Haberl *et al.*, 1995). The Mean Bias Error (MBE) approach has been used as a general statistical method to evaluate the error between the predicted and the measured data. This is represented as monthly and yearly percentage differences and the corresponding statistical indices.

The disadvantage of this approach is the 'inter-neglecting' effect, which is caused by the plus and minus error compensation of monthly differences in metered and predicted data. This might lead to artificially small levels or zero interpretation of annual error difference. Therefore, more frequently used calibration metrics, the root mean squared error (RMSE) and the Coefficient of Variation of the root mean squared error (CV RMSE $_{month}$), account for how well the predicted results match the measured monthly data on an absolute basis (Draper and Smith, 1981; Bou-Saada, 1994; Kreider and Haberl, 1994). The accepted and widely used model calibration criteria are therefore based on the statistical model calibration metrics of monthly mean difference, mean bias error (MBE), root mean squared error (RMSE) and coefficient of variation of the root mean squared error (CV RMSE). The three main types of whole building calibration standards and guidelines are as follows:

1 ASHRAE Guideline 14-2002: Measurement of energy and demand saving (ASHRAE, 2002).

2 IPMVP 2007, International Performance Measurement and Verification Protocol (Efficiency Valuation Organisation, 2007).
3 M & V Guidelines: FEMP 2008, Federal Energy Management Program Measurement and Verification Guide (US DOE, 2008).

Recommended allowable differences for daily, monthly and seasonal data calibrations for internal loads (Kaplan *et al.*, 1992a) and the whole building acceptable tolerance for monthly and yearly data calibration defined in the standards and guidelines in relation to the statistical model calibration metrics are detailed in Table 2.3.4.

Testing possible retrofitting solutions

The Forward Simulation model is used to test possible retrofitting solution sets. In this process, the old technology and operational techniques are replaced with new green technologies and more up-to-date management operations. This creates a number of Data-Driven Simulation Models (DDSM) for each solution set. A retrofit solution generates a DDSM for summer and winter. Thus the retrofit scenario of the critical case contains can contain multiple DDSMs. They can be run in sequence from the least costly to the most costly, from the non-technical to the technical, or in an order dictated by requirements. This process provides a view of a possible progressive improvement of the retrofitting project, which can give information for the staging of the project over time. Moreover, this approach, when linked to existing targets or standards through the section of energy efficiency metrics, further improves the credibility of the approach.

THE ENERGY EFFICIENCY METRIC AND ITS APPLICATION

Two further questions were considered in developing the evidence-based approach:

1 Which energy efficiency metric should be used to assess the retrofitting strategies?
2 How could this be applied in the retrofitting context?

Table 2.3.4 Acceptable tolerance for daily, monthly and seasonal data calibration

Index	Whole building (%)			Index	Internal loads (%)	
	ASHRAE 14	IPMVP	FEMP		Tenants	HVAC
$MBE_{monthly}$	± 5	± 20	± 15	MBE_{daily}	± 15	±15–25
MBE_{annual}	–	–	± 10	$MBE_{monthly}$	± 5	± 25
$CV(RMSE_{month})$	± 15	± 5	± 10	$MBE_{seasonal}$	± 25–35	

Selecting an energy efficiency metric

The existing office stock with business-as-usual practices is a significant contributor to the present increase in greenhouse gas emissions and hence retaining these ageing offices without retrofitting is problematic for meeting national and international targets for greenhouse gas (GHG) pollution reduction. Operational inefficiencies are expected to double year 2000 CO_2 levels by 2030, if action is not taken to intervene. This applies worldwide; however, responses occur at a national or even more local level. In Australia, the metric for assessing operational performance is the National Australian Built Environment Rating System (NABERS), which replaced the previous Australian Green Building Rating scheme (AGBR). This is a performance-based energy rating system and defines a star rating for energy consumption and greenhouse gas emissions for existing office buildings based on actual performance data (NABERS, 2010).

The ratings are available for three types of energy consumption patterns in office buildings, namely, tenancy, base building and whole building. The tenancy rating is limited to tenant lighting and equipment energy consumption. The base building rating considers centrally serviced energy consumption such as common area, exterior lighting and power (including parking areas), lifts and HVAC systems. Total energy consumption by tenants and the central services contributes to the whole building rating (Figure 2.3.2).

Application of the metric

Applying this metric to retrofitting requires some changes to the normal performance approach. The base case is the existing building, and this is called the baseline performance scenario. Development of a baseline scenario is the first step in the critical case methodology for assessing energy performance improvements. The baseline scenario establishes the current energy performance status according to the given metric. The baseline in a retrofit project is usually the performance of the building or system prior to modification.

Retrofitting solution sets can then be applied to the base case to create improved performance scenarios and assess them against the performance targets desired. The best case is the performance scenario, with the application of all appropriate retrofitted solution sets that provides the best performance against the performance targets. Following this kind of approach, simulation modelling of different solution sets can be applied over time and the outcomes validated through further monitoring of the building's performance.

CONCLUSION

In this chapter an evidence-based design approach to developing retrofitting solution sets was developed for energy performance. First, an important part of the EBD approach is fact-finding and identification of the sources of

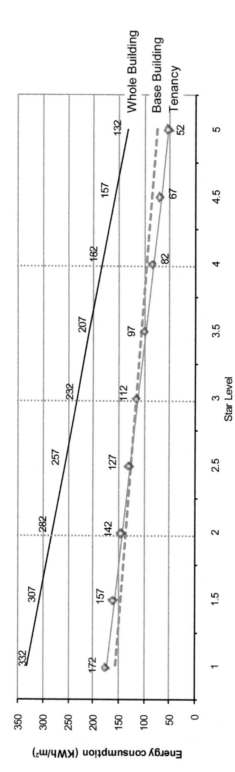

Figure 2.3.3 NABERS star rating and corresponding electricity consumption per unit area

evidence. For retrofitting, the sources of evidence are focused on the existing building. Data is needed in terms of the physical parameters of the project; inputs such as the weather data, mechanisms such as its form and fabric and ongoing monitoring of the building's performance. Second, it is important to identify a reference project or projects as a source of evidence. In this case, the existing building is identified as a critical case, i.e. its characteristics can be closely equated to other buildings within its type so that generalisations concerning the application of retrofitting strategies can be made. Third, a process is developed for the design to use results from the fact finding exercise using the critical case, and the monitoring data is used to calibrate simulation models and thus enhance confidence in this model. This model is then used to test retrofitting strategies through parametric changes in the model. It argues this increases the trust and credibility in the design decision-making process.

REFERENCES

ASHRAE (2002) *ASHRAE Guideline 14-2002: Measurement of Energy and Demand Savings*, Atlanta, GA: ASHRAE.

Bannister, P. and Duponchel, H. (2006) *160 Ann Street: Energy Audit Report*, Exergy, Australia Pty Limited.

Baumann, O. (2004) 'Operation diagnostics: use of operation patterns to verify and optimize building and system operation', in *Proceedings of ICEBO*, Paris, France.

Bou-Saada, T.E. (1994) *Advanced DOE-2 Calibration Procedures: A Technical Reference Manual*, Energy Systems Laboratory Report No. ESL-TR-94/12-01, College Station, TX: Texas A&M University.

Diamond, S.C. and Hunn, B.D. (1981) 'Comparison of DOE-2 computer program simulations to metered data for seven commercial buildings', *ASHRAE Transactions*, 87(1): 1222–31.

Doty, S. and Turner, W.C. (2009) *Energy Management Handbook*, 7th edn, Lilburn, GA: The Fairmount Press Inc.

Draper, N. and Smith, H. (1981) *Applied Regression Analysis*, 2nd edn, New York: John Wiley & Sons, Inc.

Efficiency Valuation Organisation (2007) *International Performance Measurement and Verification Protocol*. Available at: http://www.evo-world.org (accessed 3 January 2012).

EnergyPlus 2010. Available at: http://apps1.eere.energy.gov/buildings.energyplu/ (accessed 8 May 2012).

Fels, M. (1986) 'Prism: an introduction', *Energy and Buildings*, 9: 5–18.

Flyvjberg, B. (2006) 'Five misunderstandings about a case study approach', *Qualitative Inquiry*, 12: 219.

Haberl, J.B.S., Bronson, J.D. and O'Neal, D.L. (1995) 'An evaluation of the impact of using measured weather data versus TMY weather data in a DOE-2 simulation of an existing building in central Texas', *ASHRAE Transactions*, 101(1). ESL Report No. ESL-TR-93/09-02, College Station, TX.

Hamilton, D.K. and Watkins, D.H. (2009) *Evidence Based Design for Multiple Building Types*, New York: John Wiley & Sons.

Hyde, R.A., Rajapaksha, I., Groenhoet, N., Barram, F., Rajapaksha, U., Candido, C. and Abu, S.M.(2009) 'Towards a methodology for retrofitting commercial buildings using bio-climatic principles', in the *Proceedings of the 43rd ANZAScA Conference*, University of Tasmania.

Iqbal, I. and Al-Homoud, M. S. (2007) 'Parametric analysis of alternative energy conservation measures in an office building in hot and humid climate', *Building and Environment* 42(5): 2166–77.

Kaplan, M.B., Caner, P. and Vincent, G.W. (1992a) 'Guidelines for energy simulation of commercial buildings', in *Proceedings of the ACEEE 1992 Summer Study on Energy Efficiency in Buildings*, 1: 137–47.

Kaplan, M.B., McFerran, J., Jansen, J. and Pratt, R. (1992b) 'Reconciliation of a DOE 2.1C model with monitored end-use data from a small office building', *ASHRAE Transactions*, 96(1): 981–93.

Kissock, J.K., Haberl, J.S. and Claridge, D.E. (2002) 'Development of a toolkit for calculating linear, change -point linear and multiple-linear inverse building energy analysis models', *Final Report ASHRAE 1050-RP*.

Klein, S.A. (2007) *TRNSYS 16 Program Manual*, Madison, WI: Solar Energy Laboratory, University of Wisconsin, Madison.

Kreider, J.F. and Haberl, J.S. (1994) 'Predicting hourly building energy usage', *ASHRAE Journal*, 34: 72–81.

LaSalle (2004) *Existing Property Going Green*. Available at: www.agdf.org.au (accessed 8 May 2012).

LaSalle (2010) *Results of the 2009 CoreNet Global and Jones Lang LaSalle Global Survey on Real Estate and Sustainability*. Available at: wwwjoneslasalle (accessed 8 May 2012).

LBL (1990) *DOE-2 Use Guide, Version 2.1*, Lawrence Berkeley Laboratory and Los Alamos National Laboratory, LBL Report No. LBL-8689 Rev. 2, Berkeley, CA: DOE-2 User Coordination Office, LBL.

McLain, H.A., Leigh, S.B. and MacDonald, J.M. (1993) 'Analysis of savings due to multiple energy retrofits in a large office building', Report No. ORNL/CON-363, Oak Ridge, TN: Oak Ridge National Laboratory.

NABERS (2010) The National Australian Built Environment Rating System. Available at: http://www.nabers.com.au (accessed 8 May 2010).

Norford, L.K., Socolow, R.H., Hsieh, E.S. and Spadaro, G.V. (1994) 'Two-to-one discrepancy between measured and predicted performance of a "low energy" office building: insights from a reconciliation based on the DOE-2 model', *Energy and Buildings*, 21(2): 121–31.

Pan, Y., Huang, Z., and Wu, G. (2007) 'Calibrated building energy simulation and its application in a high-rise commercial building in Shanghai', *Energy and Buildings*, 39(6): 651–7.

Pedrini, A., Westphal, F.S. and Lamberts, R. (2002) 'A methodology for building energy modelling and calibration in warm climates', *Building and Environment*, 37(8–9): 903–12.

Pratt, R.G. (1990) 'Errors in audit predictions of commercial lighting and equipment loads and their impacts on heating and cooling estimates', *ASHRAE Transactions*, 11(2).

Queensland Government (2007) *Climate Smart 2050: Queensland Climate Change Strategy 2007: A Low-Carbon Future*, Brisbane: Queensland Government. Available at: http://www.climatesmart.qld.gov.au (accessed 14 August 2009).

Raftery, P., Keane, M. and Costa, A. (2009) 'Calibration of a detailed simulation model to energy monitoring system data: a methodology and case study', in *Proceedings of 11th International IBPSA Conference 2009*, Glasgow, Scotland, pp. 1199–206.

Reddy, T.A., Deng, S. and Claridge, D.E. (1999) 'Development of an inverse method to estimate overall building and ventilation parameters of large commercial buildings', *Journal of Solar Energy Engineering*, 121: 40–6.

Roussac, C. (2009) 'What is the business case of retrofit?' Paper presented at Mini-conference on Bioclimatic Strategies for Retrofitting Commercial Buildings in Warm Climates, the University of Sydney, 3 August. Available at: http://sydney.edu.au/architecture/research/archdessci_advanced_rennovation_project.shtml (accessed 3 January 2012).

Seem, J.E. (2007) 'Using intelligent data analysis to detect abnormal energy consumption in buildings', *Energy and Buildings*, 39(1): 52–8.

US DOE (2008) *M&V Guidelines*: Measurement and Verification for Federal Energy Projects Version 3.0. Washington, DC: US Department of Energy. Available at: http://apps1.eere.energy.gov/buildings/.

US DOE (2009) EnergyPlus Energy Simulation Software, US Department of Energy. Available at: http://apps1.eere.energy.gov/buildings/energyplus.

Warren Centre (2009) *LEHR, Low Energy High Rise Building Energy Research Study: Final Research Survey Report*, March.

2.4
BUILDING SIMULATION METHODOLOGY TO EVALUATE PERFORMANCE IMPROVEMENTS FOR RETROFITTING

Indrika Rajapaksha

INTRODUCTION

With the development of building simulation tools it is now possible to model the energy performance of possible solution sets for the sustainable retrofit of buildings. In Chapter 2.3 an evidence-based design (EBD) approach was proposed. In this chapter, the approach is applied to a particular building project. The choice of building project is considered to be a 'critical case' building and the evidence from it can be used not only to develop an operational plan for the existing building, but also can be generalised to a vast number of buildings with similar characteristic levels of obsolescence.

Using the existing building as a base case, a calibrated building energy simulation model was developed to assess the predicted energy performance improvements of the proposed retrofitting solutions. Performance predictions using simulation are dependent on two main modelling approaches, namely, Forward Simulation Modelling (FSM) and Data-Driven Simulation Modelling

(DDSM) (ASHRAE, 2005). In this case, the Forward Simulation Modelling was supported by two input models.

First, a Forward Building Input Model (FBIM) was developed to represent the actual physical and operational conditions of the existing building using physical and mechanical characteristics, occupancy, operating schedules, lighting, plug load densities and weather data. Then, a Critical Case Input Model (CCIM) using energy load data from monitoring of the building was developed and analysed.

Both the CCIM and the FBIM are used to develop and calibrate a Forward Simulation Input Model (FSIM) which closely represents the real operation of the building. The simulated energy performance data can then be used to prioritise case-specific retrofit solutions and the data-driven modelling can be used to evaluate the retrofit solutions in terms of energy performance optimisation.

This chapter explains the methodology and how it was tested and applied to the development of retrofitting options for a 23-storey, ageing office building in the central business district (CBD) of Brisbane, Australia. It characterises a 'critical case for retrofitting'. Two steps are used, first, defining and establishing the baseline for technological obsolescence of the critical case, and, second, applying the modelling methodology to test the best case retrofitting strategies.

BASE-CASE DEVELOPMENT: BUILDING INFORMATION

Site and climate

The critical case of this study is a 23-storey urban high-rise office building located in the CBD of Brisbane, Australia (Figure 2.4.1). Brisbane lies a few kilometres inland on the east coast of Australia at approximately latitude 27° South and has a warm, sub-tropical climate. The CBD is relatively densely populated with medium to high-rise office towers located within a relatively small footprint adjacent to the Brisbane River.

Building characteristics

The building in question has a total building area of approximately 19,600 m^2, of which the net lettable air-conditioned office and commercial areas are approximately 15,877 m^2. All the floors are above ground, 19 floors (namely the 1st, and 4th to 21st) consist typically of air-conditioned office spaces of 820 m^2 floor plates. The floor-to-floor height is 3.43 m. A typical floor consists of a service core at the centre of the rear façade and open plan office spaces at the front and sides. The front area of the ground floor consists of a coffee shop and reception, and the rear half is occupied by the mechanical plant room and service passages. Two naturally ventilated parking floors are located on the second and third floors with a floor-to-floor height of 2.6m. The

22nd floor contains plant and control rooms. Male and female amenities, fire stairs, lifts and the main lobby are contained in a core located at the centre rear on all levels of the building (see Figure 2.4.1).

The street façade is 39m in length, facing south-east (135°N) and has a wall-to-window ratio of 40 per cent. Windows are double-glazed and are fitted with mid-pane blinds. Concrete vertical fins with horizontal spandrel panels of varying overhangs, 900mm (at level 4) to 500mm (at level 22) act

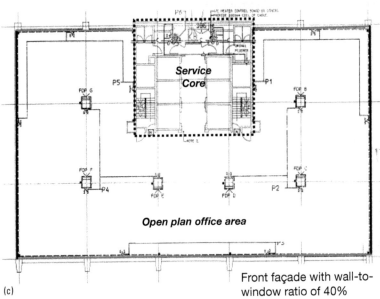

Front façade with wall-to-window ratio of 40%

(c)

Figure 2.4.1 (a) Critical case study building showing solid façade to the north-east, (b) street façade to the south-east, (c) typical office floor

as local shading devices for the glazed window bays of the front façade. The rear façade is composed of the same window types, a central solid wall and horizontal spandrel panel shading devices with 280mm overhang. The side façades are of solid textured concrete, facing towards the north-east and south-west directions. The building has a high volume-to-envelope surface area ratio of 7, confirming the deep plan configuration of the typical office spaces. Table 2.4.1 summarises the physical characteristics of the building.

Neighbouring buildings in front of the site and adjacent to two sides are typically 7–9 storeys and the rear façade is overshadowed by a taller 25-storey building.

Services

The HVAC system in the building is a variable air volume (VAV) system with terminal reheat. Each floor consists of 7–10 VAV boxes with 4.2–6kW capacity of electric reheat. Cooling of the building is provided by a central plant consisting of two water-cooled chillers of approximately 1.2MW capacity each. The chillers are combined with a common condenser and two roof-mounted cooling towers. The cooling towers are fitted with variable speed fans. The VAV terminals are programmed at a comfort set-point between 22 and 23°C for summer (December–February) and 21.5 and 22°C for winter (June–August).

Table 2.4.1 Physical characteristics of the Critical Case Input Model (CCIM)

Components	Description for CCIM
Location	Brisbane, Australia (latitude 27.4°S and longitude 153°E)
Weather data	Brisbane, IWEC, EPW format
Wind environment	Sheltered
Envelope	
External walls	200mm dense concrete with 10mm gypsum plaster inside and outside layers, $U=2.7W/m^2K$
Internal core walls	200mm dense concrete with 10mm gypsum plaster, inside and outside layers, $U=2.6W/m^2K$
Internal partitions	200mm dense concrete with 10mm gypsum plaster board, $U=2.6 W/m^2K$
Internal floors	200mm dense concrete with 20mm cement sand rendering, $U=2.3 W/m^2K$
Roof	250mm RCC dense with 20 mm Bitumen pure outside layer, $U=2.3 W/m^2K$
Ceiling	15mm suspended gypsum tile ceiling $U=2.5 W/m^2K$
Glazing	Bronze tinted outer pane, 50mm cavity with mid pane slatted blinds and 6mm clear glass inner pane. Total $U=2.7 W/m^2K$; SHGC=0.6; Solar transmittance (SC)=0.5
Local shading	Front façade (SE direction) – 0.9m overhang of aluminium-clad spandrel panels
	Rear façade (NW direction) – 0.3m overhang of aluminium-clad spandrel panels
Doors	External metal doors, $U=2.94 W/m^2K$
Infiltration rate	Leaky structure, 1.5 ac/h

Energy monitoring

Comparison of the electricity bills and aggregated sub-metering data for the building revealed an annual average electricity end-use breakdown for the HVAC system, tenant power usage (lighting and plug loads), common lifts and other miscellaneous energy uses as 46 per cent, 47 per cent and 7 per cent respectively (Figure 2.4.2). The total building annual electricity consumption has an energy footprint of 254 kWh/m^2 per annum. Detailed investigation of the energy usage pattern reveals a clear seasonal variation in the total energy consumption during the spring/summer (October–March) and autumn/winter (April–September) periods. The HVAC electricity consumption is higher in the summer months, accounting for 51 per cent of the total annual HVAC electricity consumption. In contrast, the lighting and plug loads are higher in the winter period with 52 per cent of the annual tenant electricity consumption attributed to this period. These results are intuitive and, as shown in Figure 2.4.2, the hourly HVAC and tenant electricity

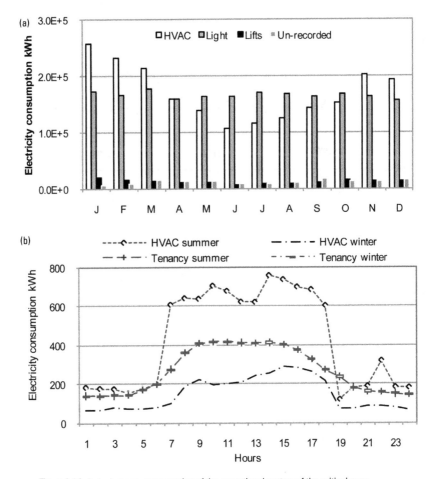

Figure 2.4.2 Actual energy consumption of the operational system of the critical case

consumption increases during the daytime from 7:00a.m. to 6:00p.m. corresponding to the typical weekday occupancy profile of the office building with a greater demand for cooling in summer, while greater need for lighting and small desk heaters account for the rise in tenant energy use in winter.

BASELINE ENERGY PERFORMANCE ASSESSMENT

Figures 2.4.3 (a) and (b) show the actual monthly electricity consumption and corresponding energy footprint corresponding to the building's NABERS ratings for three consecutive years from 2007 to 2009. The base building star rating of 0 in year 2007 gradually increases to a rating within 1 to 1.5 stars in 2009, while achieving only 1 to 2.5 stars for the whole building within the

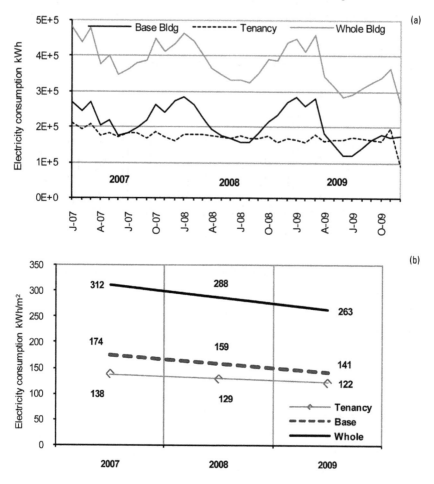

Figure 2.4.3 (a) Actual monthly electricity consumption and (b) corresponding energy footprint for NABERS rating for 2007–09

period of three years. The operational inefficiency and obsolete condition of the critical case are further established in respect to prime energy performance standards on greenhouse gas emission targets for Brisbane.

Baseline scenario

As discussed previously, the development of a baseline scenario is the first step of any retrofit project seeking energy performance improvements. The baseline scenario establishes the current energy performance status of the building and initiates the pathway for a calibrated simulation plan. The Base Building, Tenancy and Whole Building energy usage of 147 kWh/m²/annum, 126 kWh/m²/annum and 273 kWh/m²/annum define the physical baseline scenario of the critical case with a whole building energy rating of 2 to 2.5 stars (Figure 2.4.4(a)).

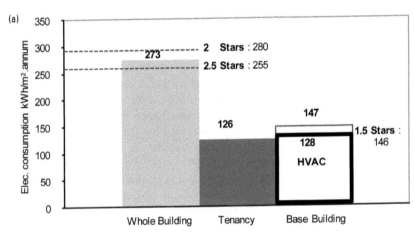

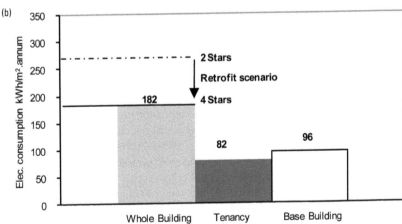

Figure 2.4.4 Energy performance and improvement scenarios of the critical case, (a) baseline, (b) retrofit

Retrofit scenario

The Section J of the Building Code Australia (BCA, 2009) recommends an annual electricity consumption of 163 kWh/m² for an office building in Brisbane and the Queensland Government's emission target initiative specifies a minimum energy efficiency rating of 4 stars for all commercial buildings by 2010 (Queensland Government, 2007). The high-energy footprint of the critical case provides evidence for retrofitting and requires energy performance improvements to sustain its future existence.

MODEL DEVELOPMENT

The steps in the model development are shown in Figure 2.4.5.

Critical case input model (CCIM)

The physical characteristics of the critical case are used as the input data for the development of the three-dimensional Critical Case Input Model (CCIM) of the building. The simulation modelling software uses the input data at different levels represented by building (site), blocks, zones and surfaces. The building level requires all site-specific information including the location, orientation, outside wind environment and weather data. The weather data file for Brisbane using the International Weather for Energy Calculations (IWEC) data in Energy Plus Weather (EPW) format was used for the simulation. The original weather data file of the World Meteorological Organisation (WMO) monitoring station at Brisbane Airport was used to generate an ASHRAE (American Society of Heating, Refrigerating & Air-conditioning Engineers) and IWEC weather data file (US DOE, 2009). Table 2.4.1 represents the detailed input data of the Critical Case Input Model.

The simulation model is generated based on the actual physical dimensions of the critical case building and each floor is represented as a block in the CCIM, and the model contains a total of 23 blocks (Figure 2.4.6). All typical office blocks (4th to 21st levels) contain an open office zone with core and perimeter thermal zones. The service core is a separate zone in each block and consists of different zones for lifts, amenities and fire stairs. The surfaces within each zone in the CCIM represent the same construction details and structural components of the actual critical case building. The 'layers method' was used in the creation of the building envelope of the CCIM for accuracy. Openings in the building are modelled at the surface level. All openings correspond to the physical construction details given in the architectural drawings and consist of double-glazed windows with mid-pane blinds. A typical office zone is composed of 18 surfaces and 50 openings.

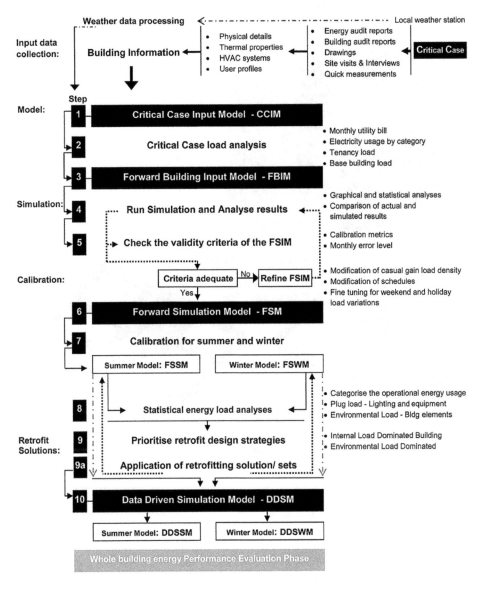

Figure 2.4.5 Whole building calibration simulation methodology

Forward building input model (FBIM)

The Forward Building Input Model (FBIM) is the operational phase of the Critical Case Input Model (CCIM). The main input data of the FBIM comprises activity, lighting, HVAC system, occupancy, and operational schedules. Table 2.4.2 represents the occupancy and operational characteristics of a typical office space of the FBIM.

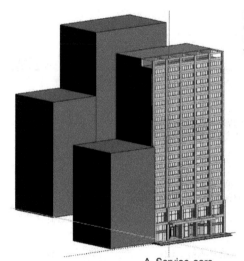

Figure 2.4.6 Critical case input model:
(a) view of the building façade and adjacent
buildings; (b) ground floor; (c) typical office
floor

A. Service-core
B. Reception & Display area
C. Coffee shot
D. Mechanical plant room
E. Open plan office

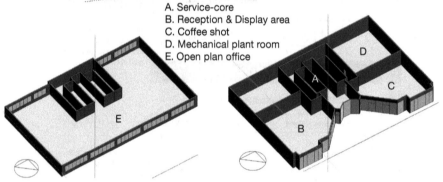

Table 2.4.2 Operational characteristics of a typical open plan office of the FBIM

Components	Description
Activity	
Metabolic rate	Light office work
Occupancy	10m²/person
Internal loads	
Lighting	14 W/m² in typical office and entrance lobby, 6W/m² in parking floors
Type of lighting	T8 Fluorescent, Triphosphor, high frequency
Target illuminance	320 lux
Computers	15 W/m²
Office appliances	5 W/m²
Ventilation rate	10/l/s per person
HVAC system	VAV system with terminal reheat
Chillers	2 of; each with COP 3.9; Capacity 1.2MWr, 4 AHUs each running at 700Pa pressure, each floor contains 7–10 VAV boxes
Ventilation rate	6.0 ac/h
Set-point temperature	23°C summer
	21.5°C winter

Each block within the model consists of different zone types, dependent on loads and the space conditioning method. The load dependent zones are based on the space function and space load characteristics. The conditioning method used for the zone types follows the HVAC drawings of the actual building. All typical offices (1st, 4th– 21st levels) and the entrance lobby, restaurant and front reception of the ground floor are modelled as conditioned zones.

Mechanical plant rooms, amenity and staircase areas, and parking floors (the 2nd and 3rd levels) represent the unconditioned zones. As opposed to commonly used core and perimeter thermal zoning types, broader categorisation of zoning with space loads and the conditioning method results in a more detailed FBIM which closely represents the operation of the critical case building. Operational schedules play a major role in the modelling of an existing building. An on-site survey of a mid-week workday and weekend was used in generation of the occupancy schedules. The basic operational schedules were obtained from the energy audit report. These were compared with the Building Management System data and on-site survey data. The updated profiles of hourly operational schedules for occupancy, lighting, appliances and HVAC for a typical weekday and weekend of the FBIM are detailed in Figure 2.4.7. An hourly, whole building simulation of the FBIM was performed for a reference year using the Brisbane IWEC data.

SIMULATION AND CALIBRATION

Comparison of the initial simulation results from the FBIM with the baseline energy consumption data is required to determine the accuracy of the model and to calibrate it. The simulated electricity consumption results from the FBIM were aggregated to represent the tenancy and building HVAC monthly electricity consumption. Figures 2.4.8 (a) and (b) show the simulated results of the FBIM in relation to the baseline monthly HVAC and tenancy energy consumption with an error range of ±20 per cent and ±5 per cent respectively.

The monthly Mean Bias Error (MBE) of HVAC energy consumption in the initial model was very high and varied from 25–48 per cent with an annual MBE of 37 per cent. The MBE of all months are therefore beyond the acceptable tolerance of the ASHRAE-14 guideline and the majority of the monthly error percentages are above the highest acceptable tolerance of calibration specified in the IPMVP. The monthly simulated electricity consumption for HVAC is lower than the actual electricity consumption and this highlighted the need to modify the estimated plant efficiency and simplified operational profiles of the HVAC system within the FBIM with more accurate data. The high percentage errors in the summer months, with a clear difference in the pattern of consumption between the model and the actual data, confirm the relevance of seasonal FBIMs with specific operational profiles and HVAC characteristics to represent the summer (October–March) and

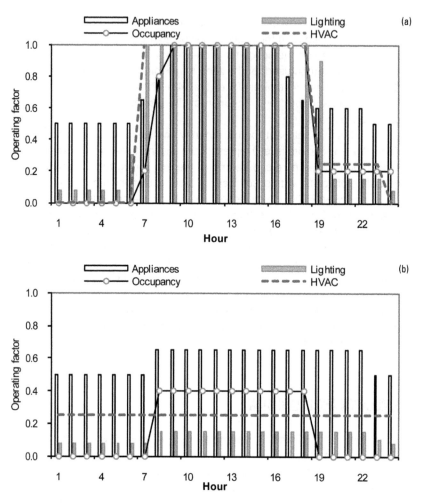

Figure 2.4.7 Hourly operational schedules of occupancy, lighting, appliances and HVAC for (a) a typical day in the week and (b) at the weekend

winter (April–September) periods. As shown in Figure 2.4.8, the simulated tenancy energy consumption shows less difference with the actual usage and the monthly error differences are within the range of 4–16 per cent with a mean annual error of 8 per cent. Higher simulated tenancy energy consumption verifies an overestimation of the plug loads and lighting levels with the simplified operational profiles.

The CV(RMSE) of 18 per cent for the simulated whole building energy consumption confirms that the initial FBIM is inappropriate to assess energy performance improvements and requires a case-specific and systematic approach, with a series of fine-tuning steps, to generate a highly calibrated simulation model with lower MBE and CV(RMSE) values. The error analysis reveals the need for fine tuning throughout the modelling process. The case-specific calibration approach prioritises the modifications to input data of the

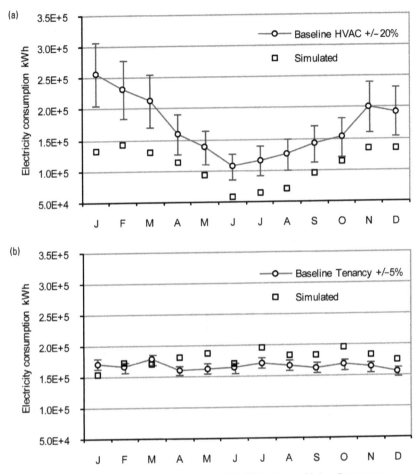

Figure 2.4.8 Comparison of initial simulated results of the FBIM with monthly baseline energy consumption and error ranges of (a) ±20 per cent for HVAC, (b) ±5 per cent for tenancy

HVAC system which will result in the development of a seasonal Forward Simulation Input Model (FSIM) for the summer and winter periods.

Forward simulation input models (FSIM)

Initial simulation results reveal the need for an iterative approach to fine tune and calibrate the Forward Simulation Input Model (FSIM). The input data for weather, lighting power density, the coefficient of performance (COP) of the HVAC system, set-point temperatures, mechanical ventilation rates, infiltration and operational schedules are adjusted and refined as an approach for calibration.

Several simulations were undertaken for sensitivity analysis and dozens of iterations of the FSIM were generated. The calibration effort provides evidence for significant diminution of monthly error values achieved through modification of overestimated system components, lighting power density, oversimplified operational schedules with incorrectly estimated off-time profiles and specific HVAC profiles for summer and winter periods. The finalised input data of the calibrated FSIM is:

- Chiller COP of 2.5
- Infiltration rate of 1.0 ac/h
- Lighting power density of 12 W/m^2
- Heating set-point of 21.5°C.

Figures 2.4.9 (a) and (b) present more detailed hourly workday operational schedules of typical offices with a modified HVAC profile to represent cooling and heating schedules.

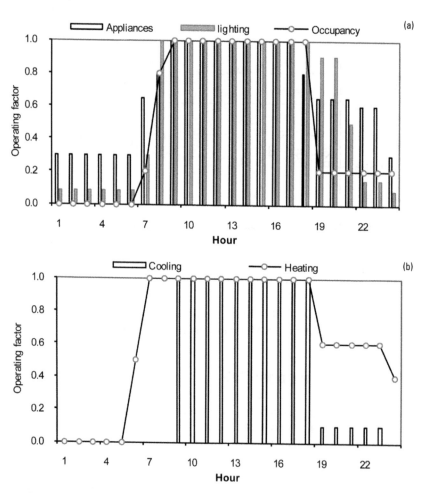

Figure 2.4.9 (a) More detailed hourly workday operational schedules for (a) occupancy, lighting and appliances, and (b) cooling and heating

Figure 2.4.10 shows the predicted results of calibrated FSIM in relation to the baseline monthly HVAC and tenancy energy consumption with an error range of ±10 per cent and ±5 per cent respectively. The overall MBE and CV(RMSE) for the building energy consumption were calculated and these values were within the acceptable tolerance levels and thus the FSIM model was considered calibrated for use in subsequent analysis.

Figures 2.4.11 (a) and (b) present seasonal models for the baseline and simulated values indicating the accuracy of the calibration.

Indicators of thermal behaviour

As an indicator of the thermal behaviour of the building, the percentage contribution of the different thermal heat loads to the overall building load is used in this study. The predicted thermal load values of the Forward

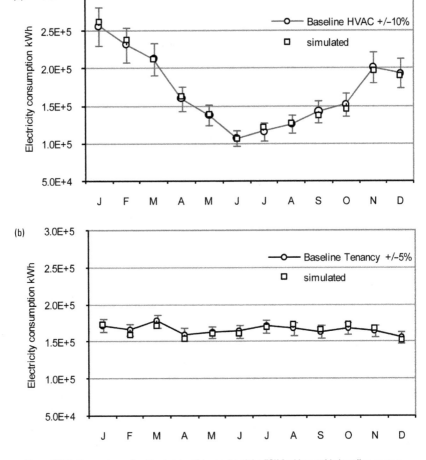

Figure 2.4 10 Comparison of calibrated monthly results of the FSIM with monthly baseline energy consumption and error ranges of (a) ±10 per cent for HVAC and (b) ±5 per cent for tenancy

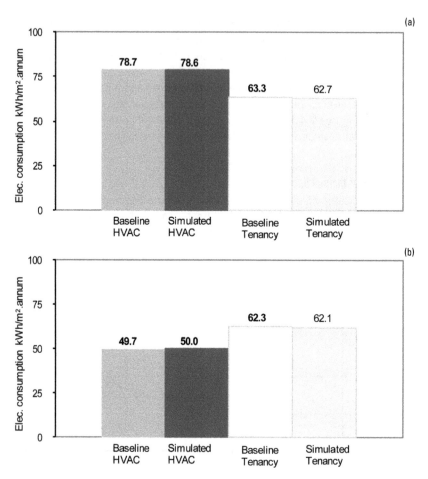

Figure 2.4.11 Comparison of calibrated results with baseline energy consumption of seasonal models, (a) the Forward Simulation Summer Model (FSSM) and (b) the Forward Simulation Winter Model (FSWM)

Simulation Model were used to determine the percentage of heat load contribution from internal and environmental loads for both summer and winter periods. There is a clear indication, as shown in Figure 2.4.12 (a), that in summer the internal loads due to internalities account for nearly 81 per cent of the thermal loads, while environmental loads due to externalities account for nearly 19 per cent of the total heat gain. While the environmental load is of less significance for heat gain in summer, the heat transfer through the total building envelope is 13 per cent and the heat transfer due to infiltration is 6 per cent. The heat gain through the walls is 8 per cent of the total load through the building envelope and 4 per cent is from windows and glazing. Thus the internal heat gain due to occupancy, lighting and appliances are responsible for the largest component of operational electricity consumption and the main contributor for cooling load in summer. The heat gain through

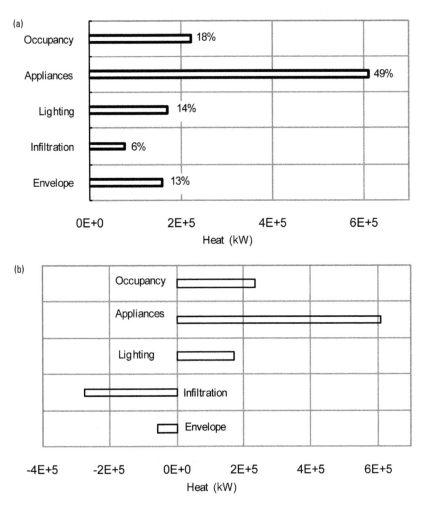

Figure 2.4.12 Thermal behaviour of the building: (a) Forward Simulation Summer Model; (b) Forward Simulation Winter Model

internal loads could promote energy conservation measures in winter if harnessed appropriately; however, heat loss through the building envelope and infiltration needs to be addressed as a retrofitting strategy. The higher cooling loads and consequent energy consumption characterise the critical case as an internal load-dominated, high-rise office for retrofitting.

BEST CASE AND PRIORITISATION OF RETROFIT SOLUTIONS

Critical case energy load profile

Office buildings such as the critical case study building with high occupant density and operational profiles that are above the benchmark standards possess a higher potential for being dominated by internal heat gains from occupants, appliances and lighting. The existing operational profile of the building has led to increased energy consumption for cooling. This places the building at the risk of becoming technologically obsolete long before its physical life has come to an end.

As an internal load-dominated building, this example signifies the importance of retrofit solutions to control internal heat gains, which demands the case-specific prioritisation of the selection of energy conservation measures. The roadmap for energy performance improvement consists of a hierarchy for retrofitting opportunities such as:

- First order decisions: non-technological and technological strategies to control internal heat loads.
- Second order decisions: strategies to control environmental heat loads.
- System improvements: non-technological and technological strategies to upgrade HVAC systems.

Figure 2.4.13 shows the retrofit scenario for performance improvement in relation to the extent of heat load types. The load type, cost effectiveness and practicality in implementation, ultimately determine the sequence of the application of the retrofit solutions.

The first order decisions to control the internal heat load start off the retrofit scenario with the application of non-technological solutions followed by several technological solutions. The bottom-up approach in application of the retrofit solution results in a cumulative retrofit energy saving at each step, which is represented by a retrofit solution set. The first order decisions in this example consist of five retrofit solutions that make up four retrofit solution sets. The second order decisions, which are less important in an internal heat load-dominated building, are primarily focused on heat gain controls to manage the environmental load entering through the building fabric. The four technological solutions proposed for this example represent the implementation of technological changes that comply with the Building Code of Australia (BCA, 2009) minimum performance standards.

The final stage in implementing energy conservation measures is focused on HVAC system improvements. These retrofit solutions are focused on cost-effective non-technological retrofit solutions such as increasing the cooling set-point temperature by 1°C and implementing a no-heating mode in winter periods. The technological solutions are focused on

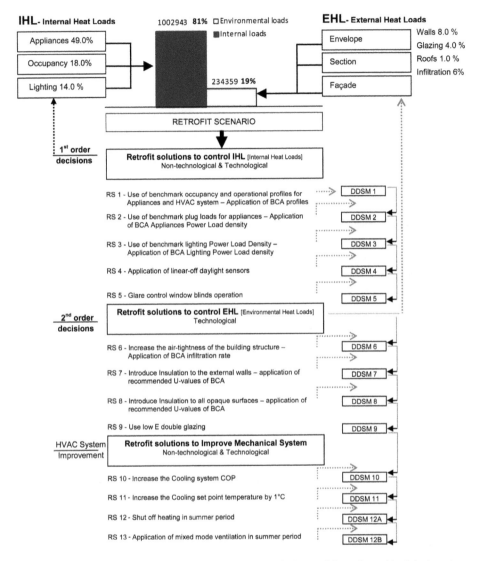

IHL- Internal Heat Loads

Appliances 49.0%

Occupancy 18.0%

Lighting 14.0 %

1002943 **81%** ☐ Environmental loads
■ Internal loads

234359 **19%**

EHL- External Heat Loads

Envelope

Section

Façade

Walls 8.0 %
Glazing 4.0 %
Roofs 1.0 %
Infiltration 6%

RETROFIT SCENARIO

1st order decisions

Retrofit solutions to control IHL [Internal Heat Loads]
Non-technological & Technological

RS 1 - Use of benchmark occupancy and operational profiles for
Appliances and HVAC system – Application of BCA profiles   DDSM 1

RS 2 - Use of benchmark plug loads for appliances – Application
of BCA Appliances Power Load density   DDSM 2

RS 3 - Use of benchmark lighting Power Load Density –
Application of BCA Lighting Power Load density   DDSM 3

RS 4 - Application of linear-off daylight sensors   DDSM 4

RS 5 - Glare control window blinds operation   DDSM 5

2nd order decisions

Retrofit solutions to control EHL [Environmental Heat Loads]
Technological

RS 6 - Increase the air-tightness of the building structure –
Application of BCA infiltration rate   DDSM 6

RS 7 - Introduce Insulation to the external walls – application of
recommended U-values of BCA   DDSM 7

RS 8 - Introduce Insulation to all opaque surfaces – application of
recommended U-values of BCA   DDSM 8

RS 9 - Use low E double glazing   DDSM 9

HVAC System Improvement

Retrofit solutions to Improve Mechanical System
Non-technological & Technological

RS 10 - Increase the Cooling system COP   DDSM 10

RS 11 - Increase the Cooling set point temperature by 1°C   DDSM 11

RS 12 - Shut off heating in summer period   DDSM 12A

RS 13 - Application of mixed mode ventilation in summer period   DDSM 12B

Figure 2.4.13 Hierarchy of energy performance improvement strategies in retrofitting an internal load-dominated building

HVAC system upgrades with more efficient components and incorporation of a mixed mode ventilation system in the summer period.

Data-driven simulation models (DDSM)

Each retrofit solution generates a Data-Driven Simulation Model (DDSM) for both the summer and winter periods. Thus, the retrofit scenario of the critical case presented here contains 24 DDSMs. The details of the DDSMs of the retrofit solutions to control the internal heat loads are as follows:

DDSM1 Operational profiles (occupancy, lighting and appliances)	RS1
DDSM2 – DDSM1 + Power load densities for appliances	RS 1+2
DDSM3 – DDSM2 + Power load densities for lighting	RS 1+2+3
DDSM4 – DDSM3 + Linear-off daylight sensors	RS 1+2+3+4
DDSM5 – DDSM4 + Glare control window blinds operation	RS 1+2+3+4

First order decisions – control of internal heat load

The first order decisions to control the internal heat load of the building characterise both non-technological and technological solutions. The retrofit solutions incorporate operational profiles and power load densities that are based on benchmark good practice performance standards as defined in Section J of the Building Code of Australia (BCA, 2009). Thus the simplest non-technological retrofitting opportunity is the use of benchmark standards for operational profiles. Figure 2.4.14 shows the weekday operational profiles for appliances, lighting, occupancy, and HVAC systems for the office building. During the weekend, lighting and appliances are assumed to be continuously operating for 10 per cent of a given 24-hour period (factor of 0.1), and the building is assumed to have no occupancy or HVAC system operating.

The sequence of technological solutions for internal heat load control is focused on improving the performance of the installed artificial lighting and appliances (plug loads), maximising the use of natural light and making effective use of window blinds to control glare and solar heat gain. The preliminary step is to increase the energy efficiency of lighting and appliances by reducing the power load densities to 9 W/m^2 and 12 W/m^2 respectively. It is important to maximise the use of daylight by incorporating daylight sensors, which will switch off artificial lighting when the ambient daylight levels are sufficient. The sensors, with linear off controls, dim the artificial lights continuously and linearly from maximum electric power consumption and light output to minimum consumption and output as the daylight illuminance increases and then switch off completely when the minimum dimming point is reached. As the daylight levels increase, the potential for discomfort glare increases for office work and a sensor will operate the mid-pane window blinds whenever the glare exceeds the maximum allowable levels.

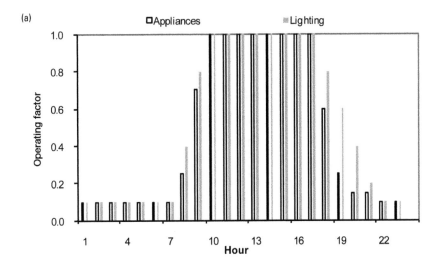

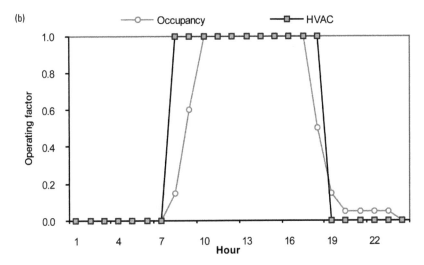

Figure 2.4.14 Weekday hourly operational profiles for office buildings in Brisbane, (a) appliances and lighting, (b) occupancy and HVAC

Performance improvements resulting from first order retrofit solutions

Figure 2.4.15 shows the energy performance improvements that resulted from the implementation of the first order decisions. The proposed first order retrofit solutions reduced the annual energy footprint of the whole building by 40 per cent. The revised whole building energy footprint of 163 kWh/m²/ annum met the set targets of the retrofit scenario by improving the NABERS star rating from less than 2 stars to more than 4 stars. This performance meets Queensland's 2010 emission target initiative for commercial buildings.

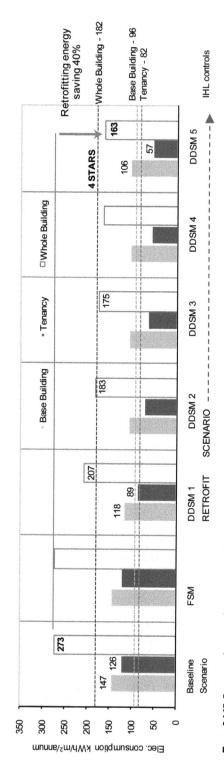

Figure 24.15 Energy performance improvements of the first order decisions: controls for internal heat load

The corresponding energy footprints of a 4 star rating whole building, base building and tenancy are 182, 96 and 82 kWh/m²/annum respectively. The first order retrofit solution set has improved the tenancy energy consumption to a rating of between 4.5 and 5 stars with an energy footprint of 57 kWh/m²/annum. The energy performance improvement confirms the potential of first order retrofit solutions in meeting minimum performance standards prescribed in Section J of the Building Code of Australia (BCA, 2009) for an office building in Brisbane.

Second order decisions: controls for environmental heat load

The second order decisions to control the environmental heat load of the building are focused on its physical structure and characterised by the components of the building envelope, section and façade. Heat gain through infiltration through the leaky structure and non-insulated envelope components represents the prime cause for environmental heat loads entering the conditioned space. Section J of the BCA (2009) recommends standards for the structure of the building to limit infiltration and environmental heat gain, and the retrofit solutions proposed are based on meeting these benchmark values. The sequence of application of control measures to address the environmental heat loads are improving the air-tightness of structure with the reduction of infiltration rate to 0.5 ach (air changes per hour), followed by the inclusion of insulation to the external walls. The R-values of the insulation included in the various models are 2.5 m²K/W. The model scenarios may then be summarised as:

DDSM6 – DDSM5 + Airtight structure
DDSM7 – DDSM6 + Insulated external walls
DDSM8 – DDSM7 + Insulated opaque surfaces
DDSM9 – DDSM8 + Low-e glazing

Figure 2.4.16 shows the energy performance improvements of that were achieved through internal and environmental heat load control. In comparison to the retrofitting energy savings achieved through the control of internal loads, the retrofitting solution set aimed at addressing environmental load accounts for only 3 per cent of the energy saving achieved through retrofitting. The energy footprint of the whole building with first and second order interventions is 155 kWh/m²/annum, which is a reduction of 10 kWh/m²/a from the solution set using only first order decisions. The base building energy consumption of 96 kWh/m²/annum achieves a NABERS 4 star rating.

HVAC system improvements

Next, interventions were aimed at improving the mechanical systems within the existing building and are summarised as follows:

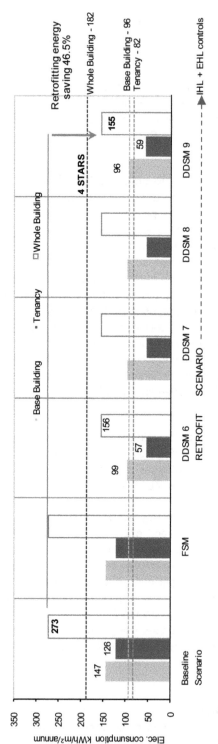

Figure 2.4.16 Energy performance improvements of the first and second order decisions: controls for internal and environmental heat loads

DDSM10 – DDSM9 + Increased cooling COP

DDSM11 – DDSM10 + 1°C increase of cooling set-point
temperature

DDSM12 – DDSM11 + no heating mode in winter

Improving the efficiency of the HVAC system to a Coefficient of Performance (COP) 5 further reduces energy consumption from 155 to 130 kWh/m^2/annum after the improved cumulative effect of the previous two sets of scenarios. Increasing the internal set-point temperature from 23 to 24°C saves another 9.9 per cent. Results for the no-heating mode in winter show a further reduction in HVAC energy consumption to 107 kWh/m^2/annum. This uses the internal loads generated by plug loads and other equipment such as lights for passive heating. However, the introduction of natural ventilation (DDS 13) has increased the energy consumption to 111 kWh/m^2/annum due to an increase in the heat loss in winter and the need for electric heating (Figure 2.4.17).

The research used a mix of thermal heat load control interventions to improve the overall energy sustainability of the building. Retrofit scenarios investigated show an interaction between the various interventions proposed and their impacts on manipulating thermal heat loads within the conditioned space. The decrease in required cooling load for all aspects of the building (including the base building and tenant areas) in the current climate are evident. It demonstrates that under current climatic conditions, the modelled building, utilising the proposed retrofit scenarios will require less cooling and heating through different stages of improvements to achieve energy efficiency, and finally that the building has the ability to move from a highly air-conditioned, energy-intensive building to a low energy, mixed-mode building.

CONCLUSION

This chapter demonstrates how the proposed retrofit approach may be used as 'a road map for energy performance improvements' in ageing office buildings utilising a whole building calibrated simulation methodology. It is essential that any ageing office building being considered for upgrading to higher energy performance levels with this methodology should have a proper diagnosis of the thermal load characteristics undertaken to confirm their behaviours, and compliance with minimum performance standards set out in building codes and regulations or other benchmarks selected by the project team. Once the thermal load characteristics are determined, retrofit scenarios and their potential contribution in manipulating heat gains and associated energy savings can be identified. Where internalities become the more dominant heat generators, the decision to look at interventions for internal heat gain control should be investigated as a priority. This chapter

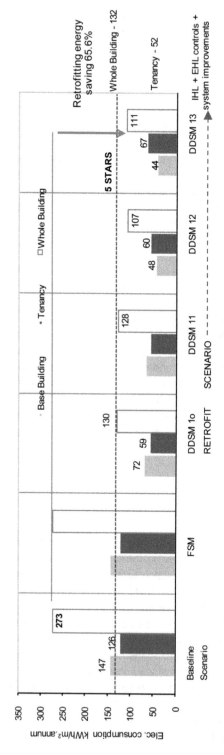

Figure 24.17 Energy performance improvements of the first and second order decisions and HVAC system improvements

addresses issues regarding providing an evidence base for decision-making in a project, particularly where local standards and benchmarks that are higher than the current performance are to be met.

ACKNOWLEDGEMENTS

The research work that underpins this chapter was supported by an Australia Research Council (ARC) Linkage grant LP0669628, 'Exploring Synergies with Innovative Green Technologies: Redefining Bioclimatic Principles for Multi Residential Buildings and Offices in Hot and Moderate Climates', initially from 2006 at the University of Queensland and subsequently at the University of Sydney from 2008 to the conclusion of this project in 2012. During this time the research work was actively carried out in collaboration with international partners Llewelyn Davies Yeang and T.R. Hamzath and Yeang, AECOM and ENSIGHT. The industry partners were joined in 2009 by INVESTA.

REFERENCES

ASHRAE Handbook (2005) *Fundamentals*, Chapter 32: Energy estimating and modeling methods, Atlanta, GA: ASHRAE.
BCA (2009) 'Energy efficiency provisions for multi-residential and commercial buildings', Building Code of Australia, The Australian Building Codes Board, retrieved from: http:www.abcb.gov.au.
Queensland Government (2007) *Climate Smart 2050: Queensland Climate Change Strategy: a low carbon future*. Available at: http://www.climatesmart.qld.gov.au (accessed 14 August 2009).
US DOE (2009) EnergyPlus Energy Simulation Software, US Department of Energy. Available at: http://apps1.eere.energy.gov/buildings/energyplus.

2.5
THE ECONOMIC CASE FOR RETROFITTING USING BIOCLIMATIC PRINCIPLES

How can existing buildings be designed for retrofitting around economic constraints?

Francis Barram

INTRODUCTION

Interest in the energy efficiency of existing building stock is high as it generates over 23 per cent of Australia's total greenhouse gas emissions with the commercial building's share generating over 10 per cent of all stationary greenhouse gas emissions (ABARE, 2006b). Since energy costs only contribute to 10 per cent of a building's operational costs and consume 2.5–3.9 per cent of a building's revenue, it is little wonder not a lot has been done to improve the energy efficiency of non-residential buildings (ABARE, 2006a). Yet due to obsolescence, a significant proportion of Australia's existing non-residential buildings will need to be renovated over the coming decades. It is estimated that at any one time 10 per cent of the building stock is undergoing renovation and it is this phenomenon that provides a point of leverage for introducing environmental renovations, that is, to use the opportunity to improve its environmental performance. Such renovations are called bioclimatic retrofitting, which not only includes the mitigation of climatic effects on energy use, but also the effects of the buildings' energy use on the climate. One barrier to this is the economic consequences of such strategies (Hyde *et al.*, 2006).

    This chapter aims to provide a useful input to considering the correct economic approach to the evaluation of improving the energy efficiency of this building stock. This case study demonstrates the benefits of using different economic analyses to achieve deep energy cuts in the energy use of existing non-residential buildings as a result of the implementation of a major renovation (see Chapter 2.4). This involves a comparative study to examine some possible energy efficiency scenarios that can be used to retrofit the case study building. Three methods are used to demonstrate some useful ways the data can be examined to facilitate decision-making:

1   the payback method;
2   the life-cycle cost approach;
3   the rental income approach.

The first section of the chapter discusses the background to economic considerations as a driver for bioclimatic retrofitting, including how economics can be modelled using the three approaches and discussion of the tool that is used to provide data for the modelling work. The second section focuses on the case study, a deep plan building located in Brisbane, Australia. This 30-year-old building is located in a subtropical climate and requiring significant cooling all year round.

ECONOMICS AND ENERGY EFFICIENCY

There are complex and interrelated economic considerations that influence the depth and extent of the implementation of energy efficiency in existing non-residential buildings. The existing Australian building stock was examined in a study by the Centre for International Economics, Canberra, and identified that the building sector as a whole could reduce its share of greenhouse gas (GHG) emissions by 30–35 per cent while accommodating growth in the overall number of buildings by 2050 (Centre for International Economics, 2007). This work built on a previous study by the Australian Business Roundtable on Climate Change (BRCC) on an economy-wide GHG emissions target of at least 60 per cent below 2000 levels by 2050 (BRCC, 2006). The majority of the predicted energy savings will be delivered by the existing building stock, when major building renovation occurs.

ABARE categorises the data of energy use and impact on the economy of the non-residential building stock under five headings, as detailed below (ABARE, 2006a):

1   Wholesale and retail
2   Communications
3   Finance & insurance, plus property & business services
4   Government, education, health & community services
5   Accommodation, cafés & restaurants.

This study focuses on office buildings, which are included in the categories 3 and 4. With almost a quarter (23 per cent) of Australia's total greenhouse gas emissions as a result of energy demand in the building sector, driven primarily by its end use of, or demand for, electricity (ABARE 2006b), this is a worthwhile project.

The energy savings in the case study office building are examined with 12 energy efficiency scenarios divided into three Sustainable Solution Sets: (1) technological and non-technological operational interventions; (2) internal heat load and environmental heat load control; and (3) direct savings on energy usage through efficient technology. The first targets improved the operational management of the building, which can simply be achieved with no or little capital investment; the second relates to improvements of the building envelope, which requires significant capital investment; and the third set of initiatives relates to capital investments in energy-efficient technologies.

However, what is 'energy efficient' is not just about reducing energy use without the context of the building being taken into account. Meier *et al.* (2002) state that there are three basic criteria for an energy-efficient building:

1   It must be equipped with efficient equipment and materials appropriate for the location and conditions.
2   It must provide amenities and services appropriate to the building's intended use.
3   It must be operated in such a manner as to have a low energy use compared to other, similar, buildings.

There is a significant push by all levels of government to address the failure in private markets to implement cost-effective energy-saving initiatives. The Federal government in Australia has passed legislation requiring all buildings over 2,000 m^2 that are leased or sold to provide a Building Energy Efficiency Certificate to the prospective leasers or buyers under the Building Disclosure Act. State and Federal governments are requiring agencies to buy or lease office buildings that have achieved minimum energy performance (Department of Environment, Climate Change and Water, 2011). Offices over 1,000 m^2 must meet a minimum 4 star and up to a 5 star NABERS office energy rating (ibid.).

These two initiatives aim to address the market failure of imperfect information to property owners and tenants. National companies and multinationals also have policies to only enter into leases that meet 4, 4.5 and 5 star ratings, depending on the business's policies. What also must be taken into consideration is the recent rapid increase in Australia's electricity costs and the further impact on electricity prices from the introduction of a carbon tax, and information market starts to get crowded.

Added to this, tighter government regulations and higher tenant expectations of sustainable energy use, there is now pressure on developers and building owners to significantly improve the energy performance of the existing building stock when it is renovated. These factors support the argument that energy-efficient commercial property has improved valuations over non-energy efficient property (Lutzkendorf and Lorenz, 2005). This suggests that energy efficiency initiatives that have a positive financial return can provide abatement benefits (Centre for International Economics, 2007).

ECONOMIC MODELLING

The energy usage of the building has been analysed using a carbon response methodology (Barram, 2011a), and a predictive energy-modelling tool (Barram, 2011b), *Energy Analyser*™. *Energy Analyser*™ enables the gathering of information on energy use related to every specific location and room and energy service type. It produces strategic management and operational management reports that provide technical and economic analysis to generate specific ideas and action plans for decision-makers and implementation staff or contractors. The energy savings for each scenario are estimated down to individual initiatives so they can be summated into a Sustainable Solution Set and compared. The approach allows for financial and technical evaluation for the 12 energy-saving scenarios. Some buildings are designed to be predominantly free-running while others are largely condi-tioned; also some are hybrids and run on mixed mode. *Energy Analyser*™ allows the user to determine which type of mode is to be modeled. This plethora of user operational modes leads to further complexity in how to rate and assess buildings; the latter favours thermal comfort criteria, the former energy use (Hyde and Watson, 2002). Thus, the energy-efficient methodology adopted allows the comparison between different buildings, as each has to meet the three elements of what constitutes an energy-efficient building.

This approach means that the energy use at the known level of the building (i.e. the block level for electricity) is led down to the lowest common node (rooms, in this case study) by pro rating the consumption, using different factors such as meters (Figure 2.5.1). The research has identified four different energy use models (Pitt and Griffith, 1996). The models of energy use are: (1) areas for heat load and lighting; (2) volume for conditioned spaces; (3) occupied volume for fresh air; and (4) heat loss in infiltration, convection and radiation.

Three assumptions underpin the technical and economic evaluation for this case study. First, capital cost estimates of energy savings initiatives are based on Rawlinson's documented unit rates for renovation projects or they are based on the engineering costing methodology (Rawlinson, 2010). Second, energy savings are predicted by *Energy Analyser*™, which are compared with that of a computer simulation building modelling to develop outputs to validate energy saving, and these are also compared with the

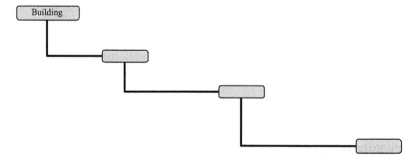

Figure 2.5.1 Area layout used by *Energy Analyser*™
Source: Barram (2011b).

meter data collected from the building. The energy cost savings are calculated by using energy savings predicted by *Energy Analyser*™ and the electricity costs paid for similar buildings. Electricity costs are based on current average costs, these are escalated at real 5 per cent a year and in 2013 electricity costs are to increase by 30 per cent to account for the application of the carbon tax.

CASE STUDY

These 12 scenarios are modelled to determine their contribution to achieving the deep energy use cuts required by the BRCC study. The analysis also goes to establish if the building can achieve a 5 star NABERS Energy rating, and at what economic cost, as this would require a cut of energy use of 65 per cent on its use in 2006. Deep energy cuts are what more and more organisations have agreed is necessary to meet the IPCC's 50 per cent cut to man-made CO_2 emissions before 2050. This is the reduction that scientists believe is necessary for the earth to avoid severe environmental problems from global warming. A consequence for the building sector is that there will be a widespread conversion of existing buildings to very low energy consumption and even zero-energy buildings (ZEB) are called for, as the necessary way forward.

Approach to modelling energy savings

The energy analysis is based on an office building located in Brisbane, which is proposed to have a major renovation (see Chapter 2.4). A major renovation is one that involves the building receiving a new façade and visible elevations; a total upgrade of lettable space; remodelling and upgrading of the core and components; modernising or new lifts; and full upgrading to state-of-the-art HVAC and energy efficiency systems. The challenge is that the building currently achieves a 1 star NABERS rating, which means that it could not be leased to government or national organisation tenants. The building owner would need to have the building achieve a 4.5 star NABERS rating or better to attract government tenants. The approach analyses 12 scenarios to determine their contribution to energy use improvements, as seen in Table 2.5.1.

Three best practice approaches are tested in this case study. One approach tested is lowest life-cycle costs for the energy efficiency level that minimises the building's lifetime costs. It is one of three approaches suggested by Meier *et al.* (2002) to create an 'energy-efficient' building. The other two are by negotiation and statistical. The negotiation approach relates to meeting an arbitrary standard determined by government or stakeholders through discussions, like the BCA or Green Star. The third approach, statistical, is based on meeting an energy efficiency standard based on the current rates of performance of existing buildings, like the NABERS rating.

This meaning of energy-efficient will be used in this case study, which means the property will not be left short to meet its energy service needs for lighting, climate control or services. Taking this broader perspective of what constitutes an 'energy-efficient' building, the 12 energy scenarios have been developed. A Scenario Title, Scenario Description of each scenario and the Scenario Design Approach used for each scenario are detailed in Table 2.5.2.

Table 2.5.1 Energy efficiency scenarios for the building in Chapter 2.4

Using technological and non-technological operational interventions

---

Scenario 1   FSM + Occupancy and operational profiles – BCA
Scenario 2   (a) Efficient appliances – MEPS
              (b) Increase of cooling set-point temperature by 1 degree
              (c) Advanced computer management system – for BCA, PH and ASHRAE

For internal heat load + environmental heat load control
Scenario 3   Infiltration – for BCA, PH and ASHRAE
Scenario 4   Insulation to external walls – for BCA, PH and ASHRAE
Scenario 5   Insulation to total opaque surfaces – for BCA, PH and ASHRAE
Scenario 6   Solar-e transmission glazing – for BCA, PH and ASHRAE

For direct saving on energy usage from technology retrofit
Scenario 7   Efficient lighting – for BCA, PH and ASHRAE
Scenario 8   (a) Daylight linear off sensors for artificial light – for BCA, PH and ASHRAE
              (b) Window blinds operation with photo sensors – for BCA, PH and ASHRAE
Scenario 9   Efficient HVAC (increased COP 5) – for BCA, PH and ASHRAE
Scenario 10   Mechanical services (Pumps and lifts) – VSD pressure and temperature control
Scenario 11   Mixed mode HVAC system (Winter no heating) – for BCA, PH and ASHRAE
Scenario 12   HVAC Fans & VAV diffuser system – for BCA, PH and ASHRAE

Table 2.5.2 Modelling approach to the 12 scenarios of the building modifications

SC	Scenario title	Scenario description	Scenario design approach
1	Occupancy and operational profile of Building Code of Australia (BCA)	Occupancy and operational profiles for appliances and lighting recommended by the BCA for office practices to the base case which is diagnosed with high cooling loads partially due to heat gain from current occupancy and operational profiles	Support efficient way of an operational management for the light and power services. Deploy Energy Management (on office areas to optimise energy consumption (i.e. lighting, administrative equipment or climate control)
2a	Efficient appliances with 15W/m² plug load density	Upgrade appliances that are identified as having poor energy efficiency and high heat generation therefore it is suggested to replace them with efficient equipment of 15W/m² as recommended in Building Code Australia	Investigate energy-efficient appliance replacement programme to lower heat gain generated from inefficient appliances (Minimum Energy Performance Standards (MEPS) Regulations in Australia used on benchmark)
2b	Increase of cooling set-point temperature by 1°C (24°C)	Increase the set-point temperature by 1°C in summer and decrease 1°C in winter.	Determine energy saving as a result of relaxing temperature set-point
2c	Advanced computer management system	Computer, printer and photocopier standby management enabled either in network or on individual machines.	Determine energy saving as a result of implementing a site-wide computer management approach with tenants.
3	0.5 ACH infiltration (airtight building envelope)	Reduce heat gain through infiltration is controlled through the improved air-tightness of the envelope reducing the infiltration from 1 air changes per hour to 0.5 air changes per hour.	Explore methods to reduce infiltration and heat gain from outside, reduce heat load by sealing windows and doors
4	Insulate external walls U = 0.3W/m²/K	Thermal performance of the external envelope, the external walls, of the building is improved through the installation of insulation with recommended U-values of BCA	Appraise the energy reduction of external wall insulation and recommend method to improve thermal performance of the building envelope.
5	Insulate all opaque surfaces.	Further improvement to the building envelope with insulation with BCA recommended U-Values to all opaque surfaces including internal walls, floor slabs and roof	Appraise reduction of internal wall, floor and roof insulation and recommend method to improve thermal performance of the building envelope

Economic outcomes of scenarios

Analysis of the lowest life-cycle cost of the energy efficiency level will be based on simple payback and net present value, to determine if the decisions are different. Simple payback is a method of evaluating investment opportunities on the basis of the time taken to recoup the investment.[1] Where net present value is the difference between the present value of the future cash savings flows from an investment and the amount of investment, a positive value indicates an acceptable investment and a negative value indicates an unacceptable investment. Present value of the expected cash flows is computed by discounting them at the required rate of return (also called minimum rate of return, which is set at 8 per cent real for this project).[2]

Then a financial and an economic method are compared. First, the financial method. It is a blunt and basic analysis of each energy efficiency initiative where it is needed to pay for itself. The second is an economic approach and is based on answering the question 'What does the building owner need to do to make the building rentable to their target market?', including analysis of the level of retrofit with and without efficiency measures. Then, separately, a broader economic question is asked, taking into account risk and marketing, asking the question, 'What level of renovation would allow the building owner to sign leases on the building after the renovation is completed as quickly as possible, and how to implement the energy efficiency?' This question is answered to determine the robustness of the questions and the investment sensitivity.

A summary of the 12 initiatives along with their predicted energy use and saving by scenario are shown in Table 2.5.3; the results of analysis of their payback and net present value (NPV) are also shown.

For payback, the basic economic decision-making criterion is simple, accept all initiatives that have a payback under 9 years or under. For NPV, accept all positive NPVs (this analysis uses a real discount rate of 8 per cent). For this case study the following scenarios in Table 2.5.4 would pass the decision criteria for the simple payback.

Based on the payback decision criteria of accepting all scenarios under 9 years, the following scenarios would fail the test (Table 2.5.5). The scenarios that have passed the economic test account for 78.7 per cent of the available energy savings identified for the building or a 51.2 per cent reduction in energy use in the building of the needed 65 per cent reduction. This would allow the building to achieve a 4.5 star NABERS Energy rating.

As can be seen from Table 2.5.4, for the financial analysis there is little difference between the outcomes of payback or NPV, as only SC-10 was rejected by the payback analysis that would have been accepted under NPV.

The economic question is 'What renovations will be undertaken on the building in a business-as-usual situation for the building owner to make the building rentable to their target market anyway?' This economic analysis is called marginal cost analysis. Marginal cost analysis is the analysis of the increase or decrease in investment from the business-as-usual decision-making framework, to a more efficient item.[3] The cost analysis looks at the

6	Solar transmission controlled glazing	Include low-e double glazing replacing existing 2 glazed panels of windows in order to control solar transmission from outside	Investigate window glazing technology and energy use reduction to reduce heat gain as a result of radiation
7	Efficient lighting with $9W/m^2$ lighting load density	Decrease the internal loads in the building by improving the efficiency of artificial lighting with BCA recommended fittings of $9W/m^2$	Explore and adopt a range of high tech lighting retrofits for the office building such as single T5 fluorescent replacements, compact fluorescent lamps (CFLs), light-emitting diode (LED) lighting where applicable to allow more efficient use of artificial lighting.
8a	Daylight linear off sensors	Assumed Scenario 3 completed, the existing lighting situation is wasteful with lighting installed in the passive zones of office floors close to the façades. Lighting energy is reduced further by the installation of photo-sensitive sensors to lighting	Explore and examine the utilisation of rewiring automated lighting controls to take advantage of natural sunlight and optimise artificial lighting, an example of this is daylight sensor control together with dimmable lighting controls on outside light rows
8b	Window blinds operation with glare-control sensors	Controlling radiation through glazing is known to be effective in controlling environmental gains. This scenario with photo-sensitive controls to blinds in window panes can promote a reduction of environmental loads	Investigate window lighting sensor control technology and energy use reduction to control heat gain as a result of radiation, improving heat both comfort and heat gain.
9	Efficient HVAC system with COP 5 with no heating	Improve the HVAC system efficiency to increase COP from 2.5 to 5.0	Investigate a water-cooled chiller product COP greater than 5
10	Mechanical services (Pumps and lifts)	On all pumps including low and high light domestic water pumps, condenser water, chilled water and condenser fan install variable speed drives, sensors and new pumps to improve efficiency	Investigate window glazing technology and energy use reduction to reduce heat gain as a result of radiation
11	Mixed mode HVAC system with winter no heating	Integrates ventilation during mornings with HVAC to work in mixed mode in summer and no heating mode in winter	Determine energy saving and as a result of mixed mode HVAC system to make use of cooler air from outside where applicable, install heat wheel AHUs
12	HVAC Fans & VAV diffuser system	Replaces the three-zone layout with new internal ductwork and create a zone with VAV diffusers and variable speed drives with pressure control on air handling fans	Investigate window glazing technology and energy use reduction to reduce heat gain as a result of radiation

Table 2.5.3 Modelling results of the 12 energy efficiency scenarios

SC	Scenario descriptions	Energy use BAU kWh p.a.	Energy use BP kWh p.a	Energy reduction (kWh p.a.)	Energy cost ($ p.a.)	Invest ($1,000)	Approach 1 Payback (years)	Approach 2 NPV ($)
SC-1	Occupancy and operational profile of BCA	355,840	142,336	213,504	25,620	100	3.9	414,089
SC-2	SC-2a Efficient appliances	1,271,960	391,040	409,588	49,151	5	0.1	967,392
	SC-2b Increase of cooling set-point temperature by 1 degree			80,292	$9,635	3	0.3	187,769
	SC-2c Advance computer management system			391,040	46,925	50	1.1	881,708
SC-3	Infiltration improvements	19,138	11,599	7,539	905	34	38.5	−14,331
SC-4	Insulate external walls	41,128	5,364	35,763	4,292	6,965	1,623.1	−6,364,846
SC-5	Insulate total opaque surfaces	84,401	53,079	31,322	3,759	11,166	2,970.8	−10,264,638
SC-6	Low emission transmission; double glazing	64,733	35,976	28,758	3,451	2,056	596.0	−1,836,197
SC-7	Efficient lighting	1,109,318	416,459	693,341	83,201	307	3.7	1,360,693
SC-8	SC-8a Daylight linear off sensors	64,680	38,808	25,872	3,105	80	25.8	−12,675
	SC-8b Window blinds operation with sensors			28,758	3,451	643	186.4	−527,309
SC9	Efficient chiller system	802,920	224,818	578,103	25,620	219	8.6	303,200
SC10	Mechanical services (Pumps and lifts)	333,000	208,300	124,700	14,964	185	12.4	555,956
SC11	Mixed mode HVAC system with winter no heating	347,000	34,700	312,300	37,476	200	5.3	124,638
SC12	HVAC Fans & VAV diffuser system	731,400	261,183	470,217	56,426	2,200	39.0	−921,133

Table 2.5.4 Energy efficiency scenarios that pass the simple payback test

SC	Scenario descriptions	Approach 1 Payback (years)	Approach 2 NPV ($)
SC-1	Occupancy and operational profile of BCA	3.9	414,089
SC-2	SC-2a Efficient appliances	0.1	967,392
	SC-2b Increase cooling set-point temperature by 1 degree	0.3	187,769
	SC-2c Advanced computer management system	1.1	881,708
SC-7	Efficient lighting	3.7	1,360,693
SC-9	Efficient chiller system	8.6	303,200
SC-11	Mixed mode HVAC system with winter no heating	5.3	124,638

Table 2.5.5 Energy efficiency scenarios that fail the simple payback test

SC	Scenario descriptions	Approach 1 Payback (years)	Approach 2 NPV ($)
SC-3	Infiltration improvements	38.5	14,331
SC-4	Insulate external walls	1,623.1	6,364,846
SC-5	Insulate total opaque surfaces	2,970.8	10,264,638
SC-6	Low emission transmission; double glazing	596.0	1,836,197
SC-8	SC-8a Daylight linear off sensors	25.8	12,675
	SC-8b Window blinds operation with sensors	186.4	527,309
SC-10	Mechanical services (Pumps and lifts)	12.4	555,956
SC-12	HVAC Fans & VAV diffuser system	39.0	921,133

additional cost of energy efficiency as the extra over cost over the business-as-usual investment (Figures 2.5.2, 2.5.3). For example, the building owner will be doing a complete HVAC upgrade, or in the case of this building $488,163 is the full cost of a new chiller, whereas if they will be purchasing a new chiller anyway, the marginal analysis would only count the extra over cost of an efficient chiller, not the cost of the total chiller. The marginal cost of an efficient chiller would be in the order of $100,000. On this basis, the following energy efficiency scenarios would pass the same payback and NPV tests (Table 2.5.6).

The scenarios in Table 2.5.6 have passed the economic test and account for 97.3 per cent of the available energy savings identified for the building or a 63.3 per cent reduction in building energy use of the available 65 per cent. This would allow the building to achieve a 5 star NABERS Energy rating. This increases the star rating from 4.5 star NABERS Energy rating for the financial approach to 5 stars, opening the market to more potential tenants. Of the energy efficiency scenarios that fail the payback and NPV shown in Table 2.5.7, only SC-8b, Window Blinds with Sensors is marginal with a slightly positive NPV, the other scenarios are all negative as well as having an excess payback period.

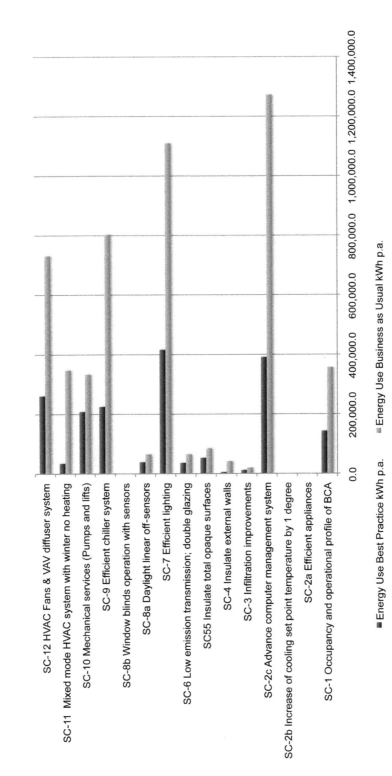

Figure 2.5.2 Energy use comparison: Business as Usual compared to Best Practice for critical case study

SC-12 HVAC Fans & VAV diffuser system

SC-11 Mixed mode HVAC system with winter no heating

SC-10 Mechanical services (Pumps and lifts)

SC-9 Efficient chiller system

SC-8b Window blinds operation with sensors

SC-8a Daylight linear off-sensors

SC-7 Efficient lighting

SC-6 Low emission transmission; double glazing

SC55 Insulate total opaque surfaces

SC-4 Insulate external walls

SC-3 Infiltration improvements

SC-2c Advance computer management system

SC-2b Increase of cooling set point temperature by 1 degree

SC-2a Efficient appliances

SC-1 Occupancy and operational profile of BCA

0.0   200,000.0   400,000.0   600,000.0   800,000.0   1,000,000.0   1,200,000.0   1,400,000.0

■ Energy Use Best Practice kWh p.a.   ■ Energy Use Business as Usual kWh p.a.

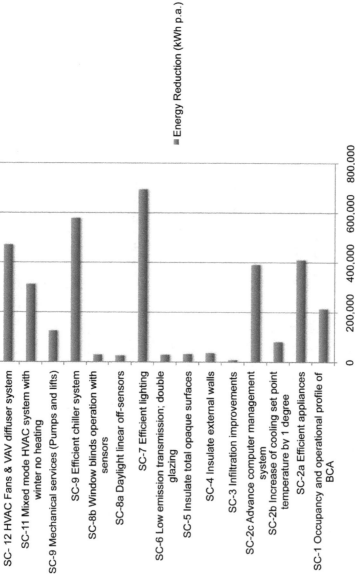

Energy Reduction (kWh p.a.)

■ Energy Reduction (kWh p.a.)

- SC- 12 HVAC Fans & VAV diffuser system
- SC-11 Mixed mode HVAC system with winter no heating
- SC-9 Mechanical services (Pumps and lifts)
- SC-9 Efficient chiller system
- SC-8b Window blinds operation with sensors
- SC-8a Daylight linear off-sensors
- SC-7 Efficient lighting
- SC-6 Low emission transmission; double glazing
- SC-5 Insulate total opaque surfaces
- SC-4 Insulate external walls
- SC-3 Infiltration improvements
- SC-2c Advance computer management system
- SC-2b Increase of cooling set point temperature by 1 degree
- SC-2a Efficient appliances
- SC-1 Occupancy and operational profile of BCA

0   200,000   400,000   600,000   800,000

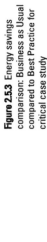

Figure 2.5.3 Energy savings comparison: Business as Usual compared to Best Practice for critical case study

Table 2.5.6 Energy efficiency scenarios that pass the economic test

SC	Scenario descriptions	Approach 1 Payback years)	Approach 2 NPV ($)
SC-1	Occupancy and operational profile of BCA	–	506,681
SC-2	SC-2a Efficient appliances	0.1	967,392
	SC-2b Increase cooling set-point temperature by 1 degree	0.3	187,769
	SC-2c Advance computer management system	1.1	881,708
SC-3	Infiltration improvements	2.8	15,577
SC-6	Low emission transmission; double glazing	1.4	63,617
SC-7	Efficient lighting	–	1,645,415
SC-9	Efficient chiller system	0.8	488,163
SC-11	Mixed mode HVAC system with winter no heating	1.7	272,786
SC-10	Mechanical services (Pumps and lifts)	0.5	722,622
SC-12	HVAC Fans & VAV diffuser system	6.2	791,830

Table 2.5.7 Energy efficiency scenarios that fail the economic test

SC	Scenario descriptions	Approach 1 Payback (years)	Approach 2 NPV ($)
SC-4	Insulate external walls	44.3	(91,054)
SC-5	Insulate total opaque surfaces	2,970.8	(10,264,638)
SC-8	SC-8a Daylight linear off sensors	25.8	(12,675)
SC-8	SC-8b Window blinds operation with sensors	18.6	8,691

SCENARIOS AND RENTAL RETURNS

Accepting the advances in technology and the current focus on sustainability in buildings, it is logical to question the impact of these changes on the financial performance of the building assets. Will the advancement in technology result in improved returns from the property assets (Boyd and Kimmet, 2005)? The answer to this question is open; however, the leasing and sale ground rules on office buildings are changing. For example, it is argued that a high energy rating (e.g. a 4–5 star NABERS office rating) on a building that otherwise conforms more or less to a standard building class gives it a market edge. There is some evidence that for public sector tenants at least, a fall in the rating during tenancy can actually trigger a diminution in rent. This suggests that a premium rent could be achieved based on an expectation of lower occupancy costs or a better working environment. These higher rents influence the capitalised value. The life-cycle cost approach seeks the efficiency levels that yield the lowest lifetime costs (taking both rental returns and energy operating costs into account). These calculations begin with a conventionally constructed building or appliance as a baseline, and the examination of trade-offs in initial investments in energy-saving technologies to energy costs (Meier *et al.*, 2002) (Figures 2.5.4, 2.5.5).

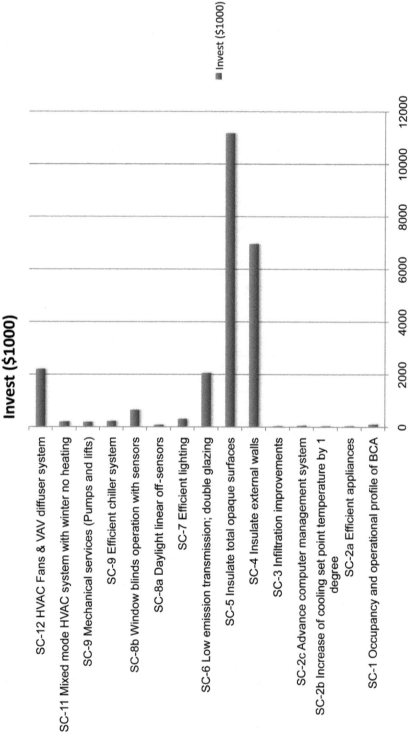

Figure 2.5.4 Investment for Best Practice retrofitting for critical case study

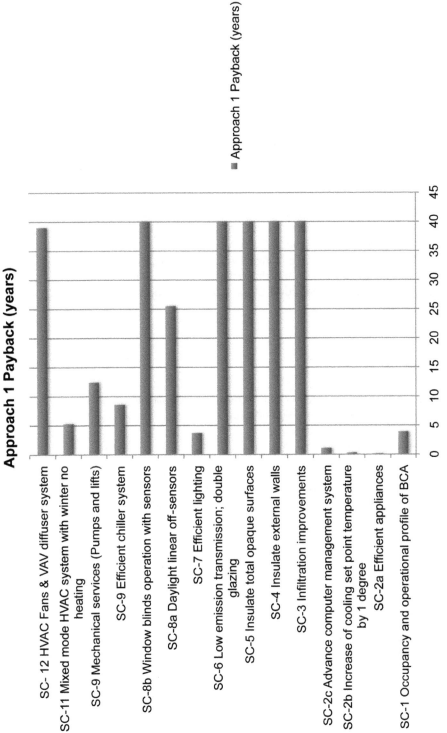

Figure 2.5.5 Payback for Best Practice retrofitting for critical case study

The CRC Construction Innovation project, 'The Evaluation of the Functional Performance of Commercial Buildings', found that a significant part of assessing functionality performance of built assets involved determining the measure of achievable sustainability relative to the market (Table 2.5.8). Technological advancement has the potential to produce high sustainability outcomes for buildings, but the cost of much of this innovation is difficult to justify to 'rational economic' investors (Boyd and Kimmet, 2005).

The question is, how big is the rent in relation to the 'energy-efficient' building we are targeting? As can be seen in Table 2.5.9, the biggest gains come from moving from a Class C building to a Class B building, with over A$30 million increase in present value income, while moving from a Class B building to a Class A building increases lifetime income by A$2.8 million. Now compare these to the investment required for the energy efficiency measures that pass the financial and the economic test. The financial test data suggest that it is viable to invest in a Class C building to achieve a Class B building but not a Class A, whereas using the economic test data suggests that it is viable to invest in a Class C building to achieve a Class A building. This is consistent with the 4.5 star NABERS rating for Class B and a 5 star NABERS rating for a Class A building (Property Council of Australia, 2009).

DISCUSSION

This case study highlights the difference between financial and economic evaluations and suggests that using an economic evaluation approach can

Table 2.5.8 Annual rental income for case study building depending on PCA Building Class obtained

Office class	Annual rent ($)	NLA m$^{2>}$	Annual gross rent ($)	Marginal annual gross rent ($)
Class A	519	15,742	8,170,098	267,614
Class B	502	15,742	7,902,484	2,817,818
Class C	323	15,742	5,084,666	

Source: Rawlinsons (2010).

Table 2.5.9 Energy efficiency classes that fail the economic test

Office class	Marginal annual gross rent ($)	Present value of rental ($)	Total investment in SC that passes ($)	
			Financial test	Economic test
Class A	267,614	2,856,720	5,361,860	480,500
Class B	2,817,818	30,079,576	5,361,860	480,500

achieve 0.5 star higher NABERS rating and 10 per cent more energy efficiency than if using the financial test approach. In addition, the study has
shown little difference in the energy efficiency outcome from using a simple
payback methodology over a more complete Net Present Value methodology. To achieve the deep energy cuts of 65 per cent in existing building
stock will require the application of economic analysis to justify these
decisions. As shown in this case study, the Australian economy will need to
make a substantial investment in converting energy-inefficient building to be
efficient and this can positively contribute to overall economic activity when
measured in terms of GDP and employment.

Based on previous work by the BRCC, the argument for a carbon-
constrained economy continues to grow, however at a slightly slower rate.
When the building sector's energy efficiency opportunities are explicitly
considered in this deep cuts scenario (that is, 60 per cent economy-wide
deep emissions abatement target), the Gross National Product (GNP) growth
rate is affected (Centre for International Economics, 2007). The case study
has shown that advanced energy efficiency retrofits in the non-residential
sector can make a significant contribution to reducing energy use and can
contribute to the growth of an economy.

CONCLUSION

This case study shows that additional research is needed on developing a
standard economic analysis test that could be used in government or asset
owner energy efficiency retrofit policy. The lack of this policy means that
energy reduction opportunities are lost. The chance to capture energy efficiency savings during major renovations in the Australian non-residential
building stock happens only very irregularly, so when the opportunity is overlooked, it is lost until the next major renovation.

NOTES

1   http://www.businessdictionary.com/definition/payback.html.
2   http://www.businessdictionary.com/definition/net-present-value.html.
3   http://www.businessdictionary.com/definition/marginal-cost.html.

REFERENCES

ABARE (Australian Bureau of Agricultural and Resource Economics) (2006a) *Australian
    Energy National and State Projections, 2029–30*, ABARE research report 06.26,
    December, Canberra: ABARE.

ABARE (Australian Bureau of Agricultural and Resource Economics) (2006b) *Energy Update: Australian Energy Consumption and Production, 1974–75 to 2004–05*, June, Canberra: ABARE.

Barram, F. (2011a) *Exploding the Carbon Myths; Secret Carbon Response Strategies to Beat the Carbon Tax*, Victoria: Global Publishing Group.

Barram, F. (2011b) *Energy Analyser: Predictive Precinct and Building Modeling Tool Manual*, Energy Coordinating Agency Training.

Boyd, T. and Kimmet, P. (2005) *The Triple Bottom Line Approach to Property Performance Evaluation*, Brisbane: School of Construction Management and Property, QUT.

BRCC (Business Roundtable on Climate Change) (2006) *The Business Case for Early Action*, Canberra: BRCC.

Centre for International Economics (2007) *Capitalising on the Building Sector's Potential To Lessen the Costs of a Broad Based GHG Emissions Cut*, Adelaide: Centre for International Economics.

Department of Environment, Climate Change and Water (2011) *How Is NABERS Being Used?* Fact Sheet. Canberra: Australian Government Printer.

Hyde, R. and Watson, S. (2002) 'A prototype E.S.D. home: towards a model for practice in solar sustainable design', in *Proceedings of ANZAScA*, Deakin University, Australia, pp. 201–8.

Hyde, R., Yeang, K., Groenhout, N. and Barram, F. (2006) *Exploring Synergies with Green Technologies for Advanced Renovation*, Canberra: Australian Research Council.

Lutzkendorf, T. and Lorenz, D. (2005) 'Sustainable property investment: valuing sustainable buildings through property performance assessment', *Building Research & Information*, 39(3): 212–34.

Meier, A., Olofsson, T. and Lamberts, R. (2002) 'What is an energy-efficient building?' In *Proceedings of ENTAC 2002-IX Meeting of Technology in the Built Environment*, Foz do Iguaçu, Brazil.

Pitt, T. and Griffith, G. (1996) 'An iterative approach to benchmarking in property', *The Cutting Edge*.

Property Council of Australia (2009) *Average Income and Operating Costs for Grade A, B and C Sydney Office Towers*. Available at: http://www.propertyoz.com.au/ (accessed 2 January 2012).

Rawlinsons (2010) *Australian Construction Cost Guide 2010*. Available at: http://www.rawl house.com/.

2.6
SUMMARY
Nathan Groenhout

The findings in Part II are concerned with the application of bioclimatic retro-fitting to commercial buildings. A critical case study was used to develop a possible, and validated, pathway for retrofitting within the context of meeting the impacts of climate change on our existing but ageing building stock.

We found that bioclimatic retrofitting essentially means that the retrofitting process of a building largely starts from the outside and works inwards. In simple terms, the biggest gains can be made via changes in the façade, while changes to the building services, the heart, lungs and nervous system of the building, are much cheaper but may have much lower long-term impact. However, pursuing this line of attack, it became apparent that a building has its own 'lifeline' or line of obsolescence, as it can be called. We found that the physical and functional obsolescence of commercial buildings starts from the inside, working outward to the façade and finally back inside to the structure. This is an obvious conclusion when we consider the service life of the elements that constitute the building. The building services elements typically have a service life of between 10 and 30 years, the façade between 25 and 50 years, while the structure has a service life of 50–100 years and beyond. So the path of bioclimatic retrofitting should follow the line of physical obsolescence and occur as the cladding and other external systems decay. Yet there are other forces impacting on the line of obsolescence such as the upgrading of the building to a higher performance rating and then to zero carbon emission status. This is a force of environmental obsolescence, and how this impacted on the line of obsolescence became unclear.

So, from a bioclimatic approach, it is no longer practicable to simply renovate as the systems become physically obsolete and wear out. Upgrade is needed to meet other, more profound objectives. In order to the meet these objectives, we must have a proper diagnosis of thermal load characteristics undertaken to understand and describe the building's thermal and energy behaviour. From an environmental sustainability perspective, once thermal load profiles and the fundamental forces contributing to those profiles have been determined, then design solutions for retrofit, with their potential contribution in manipulating the thermal loads, associated energy savings and compliance with minimum performance standards can be identified.

Hence, bioclimatic retrofitting has come about, not from the drivers of physical obsolescence but from the forces of environmental obsolescence as they apply to all buildings in this current age. It is a significant contextual change to the operational parameters of existing buildings and a number of conclusions have arisen from this thinking:

- Limitations to previous studies show a lack of understanding of the solution sets for retrofitting and methods of determining the energy performance behaviour of these strategies, in particular, the *internal heat loads*.
- The solution sets, such as those proposed by this research, are seen as interventions in the building characteristics and occupancy profiles in response to climate change effects, i.e. *increase of extreme temperatures* on energy performance and on subsequent *external heat loads*.
- A bioclimatic methodology utilises the different types of loads, internal and environmental, as indicators for energy performance and helps in quantifying the actual behaviour of the building, suggesting retrofit solutions and exploring their potential benefits.
- In warm climates, the selection of strategies and solution sets to control the environmental heat loads is a major imperative if energy efficiency is to be achieved.
- The response to energy performance and efficiency targets can be phased in over a period of time in a process of continued improvement. It is estimated that 60–70 per cent of total energy usage can be reduced through this process. Initially starting with the non-technological (or behavioural-based) solution sets involving building management, occupant behaviour and interaction, and then exploring the technical solution sets involving the form, fabric and mechanical services.
- Finally, the issues of changes to the selection of energy sources for a building's energy supply should be investigated and implemented, with the aim of addressing the current, unsustainable reliance on grid sources and its dependency on fossil fuel energy. A number of solution sets, such as gas-powered co-generation and the use of renewable energy sources, can be used to offset the remaining fossil fuel energy.

The reluctance of building owners to establish a meaningful pattern of monitoring heat loads in buildings has necessitated the development of sound arguments and evidence to support solution sets, such as those proposed in this body of work. Hence, an evidence-based design approach (EbDA) to retrofitting is recommended. The conclusions that can be drawn from an EbDA approach to sustainable retrofitting include:

- An important part of EbDA is fact-finding and identification of the sources of evidence; for retrofitting the sources of evidence are focused on the existing building. Data is needed in terms of the physical parameters of the project; inputs such as the weather data, mechanisms such as its form and fabric and ongoing monitoring of building performance.
- Identify a reference project or projects as a source of evidence. In this case, the existing building is identified as a critical case, i.e. its characteristics can be closely equated with other buildings of its type so that generalisations concerning the application of retrofitting strategies can be made.
- A process is developed for design to use the outcomes from fact-finding in the critical case. This process uses the monitoring data to test and calibrate simulation models and thus enhance the confidence in the model as a test vehicle to then explore the impact of the interventions proposed by the solution sets. The model is used to test retrofitting strategies through parametric changes across a range of critical variables through further simulation. We argue that this increases trust and credibility in the design decision-making process.

Principally, two main sources of evidence are recommended: a whole building calibrated simulation methodology, and an economic analysis model. The retrofit approach for a building can be charted as 'a road map for energy performance improvements'. From the work on these two modelling process areas, we have concluded that:

- It is essential that any ageing office building being considered for upgrading to higher energy performance levels using this methodology have a proper diagnosis of the thermal load characteristics undertaken, to confirm the building behaviour, and its ability to meet or exceed minimum performance standards defined in codes or regulations. Once the thermal load characteristics are determined, retrofit scenarios and their potential contribution in manipulating heat loads for improved environmental performance and the associated energy savings can be identified.
- Where internalities become the dominant heat generators, the decision to look at interventions for internal heat gain control should be prioritised. The study addresses issues about providing an evidence base for a project, particularly where local regulations, standards and benchmarks are to be met.
- It is important to address the highest heat generators in the initial retrofitting to have a significant impact on environmental performance and address any shortcomings in the energy performance of the building.

- Further research is needed on developing a standard economic analysis test that could be used in developing government and asset owner policy for energy-efficient retrofits.
- The lack of any current policy means that energy reduction opportunities are being lost. The chance to capture significant energy efficiency savings only happens during major renovations. In the Australian non-residential building stock, these renovations happen only very irregularly, so when the opportunity is overlooked, it is lost until the next major renovation.

TECHNOLOGICAL AND BEHAVIOUR CHANGE FOR PERFORMANCE IMPROVEMENTS

3.1
INTRODUCTION
Richard Hyde

Part III examines technological and behaviour change for performance improvements. Largely it is the general belief that the task of retrofitting is to replace existing technologies with 'new' green technology to improve energy performance. However, the improvements in the building performance have largely been directed at improving profitability through greater efficiency in the machinery used. The benefits of efficiency for environmental benefits have been largely ignored. A central question addressed in Part III is, what are the different factors that should be considered when engaging in sustainable retrofitting from design through to operation to renovation?

First, an important question is examined concerning the evaluation of typologies of commercial architecture for retrofitting, examining what the potential is for retrofitting existing commercial buildings (Chapter 3.2). Many types have been tried in the evolution of the commercial office building; each has its own set of challenges and opportunities and hence it is worth looking strategically at the range of types and evaluating which strategies are appropriate when applying both technical and behavioural strategies.

Chapter 3.3 examines issues around retrofitting thermal comfort and indoor environmental quality into existing buildings. A central issue is how improved levels of occupant comfort can be achieved with existing buildings. It can be argued that the objective measures of thermal comfort, based on the premise that comfort is guaranteed, has provided one of the drivers for high energy use in conventional commercial buildings. Changing the design brief now seems an urgent alternative and we should argue that the standard of comfort provided is determined by the energy performance that is expected. Hence the next three chapters (Chapters 3.4, 3.5, and 3.6) focus on how to determine performance in buildings. This provides information on benchmarking systems for retrofitting and examines how benchmarking can be harnessed for the purposes of retrofitting in the international and national policies context, and this may provide a new rating system for retrofitting (Chapters 3.4 and 3.5). Performance modelling using computer-based software tools is playing an increasing role in informing decisions in predicting building performance (Chapter 3.6).

Two chapters provide information on an alternative to simulation modelling which involves using data from the actual performance of the building or buildings obtained through monitoring. The authors argue that monitoring building performance can examine the environmental con-sequences of behaviour and technical change on performance (Chapter 3.7). How to measure and what to measure are important considerations (Chapter 3.8).

The remaining chapters provide a discussion of the technical and behavioural drivers for retrofitting. First, renovation and retrofitting projects should maintain the existing embodied energy in a building as far as possible. The benefits to society of this are huge; we cannot use our valuable energy resources to continually demolish and rebuild the building stock. Hence, how can embodied energy be saved when retrofitting existing buildings (Chapter 3.9)? For example, a large part of the embodied energy in a building is in its structure and this is often the most difficult part to recycle. So a central driver is to save the embodied energy in the retention of primary structural systems of our existing buildings, yet how this can be done is debatable and how to adapt them to future functions and uses is a pressing issue.

Second, retrofitting of sustainability into buildings can address far larger issues within society. For example, Chapter 3.10 on reducing peak loads in buildings speaks to this message, arguing that by implementing a number of measures in buildings we can affect the peak demand for elec-tricity and hence reduce our carbon footprint. However, as explained in Chapter 3.11, this requires new design processes such as the Penalty–Reward–Pinch (PRP).

The remaining two chapters discuss both the economic (Chapter 3.12) and bioclimatic passive low energy drivers for renovation (Chapter 3.13). Harnessing these drivers for change in buildings is central to a comprehensive sustainable retrofitting approach.

Chapter 3.14 provides a summary of this Part.

3.2
EVALUATING TYPOLOGIES OF COMMERCIAL ARCHITECTURE FOR RETROFITTING

What are the design and technical characteristics of existing commercial buildings built in the recent past and their potential for environmental improvements in the face of climate change?

Michelle Nurman

INTRODUCTION

This chapter aims to investigate commercial high-rise building as a type, focusing solely on office buildings and the typological characteristics that distinguish types of office building. It will explore the historical origins and development of the type and the current issues driving typological trends in commercial architecture. Five case studies from the Asia Pacific Region have been selected to exemplify established and emerging typologies of commercial architecture with a focus on highlighting the opportunities for appropriate bioclimatic solutions for the warm temperate climate of the Asia Pacific Region. Finally, it examines some of the opportunities and limitations for the environmental retrofitting of these commercial building types.

DEFINING COMMERCIAL ARCHITECTURE, OFFICE BUILDINGS AND TYPOLOGY

Commercial architecture can be defined as the design of buildings and structures of which 50 per cent or more floor space is for use for business, business being defined as a regular occupation, profession, or trade (Oxford Dictionaries online, 2011). This chapter will focus solely on office buildings.

A good definition of an office is a building where the end product is essentially a piece of paper (in its physical or electronic form) rather than a piece of machinery, equipment or hardware (Duffy *et al.*, 1993: 45). Type is defined as form or character that distinguishes a class or group of buildings (Curl ,1999). Typology is the study or systematic classification of such types (Houghton Mifflin Company, 2006).

Historical origins and development

This section will explore the functional, technological and cultural demands that historically have led to the most common office building characteristics and hence the formation of this building type and typologies.

Banks, stock exchanges and governmental buildings already existed in the Middle Ages. Specialised office buildings, however, did not appear until the Industrial Revolution, when clustering of work outside the home and development of mass production techniques led to the need for coordination and administration of these activities (Remoy and De Jonge, 2007: 2). Several historians such as Nikolaus Pevsner have regarded the office building as a subset emerging from these older building types such as warehouses and factories (Pevsner, 1976: 213).

Factories emerged with the development of printing and Pevsner claims the earliest was built in Nuremberg in 1497. Kohn (Kohn, 2002: 26) claims factory buildings were innovative in terms of construction (steel frames, long spans, fireproofing) and social reform and urban planning (in terms of modern planning of factory towns). The use of glass and metal curtain walls, originally used in factory buildings, became integral in the development of office building types. According to Kohn and Katz, the archetypal office building emerged in New York and Chicago around 1880 and by about 1900, the principal characteristics of the office building type were emerging.

Technological development

New technological developments such as the widespread use of steel frameworks, the invention of the elevator and curtain wall technology gave rise to new building types in the twentieth century. From the 1880s onwards, steel (and earlier cast iron) frameworks became more popular and liberated the high-rise building from the structural confines of load-bearing brick walls. Steel structures made it possible to develop larger structural spans and the light and compact nature of the material meant less space was needed for the structure and more floors could be stacked on top each other (Remoy and De Jonge, 2007: 164). From the Industrial Revolution onwards, the mass production of building materials made construction cheaper and led to modular building systems, and prefabrication.

From the late 1800s, the United States pioneered glass and iron façades, heralding the concept of large office floor plates enclosed in an envelope (Collard and De Herde, 2004: 9). Technologies in steel and concrete enabled a separation to be made between the external envelope and the main building structure and the curtain wall was born.

Before the Industrial Revolution, the need for natural lighting conditioned the form of the buildings and the design of façades in order to achieve sufficient daylighting. Artificial lighting transformed this relationship. The development of fluorescent lighting, for example, gave at least a fourfold increase in lighting efficiency and reduced the imperative for the designer to design for daylighting. Together with developments in mechanical services, it made it possible to establish a generation of fully artificial and simulated internal environments, with deeper plans and building orientation and façade characteristics independent of external conditions.

The economic growth in the post-Second World War period and the growth of service industries fuelled the economisation of office spaces as property investments in the 1960s with a trend towards speculative office building development. This led to standardisation of the structure, the interior of office building and optimisation of occupant density.

Changes in office planning

At the beginning of the twentieth century, the open planning concept became popular, based on the works of Frederick Taylor, who developed a vision of administrative work based on more efficient organisational methods borrowed from factories such as division of labour and tighter hierarchical control (Myerson and Ross, 1999: 8). While Taylor's design discouraged socialising among workers, the German 'office landscape' (*Burolandschaft*), attempted to encourage it. Developed in the late 1950s by Eberhard & Wolfgang Schnelle, it asserts that internal office design should reflect working relationships among employees and the flow of paperwork, information and staff, and encouraged communication (Curl, 1999). Other office design concepts made popular in the 1960s were the action office and group and combi office layout.

According to Duffy (1997), two schools of quite contrasting office typologies can be defined in relation to the stereotypical North American and North European office building models (Table 3.2.1).

Table 3.2.1 A comparison of North American and North European office models pre-1980s

Factors	North American model pre-1980s	North European model pre-1980s
Procurement design approach	Speculative and market oriented. Efficiency in cost and construction, Standardisation of floor plates, structure and services. Independent relationship between external shell (domain of landlord) and interior (domain of tenant) Taylor's and Ford's scientific management in office layout planning	Custom design, user oriented User needs focused style and configuration to suit user Integrated shell and scenery Architecture to comply with stricter statutory regulations (which reflect social democratic climate of Northern Europe). Users are in a better position to negotiate quality of work environment. High level of amenity and environmental standard
Architectural characteristics	High rise and high density urban skyscrapers Orthogonal office grid Deep floor plate Hierarchy of open plan and cell offices (Taylorism) and later the 'action office' style of planning Hermetically sealed artificially controlled buildings	Low rise and suburban Organic free form Geometry Narrow floor plate High degree of cellularisation i.e the combi office Naturally ventilated or hybrid buildings
Experiential characteristics	Separation between inside and outside	Direct access to views and daylighting
Typical examples	The Seagram Building (1958) by Mies van Der Rohe and Philip Johnson	Central Baheer Offices, Apeldoorn, the Netherlands (1970–72) by Herman Hertzberger

Source: Duffy (1997).

EMERGING TRENDS AND DRIVERS

Dramatic changes in information technology and increasing awareness of environmental issues are changing user demands and the way work and businesses are conducted. These changes are precipitating the need for new typological models of office buildings which combine the best characteristics of the North American model (maximum efficiency) and the North European model (maximum effectiveness) and more (Duffy, 1997: 46). New office buildings need to take the best of current practices while being capable of adapting the ever-changing twenty-first-century trends such as the Information Technology Revolution as well as being flexible, healthy, amenable, cost- and space-efficient and low in energy consumption.

Information technology revolution

A key development that has affected the typological development of office buildings is the automation and computerisation of office processes and the dramatic developments in information technology in the last quarter of the twentieth century. The popularisation of mobile phones, laptops, the internet and email means workers are less dependent on office buildings as a geographical location (Remoy and De Jonge, 2007: 16).

This has several implications. The technology information revolution has created the opportunity for employees not to be bound physically to one workspace and to be able to work outside the traditional 9–5 working hours. However, equipment-intensive spaces combined with high rates of occupancy have generated the issue of major internal loads. In addition, more flexible work schedules have also meant that management of heating, cooling and lighting loads in relation to occupancy rates at different times of the day is more difficult to predict. Developments in service zoning help in isolating and keeping under control the heat problems.

The information technology revolution has opened up the opportunity for employees to enjoy more of a balance between home, work and leisure. It made working from home much more feasible. Increased automation as well as wireless communication has also allowed routine clerical tasks to be automated or exported off-site to cheaper, suburban locations (Duffy, 1997: 57). There has been a trend to remove peripheral service functions such as property and facilities management and to outsource these services. This trend enables businesses to focus on their core business and motivates them to create an environment that supports creative, open-ended work with a wider range of work settings that support team work, sociability, exchange of ideas and encourage flexibility.

Climate change, energy and climate control

The developments of mechanical, heating and cooling technologies and the abundance of cheap energy particularly in the second half of the twentieth century meant there is a generation of commercial buildings that were able to overlook factors such as environmental and climatic issues for the sake of cost effectiveness (Hyde, 2008: 8). This is particularly true of high-rise buildings, with their significant ceiling heights and large floor areas which were conventionally mechanically air conditioned in order to achieve optimal, homogenised thermal comfort for their occupants all year around. The large expanse of poor performing façades traditionally also led to considerable energy loads to keep the internal environment comfortable.

These fully controlled artificial environments minimised opportunities for occupant control and at the same time created higher expectations of comfort (Collard and De Herde, 2004: 21). The phenomenon of Sick Building Syndrome (SBS), which has become common, has been particularly associated with deep plan air-conditioned buildings and undifferentiated lighting quality.

Climate projections

Climate researchers Steve Turton and Joanne Isaacs of James Cook University in Cairns have claimed that the globe's tropical zone has expanded between 2–5 degrees latitude over the past 25–30 years in both hemispheres. To illustrate, at this rate, they claim Sydney's climate at the end of this century will be more like Brisbane is today (Salleh, 2009). The effects of climate change mean cooler days are on the decrease. Most importantly, the increased temperatures will create additional environmental loads on current buildings as higher cooling loads will be required to keep the buildings thermally comfortable. Particular strategies are needed to address this.

A bioclimatic design approach

The bioclimatic design approach has great potential to solve these issues. It uses passive low energy techniques which relate to and are borne by the site's climate and meteorological conditions. This results in a building that is climate-responsive and environmentally interactive with reduced energy consumption in operation and embodiment (Law, 2004: 20). It focuses on providing high quality passive design of buildings through a high performance envelope, form and fabric (Yeang, 1996: 22).

A bioclimatic architectural approach gives a convincing triple bottom line. Ken Yeang (ibid.) claims the most convincing justification for the bio-climatic design for high-rise buildings is perhaps an economic one. The savings from lower consumption can be as much as 30–60 per cent of the overall life-cycle energy costs of the building.

Another justification is that bioclimatic design affords a more human experience of the high-rise building. Bioclimatic design promotes better natural ventilation and awareness of place, resulting in healthier internal environments and thereby increasing overall business productivity.

A further and important justification is the ecological rationale. Through the use of passive devices to achieve thermal comfort, bioclimati-cally designed buildings have lower total carbon dioxide emissions, which in turn lower overall air pollution, which helps minimise global warming.

This philosophy necessitates knowing the climate well and addressing the changes that climate change will bring as they arise. Higher ambient temperatures due to increasing external temperatures will need appropriate passive strategies to address the increased cooling load (Hyde *et al.*, 2009: 2). Strategies that work well to harness temperate climatic conditions such as those in Sydney, Australia, will be of moderate effectiveness as Sydney's climate gets hotter. In addition, future regulation will require higher standards of energy efficiency from buildings. Therefore, the static 'best fit' approach to façade design will fall short of the optimum performance that can be achieved relative to the changing climatic conditions outside buildings. Passive design may need to be married with intelligent materials and dynamic façades to improve adaptability (Strelitz, 2005: 92). Building Management Systems (BMS) can also play a role in the efficient management of the building's operations.

New patterns of location

These trends open up possibilities in the way office buildings are designed and geographically located. The need to have major office headquarters in expensive city centres has been rendered illogical and expensive (Duffy, 1997: 97). Architectural trends in advanced economies are towards fewer, smaller, custom-designed office buildings as well as towards recycling and converting existing buildings, from an economical and ecological point of view.

The need for flexibility

Statistics show that the average firm moves to new offices every seven or eight years and modifies the spatial configuration of 25 per cent of its workstations every year (Collard and De Herde, 2004: 5) due to a range of issues, including those mentioned. This suggests the new office buildings should therefore be regarded as a framework, so that upgrades and new components can easily be achieved (Law, 2004: 33). The building shell design should be sufficiently flexible to accommodate these changing factors and be designed to facilitate sub-tenancies on all but the smallest floor plates.

THE POTENTIAL FOR ENVIRONMENTAL IMPROVEMENT

Typologies and type are well discussed in architectural circles. Eugene Kohn relates office building types to the forces of finance, plan, programme and design (Remoy and De Jonge, 2007: 163). To this we should add the variables of bioclimatic architecture, which are climate and biology. These forces often dictate whether the project is speculative or custom-designed.

These forces drive the typological characteristics of a building, including its spatial and technological features such as the envelope characteristics, building orientation, form and depth, core location and the building operational mode (whether it is Passive mode, Mixed mode, Full mode, Productive mode or Composite mode) (Yeang, 1996).

Mechanical systems have had a significant impact on the typology. Called energetic impacts, the availability of cheap energy and the rapid increase in the development of mechanical systems have changed the typology significantly. Furthermore, the majority of post-war office buildings have not completed their anticipated life-cycle and typically are beset with higher energy requirements than those defined by current standards. The tall building over and above other built typologies uses at least a third more energy and material resources to build, to operate, and eventually to demolish (Richards, 2007: 20).

There is a real opportunity to retrofit this building type, in particular, to meet emerging needs for sustainability (Hyde *et al.*, 2007: 14). In addition, there is real value and also real opportunity in rebuilding and renovating existing buildings due to rapid building obsolescence and rapid completion of invested costs in buildings. The building shell such as the structure and the façade, services and scenery all have different points of obsolescence. A typical office building shell has a lifespan of 50–75 years while the services have a lifespan of 15 years and the scenery of 5 years. This paints a strong case for upgrading elements with shorter lifespans to maximise the life-cycle of the long-term shell (Duffy, 1997: 74). Four types of commercial building are now examined with the view to assessing the opportunities and limitations of retrofitting the building types to adapt to market demands in energy efficiency as well as changes in legislation brought about by climate changes.

The bioclimatic tower

A bioclimatic approach to architecture begins with an analysis of the meteorological data and site. Understanding this, an effective bioclimatic building maximises the benefits of climate appropriate passive systems through its envelope, form and fabric. Strategies encompass optimising passive thermal, daylighting and ventilation strategies.

CASE STUDY 1

An example is the Menara UMNO (Figure 3.2.1). Designed in 1998 by T. R. Hamzah & Yeang, this 21-storey tower located in the tropical climatic zone of Kuala Lumpur is designed for one of Malaysia's major political parties. It has an optimal long narrow floor plate utilising the restricted narrow site and the climatic conditions. All office floors (although designed to be air-conditioned) can be naturally ventilated. No desk location is more than 6.5 metres away from an operable window.

The core is located at the south-eastern wall providing a buffer zone and exposed thermal mass which mitigates solar gain. The west/north façade is shielded with solar oriented horizontal shades. Vertical fins that serve as

(a)

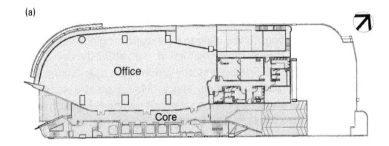

Figure 3.2.1 Bioclimatic tower: the Menara UMNO: (a) typical floor plan; and (b) view from the south-west

Source: T.R Hamzah & Yeang International (n.d.).

wind 'wing walls' on the south-east side to deflect and direct wind have been designed to capture wind data analysis. These winds are directed to balconies on every floor, which capture the wind into the building through active openings, to introduce natural ventilation into the buildings. All the lift lobbies, staircases and toilets have access to natural sunlight and ventilation, reducing operational costs.

Conventional high-rise building

CASE STUDY 2

'Stockhome' (2007) is the head office for the Stockland property company, housed in 10,000 m² over eight floors of space refurbished by Bligh Voller Nield in the Sydney CBD. The building was an opportunity for the company to highlight the benefits of retrofitting. The original 31-storey 1980s tower was a conventional speculative office building with a centrally located core elongated on the north and south axis, medium to deep floor plates with wraparound tinted glazing which had very poor light transmission at 30 per cent and no external shading, contributing to poor daylighting and high levels of heat gains (Figure 3.2.2). A major feature of the refurbishment is the insertion of a new 8.5m x 5.5m eight-storey atrium with open stairs. The atrium is staggered to maximise views and provide a greater sense of connection. Around the void's edge are a variety of communal spaces such as meeting rooms and tea benches. The façade glazing adjacent to the atrium

(a)

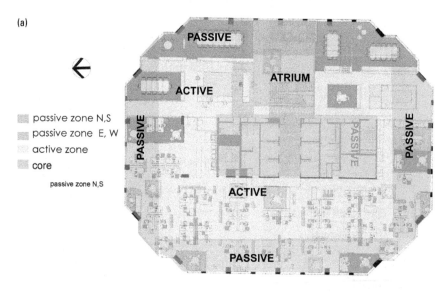

passive zone N,S
passive zone E, W
active zone
core

passive zone N,S

(b)

Figure 3.2.2 Conventional high-rise: typical floor plan

Source: Ferguson (2008).

was replaced by a high performance glass with 60 per cent visual light transmission. The services were also upgraded through installation of a trigeneration system with improved air conditioning, providing 30 per cent of the base building power consumption.

The ramified structure

The ramified structure is an alternative to the typology of the vertical high-rise office tower, characterised by its more lateral and sprawling nature (Collard and De Herde, 2004: 25). It is a groundscraper, as opposed to a skyscraper, and its origins were in the campus office parks of the 1950s such as that by Saarinen (Drake and Graham, 2005) and the Cigna building by Skidmore, Owings & Merrill. These were typically located in the suburbs, avoiding the congestion of the city and its larger surface area maximised the ability of each inhabitant to enjoy a view of the surrounding landscape. The higher degree of freedom the site often affords is an opportunity for the companies to expressly display their desired corporate image and the building is often integrated in shell and scenery construction.

The lateral nature of this typology allows for cohabitation on several levels and thus promotes a spirit of community with many intermediate areas to facilitate the transition between levels such as atriums, open stairs, shared breakout spaces, and internal 'streets' to promote equally informal and formal interactions.

CASE STUDY 3

NAB Headquarters is a seven-storey campus-style office building for 3000 people located at inner city Docklands, Melbourne (2005) designed by Bligh Voller Nield. The plan consists of four elongated floor plates running roughly north–south in a W formation. Between each pair of floor plates is an atrium, which is crossed by walkways at various angles, their connection to the floor plates defined by shared support spaces (Figure 3.2.3 (a)). Each atrium is penetrated by an open stair, encouraging vertical as well as lateral movement through the common spaces (Drake and Graham, 2005: 62). End service cores maximise the long-term flexibility of the open floor plates. A variety of sun rooms, winter gardens, meeting rooms and open plan offices and breakout areas create a variety of settings that support both formal and spontaneous work and interaction (Figure 3.2.3 (b)). The building features maximum daylighting, high performance integrated curtain and sustainable timber.

The horizontal bar

This linear office structure is often located in lower density locations such as in suburban business parks. Many of its drivers are similar to the ramified

(a)

Figure 3.2.3
The ramified structure:
(a) typical floor plan; and
(b) view towards one of the
atriums

Sources: (a) Rykwert (2005);
(b) Bligh Voller Nield.

structure. Its form is borne out of a desire to open up to the external environment, utilising the building's narrow floor plate to take advantage of the
natural light. The suburban location of such structures usually promotes the
development that is evocative of certain urban facilities with elements such
as cafés, libraries and recreation spaces, internal streets and a central plaza
(Collard and De Herde, 2004: 26).

CASE STUDY 4

The NSW Rural Fire Service Headquarters (2002) designed by Woodhead International is a three-storey custom-designed office building located in suburban Homebush Sydney, accommodating up to 400 personnel (Figure 3.2.4 (a)). It is located in a low density suburb surrounded by light industry and other free-standing office buildings. The building is linear and elongated on the east–west axis. It is designed around two internal atriums which minimise the floor plate depth to 12m. Visual connection and openness in relation to the organisation's processes and the integration of the depart- ments are promoted through the full height atrium and internal stairs and the flowing linear order of the building both in the vertical and horizontal cir- culation (Figure 3.2.4 (b)). Anticipating the 24-hour nature of their job, the

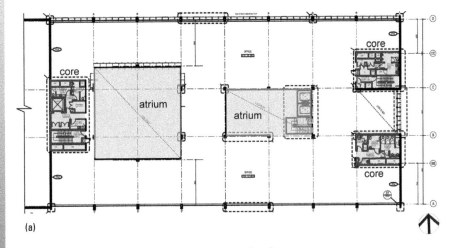

(a)

Figure 3.2.4 The horizontal bar: (a) typical floor plan; and (b) view towards atrium

Source: Woodhead (2011).

building is well equipped with amenities such as shared communal breakout spaces, kitchens with full cooking facilities, a library, and a staff café with outdoor spaces. The building has a 4.5 NABERS rating, with a high level of daylighting, integrated façades with solar-oriented shades and reduced energy and water consumption.

Adaptive re-use
The trend toward smaller core organizations and smaller, more specifically designed new office buildings, fuels the trend in converting non-office spaces into offices and recycling office shells (Duffy, 1997). Adaptive reuse makes good sense as light industry is relocating from city centres to cheaper locations, allowing office organisations to convert these centrally located spaces while avoiding the costs of renting or owning a premium large commercial building.

Different points of obsolescence in a building add further justification in converting these non-office spaces to meet current demands in commercial properties. Minimally untouched façades also promote visually continuity in the neighbourhood.

CASE STUDY 5

The Australasian Performing Rights Association (APRA) and Australasian Mechanical Copyright Owners Society (AMCOS) office headquarters (2009) by Smart Design Studio is a refurbished-to-purpose five-storey brick warehouse in Pyrmont with a high level of amenity in the form of rooftop terrace and boardroom, three levels of office and multi-purpose spaces and an active street level frontage housing a café and forum space (Figure 3.2.5 (a)). A four-storey, central atrium provides a high level of inter-relation and connectivity between these areas (Figure 3.2.5(b)). Sustainable features include stormwater retention and re-use (for toilet flushing), displacement air-conditioning system with natural ventilation option, operable louvres at the atrium rooftop to facilitate stack effect, energy-efficient lighting and water fixtures and bike facilities.

CONCLUSION

This chapter highlights the current and emerging issues, which inform the typologies of commercial architecture that are relevant as we progress in the twenty-first century. More research is needed into evaluating the opportunities and limitations of environmentally retrofitting these modern commercial building types. For example, with deep plan buildings, retrofitting for adequate

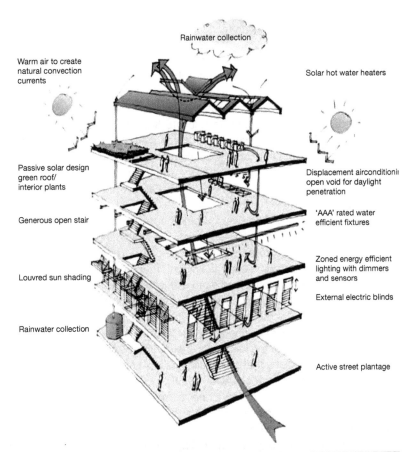

Rainwater collection

Warm air to create
natural convection
currents

Solar hot water heaters

Passive solar design
green roof/
interior plants

Displacement airconditionii
open void for daylight
penetration

Generous open stair

'AAA' rated water
efficient fixtures

Zoned energy efficient
lighting with dimmers
and sensors

Louvred sun shading

External electric blinds

Rainwater collection

Active street plantage

Figure 3.2.5 Adaptive
re-use: (a) concept
diagram; and (b) view
towards atrium

Source: Smart Design
Studio (2011).

daylight and ventilation would require major changes to the form and structure to make this possible. In addition, older buildings also often have structural constraints such as load-bearing elements or heritage restrictions which make it difficult to retrofit the shell as well as the interior. However, the case for environmental retrofitting older conventionally designed buildings or converting non-commercial building into offices far outweighs these limitations.

As highlighted in this chapter, adaptive reuse makes good sense as the trend is for light industry to relocate from city centres to cheaper locations, allowing office organisations to convert these centrally located spaces while avoiding the costs of renting or owning a premium large commercial building. Different points of obsolescence in a building add further justification in converting these non-office spaces to meet current demands in commercial properties.

The majority of post-war buildings have also not completed their anticipated life-cycle and typically are beset with higher energy requirements than those defined by current legislative standards. There is real value and also opportunity in rebuilding and renovation of these existing buildings due to rapid building obsolescence and rapid expunction of invested costs in the buildings (Hyde *et al.*, 2007: 6). The building shell such as the structure and the façade, its services and scenery all have different point of obsolescence. This paints a strong case for upgrading elements with shorter life spans to maximise the life-cycle of the long-term shell (Duffy, 1997: 74). This is further reinforced by the statistic that new buildings in Australia are only ever 1 or 2 per cent of the building stock (AGO, 1999).

Furthermore, the conventional commercial 'office' building was found to be the most significant building type in terms of energy consumption, responsible for an estimated 27 per cent of total commercial building sector emissions in 1990 (ibid.). Upgrading conventional commercial buildings, which are not yet at the end of their life-cycle, will have a substantial impact on reducing total emissions.

REFERENCES

AGO (1999) *Australian Commercial Building Sector Greenhouse Gas Emissions 1990–2010*, Canberra: Australian Department of the Environment and Heritage.
ASBEC (2007) *Capitalising on the Building Sector's Potential to Lessen the Costs of a Broad Based GHG Emissions Cut*, Sydney: Centre for International Economics.
Bligh Voller Nield (2011) *Bligh Voller Nield*, available at: http://www.bvn.com.au/ (accessed 6 August 2011).
Collard, B. and De Herde. A. (2004) *Mid Career Education: Office Building Typology*, Brussels: Architecture et Climat.
Curl, S. (1999) *Dictionary of Architecture*, Kent: Grange Books..
Drake, S. and Brawn, G. (2005) 'National Docklands', *Architecture Australia*, 94(Jan./ Feb.): 62–9.
Duffy, F. (1997) *The New Office*, London: Conran Octopus Ltd.
Duffy, F,. Laing, A. and Crisp, V. (eds) (1993) *The Responsible Work Place*, Oxford: Butterworth Architecture.
Ferguson, A. (2008) 'Sow's ear to silk purse', *Indesign*, 32(Feb.): 142–9.

Hamzah, T. R. and Yeang, K. (2011) *T.R. Hamzah and Yeang International. 'Menara Umno'*. Available at: http://www.trhamzahyeang.com/ (accessed 30 June 2011).

Houghton Mifflin Company (2006) *The American Heritage Dictionary of the English Language*, Boston: Houghton Mifflin Company.

Hyde, R.A. (2008) *Bioclimatic Housing: Innovative Designs for Warm Climates*, London: Earthscan.

Hyde, R.A. (2012) *Exploring Synergies with Innovative Green Technologies for Advance Renovation Using a Bioclimatic Approach*. Available at: http://sydney.edu.au/architecture/research/archdessci_advanced_rennovation_project.shtm (accessed 13 May 2012).

Hyde, R.A., Rajapaksha, U., Groenhout, N., Barram, F., Mohammad Shahriar, A.N. and Candido, C. (2009) *Towards a Methodology for Retrofitting Commerial Buildings Using Bioclimatic Principles*, Hobart: ANZAScA, Academic Press.

Hyde, R.A., Watson, S., Cheshire, W. and Thomson, M. (2007) *The Environmental Brief*, London: Routledge.

Kohn, A. and Katz, E. (2002) *Building Type Basics for Office Buildings*, New York: John Wiley & Sons Inc.

Law, J. (2004) *'Towards Bioclimatic High Rise Buildings*, St Lucia, Queensland: University of Queensland.

Myerson, J. and Ross, P. (1999) *The Creative Office*, London: Laurence King.

Oxford Dictionaries online (2011) Available at: http://oxforddictionaries.com/ (accessed 6 October 2011).

Parlour, R. (1997) *Building Services: Engineering for Architects*, Sydney: Integral Publishing.

Pevsner, N. (1976) *A History of Building Types*, London: Thames and Hudson.

Remoy, H. and De Jonge, H. (2007) *Transformation and Typology: Vacancy, Characteristics and Conversion Capacity*, Delft: Delft University of Technology.

Richards, I. (2007) *Ken Yeang: Eco Skyscrapers*, Melbourne: Images Publishing Group.

Rykwert, J. (2005) *Bligh Voller Nield*, Sydney: Building Press & China Architecture.

Salleh, A. (2009) 'Sydney's climate to become like Brisbane's', *ABC Science*. Available at: http://www.abc.net.au (accessed 7 June 2009).

Smart Design Studio (2011) Available at: http://www.smartdesignstudio.com/index.html (accessed 20 June 2011).

Strelitz, Z. (2005) *Tall Buildings: A Strategic Design Guide*, London: RIBA.

Woodhead International (2011) Available at: http://www.woodhead.com.au/ (accessed 26 June 2011).

Yeang, K. (1996) *The Skyscraper Bioclimatically Considered: A Design Primer*, London: Academy Editions.

3.3
RETROFITTING COMFORT AND INDOOR ENVIRONMENTAL QUALITY

Can improved levels of occupant comfort, human health, well-being and productivity be achieved from existing buildings?

Christina Candido

INTRODUCTION

With buildings accounting for up to 40 per cent of energy end-use in developed economies, regulatory and economic pressures are mounting to reduce the sector's greenhouse gas emissions (Andaloro *et al.*, 2010). One of the key lessons from the oil crisis of the 1970s is that the ultimate success or failure of a building project – in terms of its long-term viability, energy use and occupant satisfaction – depends heavily upon the quality of the indoor environment delivered to the building occupants. Since HVAC systems (Heating, Ventilating and Air Conditioning) is the single largest energy end-use in the built environment, it is inevitable that we should look critically at our dependence on mechanically cooled indoor climates. The past twenty years have witnessed a profound change in the thermal comfort field and the dialectic between conventional, or 'static', and the adaptive comfort theories became a landmark in itself. This discussion became more prominent by the end of the twentieth century with the realisation of the (unsustainable) energy carbon required to air condition indoor environments.

Cooling energy in buildings can be reduced by: (1) reducing the cooling load on the building; (2) exploiting passive design principles to meet some or the entire load; and (3) improving the efficiency of cooling equipment and thermal distribution systems. Natural ventilation reduces the need for mechanical cooling by: (1) directly removing hot air when the incoming air is cooler than the outgoing air; (2) reducing the perceived temperature due to the cooling effect of air motion; (3) providing night-time cooling for exposed thermal mass inside the building; and (4) increasing the acceptable range of temperatures through psychological adaptation where occupants have direct control of operable windows (de Dear and Brager, 2002). Even where these solutions are feasible to implement, they are also limited to a technical approach related to the building's performance, without much consideration of the potential related to behavioural change. For significant CO_2 abatement potentials to be realised, it is imperative that sustainable buildings (both newly built and retrofitted projects) meet the occupants' expectations.

Behavioural change in buildings can deliver fast and zero-cost improvements in energy efficiency and greenhouse gas emission reductions. In order to provide such behavioural opportunities, or adaptive opportunities, buildings must be designed to re-engage 'active' occupants in the achievement of comfort (Roaf, 2006) and the literature on thermal comfort can shed some light on this. The definition of thermal comfort as 'that condition of mind that expresses satisfaction with the thermal environment' (ASHRAE, 2004) begs the question of what is actually meant by 'condition of mind' and whether it is possible for that condition to change through time, but it correctly emphasises that the judgement of comfort is a cognitive process involving many inputs from physical, physiological and psychological domains. This has been challenged by an adaptive comfort model in which the indoor comfort zone tracks outdoor climatic conditions – up in warm climates and down in cool climates. The major conceptual departure of the adaptive model is its reference to thermal history, expectations and attitudes, perceived control and availability of behavioural thermoregulatory options. The adaptive model's outdoor weather-responsive comfort zone is presented as a permissible alternative to the static approach.

The recent revival of natural ventilation, as a passive design strategy, has been widening the range of adaptive opportunities available in buildings to provide comfort for occupants, both in newly built and retrofitted contexts (Brager *et al.*, 2004; Huizenga *et al.*, 2004; Toftum, 2004). This chapter reviews and discuss how changing from air conditioning to naturally ventilated buildings can assist by putting occupants at the centre of low energy (carbon) built environments, in both newly built and retrofitted projects.

THERMAL COMFORT DIFFERENCES IN AIR CONDITIONING AND NATURALLY VENTILATED BUILDINGS: MEASUREMENT AND EXPERIENCE

Most thermal comfort research to date has been premised on steady-state exposure of building occupants to indoor climates. The static theory has as its core Fanger's climate chamber experiments, which produced a comprehensive comfort index – Predicted Mean Vote (PMV). Fanger's PMV started from the premise that it is possible to define a comfortable state of the *body* in physical terms, which relate to the body rather than to the environment. Under isothermal and steady state conditions the heat balance of the human body can be defined with six basic parameters: four environmental and two relating to the occupant. The environmental parameters are air and radiant temperature (i.e. surrounding surface temperatures), humidity and air speed. The two personal comfort parameters are insulation (amount of clothing) and metabolic heat (activity rate). Steady-state comfort models such as the widely used Predicted Mean Vote (PMV) predict the average thermal sensation vote of a large sample of people on a 7-point scale (–3 cold, –2 cool, –1 slightly cool, 0 neutral, +1 slightly warm, +2 warm, +3 hot). The PMV model presumes static comfort zone boundaries. The resultant model was described as being *universally applicable*, regardless of building type, climate zone or population.

This landmark research provided the framework necessary to determine a set of design temperatures for engineering mechanically controlled indoor environments. The PMV model can also be used to assess a given room's climate, in terms of deviations from an optimal thermal comfort situation. This model has been globally applied for almost 40 years across all building types, although Fanger was quite clear that his PMV model was originally intended for application by the heating, ventilation and air-conditioning (HVAC) industry in the creation of artificial climates in controlled spaces. Apart from its controversy, this model was and still is broadly used in American and European standards (ASHRAE, 2004; CEN 2004; ISO 7730, 2005), and its influence in the thermal comfort field is widely recognised.

As with any theory, model or index, Fanger's legacy has been both widely supported and widely criticised. In his dissertation, he stated that the PMV model was derived in laboratory settings and should therefore be used with care for PMV values below –2 and above +2. Probably the most important criticism is the concept of a universal 'neutral' temperature. Regarding the inadequacies of PMV applications in naturally ventilated spaces, it has been discussed that the cool, still air philosophy of thermal comfort, which requires significant energy consumption for mechanical cooling, appears to be over-restrictive and, as such, may not be an appropriate criterion when decisions are being made whether or not to install HVAC systems (de Dear and Brager, 1998). The widely accepted 'adaptive comfort model' changed this paradigm.

The adaptive model of thermal comfort advocates the shift from statically controlled indoor environments to active naturally ventilated buildings. The posterior implementation in standards was an important step forward towards mainstreaming naturally ventilated buildings. The basic tenet of the adaptive model is that building occupants are not simply passive recipients of their thermal environment, like climate chamber experimental subjects, but rather, they play an active role in creating their own thermal preferences (ibid.). Based on an analysis of over 20,000 row set of indoor microclimatic and simultaneous occupant comfort data from buildings around the world, the database found that indoor temperatures eliciting a minimum number of requests for warmer or cooler conditions were linked to the outdoor temperature at the time of the survey. Thermal acceptability was found for 80 and 90 per cent by applying the 10 and 20 per cent PPD criteria to the thermal sensation scale recorded in the building. Buildings were separated into those that had centrally controlled heating, ventilating, and air-conditioning systems (HVAC), and naturally ventilated buildings (NV). Since the database comprised existing field experiments, the HVAC versus NV classification came largely from the original field researchers' descriptions of their buildings and their environmental control systems. The primary distinction between the building types was that NV buildings had no mechanical air conditioning, and that natural ventilation occurred through operable windows that were directly controlled by the occupants. In contrast, occupants of the HVAC buildings had little or no control over their immediate thermal environment.

The adaptive comfort theory predicts that the limit of acceptable indoor temperatures will drift up from the conventional wisdom of 23–24°C during warm weather. It also suggests that the cognitive and behavioural factors impinging on comfort in real buildings will qualitatively differ from those in laboratory environments, where conventional (PMV) thermal comfort theory was originally developed. Examples of building designs focusing on naturally ventilated or mixed-mode indoor environments are increasing. For instance, the recently completed green flagship Federal Building in San Francisco deploys a number of innovative technologies, including an integrated custom window wall, thermal mass storage, and active sun shading devices to regulate internal thermal environmental conditions within the adaptive model's seasonally adjusted comfort ranges. In this building's initial design stage, San Francisco's Typical Mean Year (TMY) of meteorological data was used to calculate month-by-month ranges of acceptable indoor temperature using the ASHRAE 55(2004) adaptive model.

RETROFITTING FOR ADAPTIVE OPPORTUNITIES

The adaptive comfort theory showed that occupants play an active role in creating their own thermal preferences, and satisfaction with an indoor environment occurs through appropriate adaptation. While the heat balance model is able to account for some degree of behavioural adaptation, such as changing one's clothing or adjusting local air velocity, it ignores the psycho-logical dimension of adaptation, which may be particularly important in contexts where people's interactions with the environment (i.e. personal thermal control), or diverse thermal experiences, may alter their expecta-tions, and thus, their thermal sensation and satisfaction. Based on this concept, two important items must be considered when designing indoor environments, both in newly and retrofitted contexts: natural ventilation enhancement and occupant control.

Natural ventilation enhancement

One of the challenges in optimising natural ventilation is to define when air movement is desirable and when it is not. Based on the argument that elevated air speeds in indoor environments could be unwelcome (draughts), air velocity limits have been skewed downwards in the standards. However, a considerable number of laboratory studies and particularly field experiments in real buildings have been providing evidence that occupants prefer the contrary (Arens et al., 2009). Indeed, occupants have been demanding 'more air movement' in numerous field studies (Zhang et al., 2007). While in cold and temperate climates, air motion might cause an unwanted 'draught', in hot humid climates, air movement enhancement is, without doubt, one of the key factors in providing occupant thermal comfort.

Many of the justifications for the shift from naturally ventilated indoor climates to HVAC during the late twentieth century emphasised the risk of local discomfort, or draught, in situations where indoor air movement relies on natural processes instead of controllable mechanical ones. As a concept, draught means any unpleasant air movement and is related not only to air temperature and air speed but also other factors such as area, variability and the part of the body that is exposed. In moderate climates, a draught is one of the main sources of complaint in regard to the workplace environ-ment, concerning up to one-third of office workers and at least two-thirds of workers in moderately cold environments. No consistent influence of thermal sensation was found in these studies, although a cool thermal sensation seemed to increase draught complaints at low air velocities and decrease draught complaints at high air velocities. One reason for the large number of draught complaints among people working in cool or cold environments is simply because they are more sensitive to draughts than people who feel thermally neutral. In situations where people are more likely to feel warmer than neutral, the scenario is qualitatively different.

In hot and humid climates, natural ventilation plays an important role in controlling indoor air quality, indoor temperature, and also prevents the risk of occupants overheating. Investigations indicate that inadequate ventilation is probably the most important reason for occupant discomfort in naturally ventilated buildings. In hot humid climates, people could live comfortably in naturally ventilated buildings as long as they were provided with appro-priate air velocities within the occupied zone. If we agree that thermal environments which are slightly warmer than preferred or 'neutral' can still be acceptable to building occupants, as the adaptive comfort model suggests, then the introduction of elevated air motion into such environments should be universally regarded as desirable because the effect will be to remove sensible and latent heat from the body, thereby restoring body temperatures to their comfort set-points. Based on this scenario and in order to define the maximum air velocity range acceptable for the occupants, many studies were carried out and it is possible to identify considerable differences between them. Relaxing the current draft limit for neutral-to-warm conditions (above 26°C) would open up opportunities for saving energy that, under current regulations and standards, is now restricted to personally controlled air movement devices. When retrofitting buildings, it is important to consider this relaxation, focusing in airflow distribution and higher air speed throughout the occupied zone.

Providing occupants with control over their immediate indoor environment

Control over air velocity is considered a form of behavioural adaptation when people are able to make the environmental adjustments themselves such as opening or closing a window, turning on a local fan, or adjusting an air diffuser. The adaptive model has long insisted that a given thermal environmental stimulus can elicit disparate thermal comfort responses, depending on the architectural context in which it is experienced. It has been noted that thermal environmental conditions perceived as unacceptable by the occupants of centrally air-conditioned buildings can be regarded as perfectly acceptable, if not preferable, in a naturally ventilated building. From a psychological perspective, studies reveal that offering personal control over the indoor environment seems to be very effective in minimising negative effects, such as stress.

The direct effect of personal control on the occupants and their satisfaction with their work environment in general are seen as acting as 'compensation' for the influence of environmental factors (Boerstra, 2010). Occupants tend to be more forgiving of daily malfunctions in their work environments, such as problems with equipment and systems, when they have greater degrees of freedom in adapting their immediate indoor conditions. More recent research confirms the importance of having some level of direct control over the environmental conditions in the workplace to occupant satisfaction. So is the challenge of new or reviewed standards to somehow include occupant control? Offering occupants control over their indoor climate results in fewer sick building symptoms, higher comfort satisfaction rates and improved performance (Boerstra, 2010). Therefore it seems logical to include the aspect of personal control over indoor climate in future (thermal) comfort standards.

CAN IMPROVED LEVELS OF OCCUPANT COMFORT, HUMAN HEALTH, WELL-BEING AND PRODUCTIVITY BE ACHIEVED FROM EXISTING BUILDINGS?

The past 20 years have witnessed major international research efforts directed towards quantifying the relationship between the quality of the indoor environment, as perceived by occupants, on the one hand, and the physical character and intensity of the indoor environmental elements, on the other. The benefits of people spending more time inside artificial and controlled environments during their daily activities in order to keep 'neutral' have been questioned.

Naturally ventilated buildings indeed provide indoor environments with higher percentages of occupants' overall satisfaction and they present enormous potential in contributing to energy conservation challenges faced by the building sector. There are important questions remaining related to allowable air velocity values (maximum) and occupants' control within the occupied zone that should be investigated in more depth. Much has been done on focusing when air movement is 'unwelcome' (i.e. a draught) but there is an enormous potential in research considering air movement enhancement in buildings as a 'welcome breeze'. Especially in hot humid climates, this research topic is pivotal in providing thermally acceptable indoor environments and occupants' satisfaction.

Architects are becoming aware that designing buildings totally disconnected from the outdoor climate and environment in which they are located is out of date. With this in mind, designers are beginning (rather slowly) to shift their attention to widening the range of opportunities available in a building to provide comfort for occupants, both in newly built and retrofitted contexts. This in turn has re-awakened an interest in the role of natural ventilation, not only in the provision of comfort but also in terms of regulations and standards.

When designed carefully, naturally ventilated indoor environments do not need to compromise occupants' comfort, well-being or productivity. Indeed, some argue it is quite the opposite – naturally ventilated buildings provide indoor environments far more stimulating and pleasurable compared to the static indoor climate achieved by centralised air conditioning. When retrofitting for comfort and indoor environmental quality, occupants must be at the centre of the design concept. Occupants' expectations and attitudes, perceived control and availability of behavioural thermoregulatory options are essential when providing not only comfort but also *satisfaction*.

REFERENCES

ASHRAE (2004) *Thermal Environmental Conditions for Human Occupancy*, Standard No. 55, Atlanta, GA: ASHRAE.

Andaloro, A.P.F., Salomone, R., Ioppolo, G. and Andaloro, L. (2010) 'Energy certification of buildings: a comparative analysis of progress towards implementation in European countries', *Energy Policy*, January.

Arens, E., Turner, S., Zhang, H. and Paliaga, G. (2009) 'Moving air for comfort', *ASHRAE Journal*, May: 18–28.

Boerstra, A. (2010) 'Personal control in future thermal comfort standards?' Paper presented to the Windsor Conference: Adapting to Change: New Thinking on Comfort, Cumberland Lodge, Windsor, UK, 2010.

Brager, G.S., Paliaga, G. and de Dear, R. (2004) 'Operable windows, personal control, and occupant comfort', *ASHRAE Transactions*, 110: 17–35.

CEN (2004) *Ventilation for Buildings: Design Criteria for the Indoor Environment, CR 1752*, Brussels: CR CEN, European Committee for Standardization.

de Dear, R.J. and Brager, G.S. (1998) 'Developing an adaptive model of thermal comfort and preference', *ASHRAE Transactions*, 104(Part 1A): 145–67.

de Dear, R.J. and Brager, G.S. (2002) 'Thermal comfort in naturally ventilated buildings: revisions to ASHRAE Standard 5', *Energy and Buildings*, 34(6): 549–61.

Huizenga, C., Zagreus L., Arens, E. and Lehrer, D. (2004) 'Listening to the occupants: a web-based indoor environmental quality survey', *Indoor Air*, 14: 65–74.

ISO (2005) *Moderate Thermal Environment: Determination of the PMV and PPD Indices and Specification of the Conditions for Thermal Comfort*, ISO 7730, London: International Organization for Standardization.

Roaf, S. (2006) 'Comfort, culture and climate change', in 2006 Windsor Conference: Comfort and Energy Use in Buildings: Getting Them Right, April 2006, Cumberland Lodge, Windsor, UK. Available at: http://nceub.org.uk.

Toftum, J. (2004) 'Air movement: good or bad?', *Indoor Air*, 14: 40–5.

Zhang, H., Arens, E., Fard, A.S., Huizenga, C., Paliaga, G., Brager, G. and Zagreus, L. (2007) 'Air movement preferences observed in office buildings', *International Journal of Biometeorology*, 51: 349–60.

3.4
REVIEWING BENCHMARKING SYSTEMS FOR RETROFITTING

How can benchmarking be harnessed for the purposes of retrofitting?

Richard Hyde

INTRODUCTION

The retrofitting process is increasingly using some form of Building Environmental Assessment (BEA) system. It has become a new pathway for green design using building performance as a basis for decision-making (Hyde *et al.*, 2007). However, it is now recognised that this approach has yet to be fully developed for retrofitting. Research has begun to fully examine the connection between BEA and retrofitting. As seen in Chapter 3.3, the challenge of developing a common approach to retrofitting is how to encompass a wide range of building types which are in varying stages of building obsolescence. The challenge for building environmental assessment is to span this wide range, all with very specific preconditions. Our current thinking is that this is not possible as the typology is so broad that it defies classification, hence a process which is more project-specific is needed.

This chapter examines the components of Building Environmental Assessment and argues that the central limitation of this approach is it is largely a tool-based rather than a principled approach to sustainable design. It argued that it is easier to generalise from sustainable principles to a broader context than to define sustainability around a particular building typology or end use. For retrofitting in building, the typology is of less importance than the benchmarking process as the latter can form the basis for a process of relative continuous improvement.

COMPONENTS OF A BUILDING ENVIRONMENTAL ASSESSMENT (BEA)

Currently two types of assessment are found: design phase assessment and operational phase assessment.

Design phase assessment typically is carried out prior to construction. The general characteristics of these systems are found in those developed by the respective Green Building Councils of the USA, the UK and Australia. For example, LEED (Leadership in Energy and Environmental Design) provides a Green Building Rating System™ and is the nationally accepted benchmark in the USA for the design, construction and operation of high performance green buildings (www.usgbc.org). BREEAM (http://www.breeam.org/) and Green Star (Australia) (http://www.gbcaus.org/) are aimed at developing environmental standards to improve the sustainability of building development. Recent research has questioned whether these systems are achieving this objective. Schendler and Udall (2005) have reported four main criticisms of the LEED system used in the USA:

1 *Assessment process*: The LEED system uses a points-based rating assessment instead of simply assessing green design. Evidence to support the rating involves crippling bureaucracy and is reportedly too costly to implement. This has raised concerns that the system is not supporting the design process but is reducing its effectiveness. Some designers are questioning the compatibility of the building assessment systems and the design process.

2 *Benchmarking methodology*: The benchmarking methodology for providing evidence of performance used in the LEED rating is complex. For example, with energy, computer simulation is used to calculate the data for benchmarking. Some suggest this process is imprecise and may not necessarily be indicative of the likely operational performance of the project.

3 *Outcomes of the assessment*: A consequence of this line of argument suggests the outcomes have led to misleading statements on the benefits of green buildings. For example, buildings which have a design benchmark of 'zero fossil fuel energy' do not achieve this when the building is in operation.

4 *Reduction in cumulative environmental impact*: The LEED system has been in operation for a number of years yet there is little evidence of the overall reduction in environmental impact due to its implementation, hence making it difficult to gauge whether the system is achieving its objectives.

Operational phase assessment is carried out after construction. Typically this is based on operational phase assessment after construction and over a particular time period. The general characteristics of these systems is represented by the Earthcheck system (http://www.ec3global.com), which can be applied to a range of tourism and non-tourism buildings and development.

These operational phase systems use a number of sustainability indicators. Data on these indicators is collected from the operation of the buildings, which is then compared to a database with the performance of similar buildings. Baseline and best practice levels are determined from the population of buildings in the database. Schianetz *et al.* (2007) report faults with these systems:

1   It provides only a '*snapshot*' of the benchmarked performance of the building at a particular place and time.
2   *Inherent bias*, as the indicators are prescribed. There is no mechanism for adjusting the indicators to particularities of the site or other circumstances, although some standards have some elective indicators.
3   *Impact assessment is lacking*: that is, there is no way of gauging whether the benchmarks will lead to significant impact reduction (ibid.).

Another scheme using the operational phase is the Australian and New South Wales Government's NABERS (the National Australian Built Environment Rating System). This is a performance-based rating system for existing buildings. NABERS rates a building on the basis of its measured operational impacts on the environment. It is aimed at residential and office buildings (http://www.nabers.com.au/).

Many changes to this approach have occurred in recent years. For example the current system for homes (NABERS HOME) has been scaled back from a comprehensive and rigorous tool for experts to a simplified tool, which is more practical for homeowners. This change demonstrates the current dilemma facing BEA: the current version of NABERS HOME may be appropriate as a guidance tool for homeowners but problems occur when it is used in conjunction with the design process since it now does not have the depth of indicators found in previous versions (Hyde, 2006).

The significance of these issues has prompted research into methods of improving tools through examination of issues of the rigour of the systems while still achieving a level of workability within the design process. To do this, an examination of the BEA process has been carried primarily through examining the underlying 'principles to indicator model' on which many of the assessment processes are based (ibid.). The first section of this chapter summarises this analysis.

PRINCIPLES TO INDICATORS

Objectives

Understanding the nature of BEA is necessary to establish the importance of benchmarking in the assessment process. BEA is not a new concept and has developed from Environmental Assessment (EA). This is a methodology that has evolved to examine the impacts of human processes on ecological systems and has been integrated into a number of systems and tools for examining aspects of the built environment. At the core of most environmental assessment methods is an examination of a 'process'. For a building, this can be the design process, prior to construction, or the operational process, post construction. The process can be examined in terms of the 'inputs' to the process and the 'outputs' as well as activities that take place within the process.

Principles are rules about what aspects of the environment we wish to protect, maintain, or improve. Agenda 21 contained a range of aspects on the environment we wish to protect. Environmental indicators can be used to examine all aspects of the process – the inputs, the activities, and the outputs – to assist with measuring the performance of the processes that impact on the environment. Commonly called auditing, this is usually carried out by someone independent of the process to ensure rigour in the audit. Indicators are usually elements of the process, which can be measured in a manageable way. For example, the energy consumption of a house is used as an indicator of environmental performance because it is recorded in the energy bills, which are relatively easy to access.

Nesting BEA in design is not an easy development as it involves linking all aspects of design. Roaf *et al.* (2004), for example, argued that 'closing the loop', that is a process involving design, construction and evaluation post construction, is needed to facilitate this process. Not only does this feed back into the beginning of the design of the next project but it also can lead to retrofit and improvement of the building over its life-cycle. Due to this complexity a number of tools have been developed to assist in steering projects to a sustainable outcome. It is important to understand the models that form the basis to the tools to see if they have such a capability.

'Principles to indicator model'

The model that addresses all three types of factors – input, process and output – in environmental assessment is commonly called the 'principles to indicator model'. For example, the assessment process consists of defining input factors such as the policy framework and sound environmental principles. Indicators are then developed and the process assessed, which involves using 'best practice' benchmarks. Mawhinney argues the chief limitation of the model is the process of determining 'best practice'. Hence the 'principle to indicator approach' has been integrated into a number of tools for use in evaluating buildings. There are some emerging criticisms of this approach (Mawhinney, 2002).

1   What is the theoretical basis to policy and principles arising from how sustainability is defined?
2   How are principles and indicators linked to policy and principles? Are there any missing indicators?

3   How is best practice defined and what is the rigour by which this is accessed?
4   How are benchmarks derived from 'best practice'?

Conclusions from this section show that there are many limitations to BEA tools. A central limitation lies in the methodological approach used for benchmarking; in fact, benchmarking is the core concept to many of these tools.

BENCHMARKING

Definition

Benchmarking is the continuous search for and adaptation of significantly better practices that leads to superior performance by investigating the performance and practices of other organisations (benchmark partners) (Kulshresth *et al.*, 2004). In addition, it can create a crisis to facilitate 'the process of change'.

Benchmarking within the building industry is seen as a central part of business management that enables organisations to structure change. There are four types of benchmarking.

Strategic benchmarking

This is concerned with collecting information on the different strategies used in the building. Two forms of analyses are used to determine the success of these strategies. First, comparing the strategies to the design intent, principles, elements, processes and technologies used in the project. Second, comparing the strategies used in the project to meet human needs and expectations of the client, user and designers. This can be achieved through surveys to measure satisfaction and the gaps between performance and standards for human needs. The main difficulty is clearly with trying to identify the strategies used. However, there is a great deal of information which can be obtained from users, environmental standards and public domain information.

An example of this approach can be found in the work of Huizenga, Laeser, and Arens, who carried out a web-based benchmarking study into air quality issues in office buildings using post-occupancy methods. Huizenga argues, 'Historically, building occupants have been underutilised as a source of information on building performance', and they have developed web-based reporting systems for collecting environmental information. The survey has been used to evaluate the performance of 22 buildings in the United States, including office buildings, laboratories, banks and courthouses. Findings from several of these studies are highlighted, demonstrating the potential of this approach. An important potential is that the approach has the ability to provide broad data across building clusters to address selected indicators. Also the approach allows a detailed investigation of key strategies, in this case, drilling down and obtaining diagnostic information on performance of selected strategies concerning technical systems (Huizenga *et al.*, 2002).

Functional benchmarking

The aim of this type of benchmarking is to investigate the performance of core processes, functions and activities in the building and benchmark these with buildings which have similar core functions. Essentially it is a typological approach. The benchmark partner needs to be in a similarly characterised building for useful comparisons to be made. The benchmarks are consistent for each function. So buildings are divided into types such as offices, hotels, domestic buildings and so on, as key functional groups for the purpose of benchmarking (Schendler and Udall, 2005). An example of this can be seen with regard to core issues such as energy and water conservation in buildings.

Governments and organisations can use this approach to address the greenhouse effect and other issues. The former state Governor of California, Arnold Schwarzenegger, developed a functional benchmarking programme to support energy conservation legislation. This involves measuring how state facilities perform in terms of energy management and tracking the building's energy profile over time in order to establish baseline data. In addition to analysing energy performance against a baseline, the state can use benchmarking to compare a building with properties of similar characteristics, such as climate, size, operations and age.

The Governor's Green Building Action Plan directs state agencies to meet a benchmark of reducing energy consumption by 20 per cent by 2015 compared to a baseline year of 2003. The Department of General Services will collect and summarise energy consumption data provided by state agencies and will report annually on the progress toward attaining the energy reduction goal including recommendations on any changes in rules or procedures to ensure the goal is met. The California State Green Building Action Plan also calls for benchmarking of all state buildings by 2007. The benchmarking process will evaluate a building's energy use by comparing it with a similar facility. This approach not only sets benchmarking targets but also sets up benchmarking partners to assist with baseline information (California State Government, 2007).

Best practices benchmarking

This approach focuses more on the processes. It breaks functions down into discrete areas and establishes the targets for benchmarking and is therefore a more focused study than functional benchmarking. This approach is largely non-typological as some building processes are the same regardless of the type of building. This approach focuses on attempts to benchmark not only green processes, but also the management practices behind them. In this case the process is consistent but the benchmarks vary with changes in parameters such as location, climate, types of buildings and other independent factors. For example, the Earthcheck system, the BEA tool for the tourism industry, has standard calculations and empirically based systems to derive the benchmarks but these vary with parameters. The parameters include factors such as climate, location and type of resort, guests and area under roof. 'Best practice' benchmarking has been commercialised by industry organisations seeking to be identified with green development and market their developments as 'best practice'.

Limitations with this approach are trying to define what constitutes the 'best practice' benchmarks, and the rigour with which these benchmarks are obtained. One method is to adopt additional approaches to benchmarking in order to enable additional information regarding the products and/or subsystems into the process.

Product or sub-systems benchmarking

This is commonly known as 'reverse engineering' or competitive product analysis. In the building context this comprises elements of strategic benchmarking and precedent analysis, i.e. stripping down and analysing similar existing high performance buildings in terms of design intent, strategies principles, technologies and costs. These are reworked into a new building. Also BEA tools can be used as a precedent for new design. The building is invariably designed around the BEA standard. The limitation of this approach is that it creates what is called 'the LEED brain', i.e. the purpose of design is to meet the benchmarks at the lowest cost. Points in the rating system are selected on the basis of getting maximum score for the least cost. The building is reverse engineered to meet the benchmarking standard (Schendler and Udall, 2005).

In summary, many BEA tools use one or more of these approaches to benchmarking in an attempt to create more rigour in their systems. One major indicator of rigour is the extent to which these tools constantly follow the approach adopted. So if a functional benchmarking approach is used, a central question is the level of similarity between benchmarking partners. A further opportunity for improving rigour can be found within the steps used to carry out benchmarking. Davies argues that many benchmarking systems are unclear about these steps (Davies *et al.*, 2000). Davies has identified a basic approach to benchmarking containing four steps:

1   *Scoping*: Identify what to benchmark; in the case of 'best practice' benchmarking, what is the level of benchmarking, ensure management support and involve all stakeholders, select the benchmarking team, analysis of internal processes, identify projects to be benchmarked.

2   *Data collection*: Decide on method(s) of data collection, collect public domain information, analyse collected information to establish what other information needs to be collected, establish contacts with benchmark projects, plan the actual visits, and conduct the benchmarking visits.

3   *Analysis*: Establish whether a performance gap exists, predict future performance levels, communicate benchmark findings, and establish action plans with appropriate targets.

4   *Implementation*: Gain support and ownership for the benchmarking action plans and goals. Implement the action plans, measure performance and communicate progress, re-calibrate benchmarks, adopt benchmarking on a broad scale (Davies *et al.*, 2000).

In the context of BEA tools, the last three steps are accommodated within the rubric of the tool. Interestingly, it is with these three steps that some of the methodological problems with benchmarking lie. So a further way of establishing the rigour of the system is to examine compliance with the steps shown above. For example, poor data acquisition will erode the rigour of the systems.

Yet, of all the steps that are critical to the performance of the tool, it is perhaps analysis which established best practice levels for particular indicators. The methods of validating performance are a crucial indicator of rigour since it establishes the credibility of the standard.

METHODOLOGICAL APPROACHES TO VALIDATE BENCHMARKING INFORMATION

The aim of this section is to examine issues concerning the methods used to derive benchmarks and how the best practice and baseline level are achieved is complex. In the building context it is largely about using principles of green design and environmental technologies well integrated within the building siting, form and fabric. Benchmarks are best derived from examples, which are comprehensively resolved. Deriving benchmarks is largely centred on working from these examples to establish 'best practice' case base performance. In recent years as the number of green buildings has increased it has been possible to carry out cumulative studies, which draw together the case base performance of a number of buildings.

Called cumulative case base performance, the approach has the potential to yield better understanding of the benchmarking data as the sample size increases. Furthermore, systems analysis works on the performance of the management systems and technical systems provides breakdowns of the overall performance measures within the case study. This can help with validation.

Validation

Questions have arisen as to the validity and robustness of the approaches to define benchmarks. It is argued that 'triangulation' in the methodology can address this problem. Triangulation in the context of this chapter is defined as a research method for establishing the reliability of research findings. This is achieved by examining a phenomenon using a number of perspectives or techniques. It is argued that by using a combination of data sources to derive information to define benchmarking, a more robust methodology might prevail.

Examining the benefits of all three approaches to benchmarking was carried out by testing this approach. This can be seen in detail in the case study information on the Couran Cove Resort (CCR), Surfers Paradise, Queensland (Hair, 2005) (Figure 3.4.1). The initial work was carried out by Barram involving systems design for energy use in the design phase of the project and later by Hair on the operational phase (Barram, 2004; Hair, 2005). Functional benchmarking has been carried out into energy systems for this type of island resort to try to continually improve its environmental performance.

Design phase benchmarking

The initial work was carried out using a systems approach for setting not only environmental but also economic benchmarks. The design process (Figure 3.4.2) involves defining the service requirements for the project, modelling energy demand and supply options and also the financial costs.

The design of the energy systems involved end use analysis, which compared standard solutions to efficient solutions for demand management, supply efficiency and utilisation of renewable energy. Outcomes led to an

Figure 3.4.1 Couran Cove Resort, Surfers Paradise, Queensland
Source: Barram (2004).

Figure 3.4.2
Design process

Source: Barram (2004).

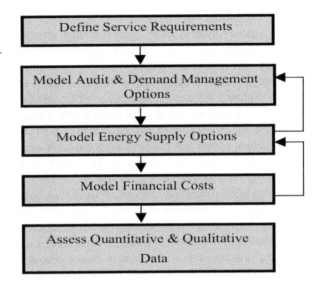

annual predicted electrical energy reduction of 5GWh, that is, from 7GWh to 2GWh. This is achieved by using renewable energy through a reduction of 400MWh(e) by using solar water heating, and by utilising waste heat from the power station to provide 1.4GWh(e) for pool heating (Barram,2004).

The consequences of this design process produce synergies in supply efficiency and demand conservation to create a reduction in power station capacity from 3.9MW to 920kW. In addition, the use of gas as a power source further reduced the CO_2 emissions.

The benchmarks for the CCR case were achieved as follows:

- Primary energy use down 60 per cent.
- Greenhouse gas emissions down 70 per cent on original design.
- Fossil fuel reduced by 1,500 tonnes per annum.
- 33 per cent of primary energy supplied from renewable energy.
- $2.5M upfront capital cost saving.
- $0.8M annual operating cost saving (Barram, 2004).

Operational phase benchmarking

The project is now completed and has been in operation for a number of years and operational data is now available to assist with cumulative benchmarking through the Earthcheck database. This database, which supports the Earthcheck's environmental standards, is a key part of the BEA tool for tourism. From this, for particular types of vacation hotels like CCR, best practice levels of energy consumption were found to be an average of 145.8 MJ per guest night annually (this is a measure of occupancy rate). The main question for BEA to address is which physical aspects of the building should be retrofitted and what will be the performance improvements of the retrofit?

A study by Hair into benchmarking at the Couran Cove Resort found energy consumption at CCR is greater than energy consumed at best practice resorts by 57.06 MJ per guest night or 39.14 per cent. The difference between CCR energy use and best practice median is even greater, being 66.86 mega joules per guest night or 49.16 per cent greater than the median.

Figure 3.4.3 shows the position of Couran Cove, in relation to the distribution of resorts with best practice energy consumption, sits roughly within the top 25 per cent of the distribution of best practice resorts. These are the largest consumers of energy among resorts with best practice energy use (Hair, 2005). Work is still under way to fine-tune the case study to compare findings from the design phase benchmarking and the operational phase. It is interesting to also relate the methods of benchmarking used in the case study with the definition and types of benchmarking earlier. The design phase work seems to involve an approach similar to product benchmarking which compares the performance of management systems and technical systems, while the operational benchmarking is clearly best practice benchmarking. Using combinations of approaches to benchmarking seems to yield better understanding of the design and operation of the buildings and this may be more use in achieving the overall objective of benchmarking which is to create a framework of change within organisations to achieve sustainability. Yet using a combination of approaches to benchmarking may increase the rigours through triangulation. It may also reduce the practicality of the benchmarking approach, making it a more costly and time-consuming process. This leads into the next section, which examines the extent to which BEA tools have adopted rigour in their benchmarking processes.

RIGOUR VERSUS PRACTICALITY

The aim of this section is to examine the benchmarking methods used in some of the current tools. The examination of benchmarking within BEA tools involves addressing three questions:

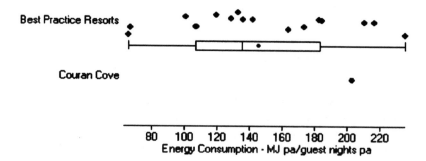

Figure 3.4.3 Comparison of Couran Cove Resort and resorts with best practice energy consumption
Source: Earthcheck and Hair (2005).

196

1   To what extent is the benchmarking approach used in the tool consistent with its benchmarking type?
2   How comprehensive is the benchmarking process used in the tool? Does it follow the steps to achieving benchmarking?
3   What level of triangulation is used in the methodology for creating valid benchmarks?

It is beyond the scope of this chapter to address all these questions in adequate depth. A review of selected tools follows which sets out some issues concerning the issue of rigour.

Benchmarking type

First, though, it is important to identify which of the benchmarking approaches is used in the selected tools. Table 3.4.1 shows an approximate assessment of the type of approach to benchmarking for selected BEA tools. These have been separated into tools for benchmarking the design phase, which are using performance evidence from the design process, and those that apply to the operational phase, post construction, which use evidence from the facilities management phase.

Rating tools such as LEED, BREEAM and Green Star are similar tools, which give points for each indicator; the building's environmental performance is an aggregation of points to give one score, a rating. A specific number of points are counted as the 'best practice' benchmark. This system is largely a functional benchmarking system, looking at key processes across similar building types. Yet one of the problems that arises is that many of the buildings within the types are different. This means that the benchmarking process is not comparing like with like.

Tools such the NABERS HOME, the 2006 version, have the potential to be more comprehensive. As seen in Table 3.4.1, the system covers a wide range of benchmarking types. It can be used for strategic purposes; it has a post-occupancy questionnaire for house-owners to complete; it can address functional benchmarking through a number of key

Table 3.4.1 Comparison of benchmarking approaches used in selected BEA tools

BEA tools	Strategic bench- marking	Functional bench- marking	Best practice bench- marking	Product bench- marking
Building design phase				
LEED/BREEAM/Green Star		✓		
Earthcheck Design and Construct		✓	✓	✓
Operational phase				
Earthcheck Company standard			✓	
NABERS (2006)	✓	✓	✓	

performance indicators, through 'best practice' benchmarking since individual performance measures can be examined separately through the collection of operational data on houses. This yields a wide range of data. Use of this tool for the design phase has been tested for the renovation of houses. The tool is used on existing buildings to obtain feedback from users as well as operational performance data as diagnostic information. Benchmarking is used to inform the design brief of the renovation project (Hyde, 2006). Recent developments such as NABERS HOME have resulted in a simplified approach for laypersons through a web-based system (http://www.nabers.com.au/). Changes to NABERS HOME draw attention to the way the benchmarking process is developing, namely that of working from a self-reporting system with the implicit goal that this will change occupants' behaviour and lead them to adopt more energy- or water-efficiency technologies and practices. However, it is a questionable strategy. Questions remain as to how this approach will be integrated with the design process for retrofitting. Hence its evolution may be criticised for having sacrificed rigour for practicality.

A key question is whether the tools consistently follow the benchmarking steps outlined earlier. More importantly, does the benchmarking process assist the user with change within their organisation?

Steps in the benchmarking process

Table 3.4.2 shows some key issues regarding the benchmarking process. Research is under way to address these key questions. For example, at the scoping stage it is important that the process for determining benchmarks is transparent so the user has a sense of whether the benchmarks are achievable. Benchmarks can be derived theoretically, empirically and by expert opinion (Davies et al., 2000). Systems like LEED and BREEAM use case studies of high performance buildings to derive the best practice benchmarks. This is complemented by an expert committee that decides the rating scale, yet are these benchmarks realisable within current practice?

Further questions include:

1   Have benchmarking targets been identified?
2   Is the data collection process manageable?
3   Are performance gaps between project data and benchmark data identified?
4   Are benchmarks recalibrated with new project data?
5   Is the project data validated?

Of primary importance is the issue of validity.

Table 3.4.2 Emphasis placed on the steps used to achieve benchmarking in selected BEA systems

Steps	LEED/BREEAM/ Green Star	Earthcheck Design and Construct	Earthcheck Company standard	NABERS HOME 2006
Step 1 Scoping Is the process of determining benchmarks transparent?	Best practice case studies and expert opinion Rating levels provided	Best practice case studies and expert opinion Pre-commitment questionnaire	Principle of continuous improvement	Expert opinion
Are benchmarking targets identified?	Checklist for initial rating			
Step 2 Data collection Is the data collection process manageable?	Extensive data collection	Integrated with the design process	Yes	Yes
Step 3 Analysis Are performance gaps between project data and benchmarks identified?	Yes	Yes	Yes	Yes
Step 4 Implementation Are benchmarks recalibrated with new project data? Is project data validated?	No Extensive documentation for certification	Yes Extensive documentation for certification	Yes Ongoing monitoring	No No

Benchmarking process and rigour

As discussed earlier, validity is a research methodology designed to test the logic to conclusions through ensuring the premises to the argument are true. If the premises are false, the conclusions are false. In the context of BEA tools this means ensuring the benchmarking process is sufficient to support the benchmark. One way to achieve this is to use triangulation; this uses information from a number of sources to support or negate the benchmark. The use of triangulation can help with validating benchmarks (Sallam, 2005).

Work is under way to assist validity; Bannister argues that combining design phase tools with ongoing monitoring in the operational phase is crucial to address this question. Sustainable benchmarking system are well placed to do this with both design and operational standards (Bannister, 2006).

The reason for Bannister's argument is found in the work of Cohen *et al.* (2007) and is centred on their work concerning the EU Energy Performance of Buildings Directive Article 7.3. They argue that during the design phase, computer modelling is used to create benchmarks. They argue there is a large gap in science between the benchmarks derived theoretically in this way and reality. Empirical benchmarking is needed to draw the two together.

Furthermore, empirical benchmarks lack detailed information and also a larger sample of buildings to gain meaningful information. Cohen *et al.* suggest using both fixed and customised benchmarks to improve validity. Fixed benchmarks are derived from statistics on a large cohort of buildings, but tailored benchmarks are customised to take account of special circumstances in buildings and to assist with finding more detailed information on the buildings (Cohen *et al.*, 2007).

IMPORTANCE OF PROVING BENCHMARKING

Benchmarking in the context of BEA tools is still a weak science. For example benchmarks created through computer simulation are yet to be effectively reconciled with operational performance, particularly in the energy area. Davies argues that the real aim of benchmarking is to open up organisations to new ideas, concepts and techniques. In the case of the BEA, this means identifying creative ways of achieving sustainability and hence improved performance (Davies *et al.*, 2000). Cohen *et al.* (2007) argue that increasingly the use of benchmarking to inform design decisions about building use is needed to meet sustainability criteria. Also they argue that benchmarking is a much needed process to inform government policy on building design and use. Large-scale studies in the USA and the EU are elevating the importance of benchmarking in the creation of new policy and design guidance.

This emphasis is on the importance of improving the benchmarking process through a clearer understanding of the type of benchmarking used, the steps taken and the validity of the benchmarks used. Moreover, greater transparency in benchmarking can help with addressing the limitations experienced in using some of the rating tools by publishing better specifications that achieve specific benchmarks.

THE RETROFITTING PROCESS

Placing these systems in the context of retrofitting we see that this separation of design and operations is largely artificial and creates in itself a barrier to the improvement of projects. The main reason for this is that in design phase assessment the scope of interventions largely work on design issues and cannot condition or prescribe future operational conditions. Similarly with operational assessment, the operational conditions are really the main focus for improvement and not the redesign of the building. Furthermore, when the building is in operation, the design interventions should consider the operational conditions, which are continuing. Unless the building occupants are decanted into another building, then the process of retrofitting should work in the context of its continued operation. Hence, for retrofitting, the context is virtually intractable as the number of complex factors pertaining to the assessment increase significantly over new build projects. The main question for BEA to address is which physical aspects of the building should be retrofitted and what will be the performance improvements of the retrofit?

CONCLUSION

BEA tools vary in the extent of the rigour with which they apply bench-marking. Theoretically highly rigorous tools comprehensively match the principles and/or policy frameworks with the indicators, the indicators are assessed with information, which is easy to collect, but, once collected, yields valid data, and finally 'best practice standards' are not biased or based on impractical levels of performance.

Very often highly rigorous tools are data-hungry and expensive to service, making them impractical to use in the schema of some organ-isations. Lack of rigour can make the tools useless and hence here lies a conundrum for this type of approach. As rigour is often traded against practi-cality of use, so many tools usually include a checklist for pre-assessment, which allows potential users to quickly check to what extent their project engages with the environmental criteria, the process becomes an end in itself, a 'snapshot' in time. To avoid these problems, further research is needed to validate benchmarks through larger-scale studies, which look at larger populations of buildings using a range of types of benchmarking. It is important that benchmarking is more strategic, that it is based on a number of sources of information and drawn from both design and operation condi-tions. A more strategic focus to benchmarking might also engage building users to assist with facilitating change, which is the central aim of bench-marking.

ACKNOWLEDGEMENTS

This chapter is based on the paper by Hyde, R.A, Prasad, D., Blair, J., Moore, R., Kavanagh, L., Watt, M. and Schianetz, K., 'Reviewing benchmarking approaches for Building Environmental Assessment tools (BEA): rigour versus practicality', in the Proceedings of 24th international Conference on Passive and Low Energy Architecture (PLEA 2007), Singapore, 22–24 November 2007, pp. 58–67.

This chapter is based on research developed and supported by the Sustainable Tourism Cooperative Research Centre, Australia, which has been working on developing planning and design standards for sustainable development using Building Environmental Assessment (BEA) tools. EC3 Global, through the support from the Sustainable Tourism Cooperative Research Centre, developed systems and standards that connect both design and operational phase to address some of these limitations. The Design and Construct Standard provides guidance for the design phase, while the Precinct Planning and Design Standard addresses planning issues associated with sustainable development.

REFERENCES

Bannister, P. (2006) 'Benchmarking building performance and the Australian building greenhouse rating scheme', Seminar, 21 August 2006. Available at: http://eetd seminars.lbl.gov/seminar.php?seminar=8 (accessed 1 July 2007).

Barram, F. (2004) 'Economic case for sustainable energy design in resorts', in the Proceedings of the Green Globe Training Course, Coolangatta, Australia.

California State Government (2007) *Green Building Action Plan*. Available at: http://www.green.ca.gov/GreenBuildings/benchmark.htm (accessed 1 July 2007).

Cohen, R., Bordass, W. and Field, J. (2007) *Fixed and Customised Benchmarks for Building Energy Performance Certificates Based on Operational Ratings*. Available at: http://energyprojects.net/links/EPLabel_EPIC_paper_final_03Jul06_corr.pdf (accessed 1 July 2007).

Davies, A., Kennerley, M., Kochhar, A.K., Thacker, M., Oldham, K. and Lucas, O.S. (2000) *An Introduction to Benchmarking*, (c) SM Thacker & Associates.

Hair, A. (2005) 'Examining sustainability in tourism: an assessment of environmental performance at Couran Cove Island Resort', unpublished thesis, The University of Queensland, Australia.

Huizenga, C., Laeser, K. and Arens, E. (2002) 'A web-based occupant satisfaction survey for benchmarking building quality', *Indoor Air Conference*, p 1.

Hyde, R.A. (2006) 'Teaching architectural science in context', guest lecture, The University of Sydney.

Hyde, R.A., Moore R., Kavanagh, L., Watt, M., Prasad, D., and Blair, J. (2005) 'Research report: Green Globe 21 Precinct and Planning Design Standard', Green Globe 21 Australia.

Hyde, R.A., Watson, S., Cheshire, W. and Thomson, M. (2007) *The Environmental Brief*, London: Routledge.

Kulshresth, M., Bharadwaj, U., and Mittal, A.K. (2004) 'A genetic approach to benchmarking of water and sanitation facilities', in D.K. Stevens, G. Sehlke and D.F. Hayes (eds), *Critical Transitions in Water and Environmental Resource Management: Proceedings of World Water and Environmental Resources Congress 2004*, CD ROM, American Society of Civil Engineers.

Mawhinney, M. (2002) *Sustainable Development: Understanding the Green Debates*, Oxford: Blackwell Science.

Roaf, S. *et al.* (2004) *Closing the Loop: Benchmarks for Sustainable Buildings*, London: RIBA.

Sallam, I. (2005) 'Triangulation methodology: can it be used to improve environmental assessment strategies?' in the *Proceedings of ANZAScA*, 2005.

Schendler, A. and Udall, R. (2005) *LEED Is Broken; Let's Fix It*. Available at: http://grist.org/comments/soapbox/2005/10/26/leed/index1.html (accessed 13 August 2006).

Schianetz, K., Kavanagh, L. and Lockington, D. (2007) 'Concepts and tools for comprehensive sustainability assessments for tourism destinations: a comparative review', *Journal of Sustainable Tourism*, 15(4): 369–89.

3.5
ENERGY PERFORMANCE RATING SYSTEMS

How international and national policies and rating systems leverage a methodology for retrofit

David Leifer and Alan Obrart

INTRODUCTION

Building Environmental Assessment (BEA) has become part of the construction industry lexicon as a result of the growing concern about buildings' contribution to environmental despoliation. The resources consumed in their construction and operation form a significant proportion of the total resources used by societies, and therefore a necessary step towards resource management is the ability to describe and measure their use and energy flows involved. Although there is still debate about the causalities in the model, we can safely state that a building life-cycle can be split into design, construction, operation, renovation and disposal that form discrete and sequential stages. This chapter examines work under way to look at energy rating systems. Although there is a large amount of rhetoric about energy benchmarking there is no energy retrofitting standard or rating tool. Hence we ask, if there were such a tool, what would it be?

POLICY CONTEXT

Whether man-made or natural, climate change will present challenges to society and it must be assumed that the *human contribution* is a causal factor, and the only parameter that can be manipulated for mitigation. In the mid-1970s environmental concern centred on the energy used in the instantaneous operation of buildings, and in the UK Section L of the Building

Regulations was enacted to establish minimum standards for the thermal transmission of new building envelopes and this affected their design. In Australia, the rationale for the need to target energy performance in buildings was justified by the reason that 'it is estimated that the building sector accounts for approximately 19 per cent of the country's total energy consumption and 23 per cent of greenhouse gas emissions' (NBSS 2010: 3.5, see also 3.5.1, 3.5.1(a)).

The evolution of rating systems, in particular, LEED (Leadership in Energy and Environmental Design) (in the USA), BREEAM (Building Research Establishment Environmental Assessment Method) (in the UK) and Green Star (in Australia) reflect the widened concern, but since new buildings add only a small percentage to a nation's building stock, there is a long lead-in time before these models have a significant effect. In Australia, a legal requirement for building energy assessment (Building Energy Efficiency Disclosures Act, 2010) prescribes that all office workspaces of over 2,000m^2 area disclose their energy efficiency through a Building Energy Efficiency Certificate (Australian Government, 2011) using the NABERS (National Building Energy Rating Systems) programme.

A stated intent of the NABERS Energy programme is that it is 'designed to allow building owners, managers and tenants to benchmark the environmental impacts of operational performance and get market recognition for their performance' (DOECCW, 2010). It is also a national voluntary tool for *existing* buildings incorporating the ABGR (Australian Building Greenhouse Rating). 'NABERS ratings are based on measured operational impacts on the environment, adjusted to account for climate and how the building is used – the more stars, the better the performance' (DOECCW, 2010).

ASSESSMENTS AND PREDICTIONS DURING THE DESIGN PHASE

A weakness of assessment tools for implementation during the design phase is that the least amount of data is available and its accuracy is uncertain (Leifer, 1999) when the need for design advice is greatest, i.e. when the most significant decisions are being made. These early decisions establish the form, structure and materials of the envelope that will filter the external environment and establish the building's embodied energy (Henriksen, 2007). The extent to which the envelope fails to mitigate the adverse effects of the climate has to be rectified by mechanical means. The mechanical installations are then designed to cope with the 'worst case' conditions; while these conditions will be encountered (and perhaps exceeded) during a building's lifetime, the equipment will remain underutilised for the largest portion of it. This means that the equipment will usually be operating at less than optimum energy-efficient performance.

It has been pointed out that in building energy simulations, the reliability of input data and their influence on the uncertainty of the results can be significant, particularly where physical values are based on subjective judgements, or they ignore the effects of quality of construction. These techniques are useful when comparing alternatives since the errors in each variant are similar and cancel each other out (Leifer, 1999).

Given the local climatic conditions, buildings are load- or surface-dominated depending on whether the contents and equipment lead to over-heating, or the thermal transfer through the exposed surface compromises comfort respectively. Where the external conditions and insolation fluctuations oscillate between heating to cooling need, there is an opportunity to use the building envelope's properties to dampen the fluctuations, leading to a lessened requirement for energy-consuming systems to rectify the mismatch.

Initial BEA tools are needed at this early stage to indicate the best strategies to pursue (Szokolay, 1977). During the design development more sophisticated modelling can be carried out as more precise data is made available, leading to the possible use of 'heavyweight' building energy simulation programs such as DOE and ESP. With these latter programs the time investment in developing the design and generating input data makes designers reluctant to revisit major decisions that would be necessary to significantly improve performance. Rather, these designers end up tinkering with ephemera to make minor improvements.

ASSESSMENTS DURING THE OPERATIONAL PHASE

Over time and across nations, concern has widened to embrace building consumption of water, the generation of waste, and the impact on local flora and fauna and the biosphere, in which the actions of the users are causal. Therefore the current BEA tools are concerned with actual performance as compared with that anticipated by the designers and builders.

BEAs such as LEED have come under criticism, for example, Schedler and Udall (2006) make the point that designers using LEED may aim to achieve the greatest number of 'stars' rather than the optimum performance, and note that there is a great deal of bureaucracy involved leading to high transaction costs. The results to date have provided weak correlation to show that 'good' LEED design produces the results predicted.

Energy performance rating systems are designed to measure the *actual* day-to-day usage of energy by buildings. This is different from energy performance estimation systems that are used to *predict* the annual energy usage of building designs, or indeed, energy simulations that are used to model the *dynamic energy performance* of buildings – both existing and in design. Prediction can be carried out by reference to simple benchmarks across building usage types on a per square metre basis, whereas the dynamic simulations look at the thermal transfer's impact on internal comfort conditions. Conversely, the aim of an energy rating system is to ensure that the energy performance of the building lives up to the designed energy performance, and this is made explicit in the Building Code of Australia [1], Part I2 – Energy Efficiency Installations:

> I02 The objective of this part is to reduce greenhouse gas emissions by efficiently using energy throughout the life of the building.

> IP2.1 A building's services must continue to perform to a standard of energy efficiency no less than that which they were originally required to achieve.

Ratings' purpose has been centred on generic building typologies, particularly residential and commercial offices. The variables that affect the energy usage of buildings are many; they include the local climate and internal environmental set-points; the buildings' usage patterns; and the buildings' designs. It is assumed that building clients and designers are capable of developing a 'green brief' (Hyde et al., 2007) and proposing building envelopes that are climate-responsive and mitigate the excesses of the exterior climate.

'A whole of building by energy comparison does not deliver useful information' (Obrart, 2010) for campus buildings. The utility of measuring and monitoring energy usage is limited by the degree of aggregation of the information obtainable; for example, while periodic electricity bills can show how much energy is used per period, they fail to show where it has been used unless metering has been set up on each electrical circuit or better still, each energy-using device. It may be confidently assumed that the development of 'intelligent metering' will eventually overcome this problem, perhaps by installing a reporting chip in every device. Moreover, the boundary of control of the various stakeholders, that is, the use that will be made of the information, will depend on whether the building owner is also the building user, if the building is leased to a single tenant, or leased to multiple tenants.

The separation of fields of concern of different stakeholders has led to consideration of rating either (a) the base building; (b) the individual tenancies; or (c) the whole building (base building plus tenancies). Indeed, recent legislation mandates energy performance ratings to be certified and displayed in some building types, or included in advertisements when they are sold (Building Energy Efficiency Disclosures Act, 2010).

ASSESSMENTS IN AUSTRALIA

The BEA tool favoured in Australian is Green Star that adopts NABERS as its energy rating engine. This is a performance-based rating system for existing commercial buildings using measured data including rated area, hours of operation, number of occupants, number of computers (a surrogate for occupancy numbers), postcode (climatic variation) and the annual energy consumption (from utility bills), all according to a strict protocol exercised only by accredited assessors. In its rigorous implementation it contains rules about which spaces are included or excluded from consideration. There is a self-rating and reverse calculator available on the NABERS website (www. nabers.com.au) for use by building occupants or users. There is no stringency on inputs, hence accuracy of output, but nevertheless it provides a useful guide to results although it cannot be used to make binding contractual claims. This version's inputs are much reduced and require only the area being served and the annual energy consumption to give a performance compared to 'benchmarks' for well and poorly performing examples (Hyde *et al.*, 2007).

It has been convincingly demonstrated (LEHR by the Warren Centre, 2009) that the skill and knowledge of a building's users can greatly contribute to the efficient operation of a building. Chapter 2.3 shows how a major building owning organisation's property portfolio has documented a longitudinal study showing a timeline of interventions to the building and the energy consumption improvements that resulted. There is no question that building users are rarely trained up in the use of their premises and will take ad hoc actions to modify their immediate climate, thus upsetting the controls and leading to increased energy use. The NABERS scheme has now been extended to include retail (shopping centres) and schemes for hotels and industrial buildings are also being trialled.

Input and output benchmarking

While most institutional and large property portfolio owners assess and benchmark buildings as a whole, the information does not provide useful data between the diversity of building types, operation and usages. For example, TEFMA (the Tertiary Education Facility Managers Association) have been benchmarking the performance of university campuses following the initiative of Fenn (1991). However, their energy benchmarks are reported in global terms such as kWh/m^2, kWh per student, etc. Unfortunately, university campuses and the buildings that make up these campuses vary markedly. This means that inter-campus and inter-building benchmarking cannot work effectively as the input parameters vary considerably. For example, the presence or absence of a teaching hospital or engineering faculty will make a significant difference to a campus's energy consumption profile. The unit of analysis therefore needs to be carefully scrutinised before the benchmarking process is started so that where inter-building comparisons are made, the input parameters are identified as input benchmarks and this forms the initial mode of comparison. Furthermore within buildings it is possible to break down the building into further sub-functions and benchmark these elements (Table 3.5.1).

A study following this approach is currently under way by Obrart *et al.*, who are collating functional energy consumption benchmarks for identified areas of common use within the diversified buildings that comprise

Table 3.5.1 Functional benchmarks of outputs from selected usage categories in university buildings

Campus buildings preliminary electrical energy consumption benchmarks

Usage category		Electrical energy consumption kWh/sqm/a	
		LOW	HIGH
1	office, admin or academic	150	250
2	library (staffed)	70	390
3	computer labs (over 15 desktops)	97	520
4	lecture rooms (over 15 seats)	90	160
5	commercial (staffed) kitchen	250	1000
6	workshops	120	145
7	teaching labs (undergrad)	50	220
8	research labs (post-grad)	70	80
9	IT server rooms	2100	7720
10	other		

the University of Sydney's property portfolio and those of other universities, i.e. they are attempting to obtain energy usage benchmarks for consumption by individual usages, e.g. computer rooms, lecture theatres, wet laboratories.

While most individual buildings' electrical installations are separately metered, the sub-metering, where it exists, rarely coincides with the usages. Retrofitting sub-metering in existing buildings may be difficult and expensive due to the non-availability of suitable power distribution sub-circuits Therefore it has been necessary to undertake detailed audits of every room in the building. The process utilises trained teams completing room-by-room surveys, collating areas, equipment faceplate data, occupant-operating information with corrections, for each of ten common campus usages which are then summed to provide an accuracy check against the actual building meters' energy use per annum. After surveying 16 buildings totalling approximately 100,000 m² of space to date, the flowing preliminary energy consumption benchmarks are emerging. It will be seen that the predicted energy consumption per square metre for similar usages within buildings can be compared and average and standard deviations calculated.

CASE STUDY: UNIVERSITY OF SYDNEY CAMPUS BUILDING ENERGY AUDITING AND FUNCTIONAL BENCHMARKING FOR PREDICTION AND MANAGEMENT OF ENERGY CONSUMPTION

The widely adopted campus building energy audit approach

From overseas and in Australia, there have been many and varied energy usage surveys and audits completed over the whole of the campus and individual existing buildings. The results are generally presented in terms of normalised consumption per building or group of buildings in kWh (or MJ) per square metre per annum, with sometimes additional breakdowns via type of consumption, lighting, HVAC, etc. and types of energy source, electrical, gas, and also normalised to include building occupancy or campus (faculty student numbers).

Limitations of this information

The preeminent feature of most university campus buildings is the great diversity in style, construction type and usage. This means the whole of a building survey, audit and energy consumption information is of limited use for comparison with other, very different campus buildings, or in establishing reliable, functional energy consumption benchmarks or for use in estimating and managing energy consumption.

The key is to focus on similar academic, administration and user areas within buildings, across the differing campus building stock, to establish and verify functional energy consumption benchmarks for common campus areas and usages, such as:

- administration and academic offices
- lecture theatres
- research laboratories
- computer laboratories.

A comparison of the whole of the building by energy does not deliver useful information.

Since 2009, graduate students under the supervision of Dr David Leifer and Alan Obrart have completed energy surveys in 16 campus buildings. as part of the University of Sydney Faculty of Architecture Design & Planning (FADP) Graduate Programme Work (DESC9111) Energy Survey and Audit Work (Table 3.5.2). The aims of the study are:

1   to identify input and output benchmarks, to identify cost-effective energy reduction methods for existing campus buildings;
2   to identify the existing building stock in an established campus that has greater potential for energy and greenhouse savings than new buildings which are likely to use new methods from energy-saving design and operation;
3   to enlist campus staff to assist with the development of energy efficiency benchmarks;
4   to provide campus staff with input and output benchmarks enabling them to understand the significant variables leading to energy consumption and to self-rate the outputs for budget purposes.

Table 3.5.2 Building affiliation and use area m². Preliminary Data Usyd Desc 9111 Energy Audits

Bldg no.	Building name	GFA sqm	2009/2101/2011 electrical consumption kWh/a **	Brief building description
A18	Brennan McAllum	7500	454,000	7-9 level (refurb 2000) masonry, admin & labs
G04	Wilkinson	11,313	1,699,000	5-6 level 1960 masonry, admin, lecture rooms, studios, workshops
D04	Bosch 1A	2009	333,000	2 level 1950 masonry, lecture rooms
J03	Electrical Engineering	9396	681,000	7 level 1970 masonry, admin, labs, workshops
J13	Electrical Link	3884	510,000	5 level 1990 masonry, admin, labs, lecture rooms
J02	PNR & Old School	3931	352,000	3 level 1970 masonry admin, library, lecture rooms
D02/QE2	Coppleson	1219	114,000	2 level 1950 masonry, admin & small labs
A28	Physics	8404	798,000	3-5 level 1940 masonry, admin, teaching, labs
F12	Transient	2216	262,000	2 level 1950 fibro, admin, labs
F19	Eastern Ave	3046	427,000	4 level 1990 lecture rooms only
H12	Book Repository		146,000	1 level fibro book repository warehouse
H69	Business & Accounting	7180	1,572,000	8 level 2000 masonry admin, teaching, computer labs
G09	Aquatic	6647	1,524,000	2 level 1990 masonry aquatic centre, sports facility, shops, cafe
A23	Manning	3497	937,800	5 level 1940 masonry student facilities and recreation
A09	Holme	6760	632,000	5 level 1940 masonry student facility and recreation
A29	Physics Annexe	1738	196,000	4 level 1970 light construction, admin & labs

Notes: ** Audited figure within 10 per cent of CIS electrical consumption utility data.
Gas consumption data not available. Individually building metered. ESAP 2008 audit results indicate gas consumption costs less than 10 per cent of electrical consumption in 14 major campus buildings.

Graduate programme energy survey and audit process

The process is essentially as follows:

- five-day intensive training for graduate audit teams;
- one-day building survey and audit done with access and building information from Campus Infrastructure Services (CIS) staff and from interviews with building occupants;
- completion of room by room survey and audit of all energy-consuming equipment;
- spreadsheet calculations and extensions using LLO correction factors for FEF, ELF, and OLF for all line items to estimate total annual consumption in kWh;

- reality-check the total building kWh against (CIS) provided utility bill annual data, looking for +/− 10 per cent correction to verify the spreadsheet calculation accuracy is adequate;
- group the room x room survey data, as corrected, into:
  - consumption type categories – lighting, HVAC, etc.
  - common area and usage categories 1–10.
- present preliminary findings to a meeting with CIS and building staff for comment;
- determine 'top ten' energy efficiency measures (EEM) to apply to rooms or systems, including estimates of:
  - annual energy saving (kWh);
  - annual cost saving ($);
  - emission reduction ($kgCO_2$);
  - cost of measure (supply, install and maintain) ($);
  - simple payback period.
- determine functional energy consumption benchmarks for the 10 categories of faculty usage within the building.

The ten usage categories are:

1 offices (administration or academic)
2 library (staffed)
3 computer lab (over 15 desktops)
4 lecture room (over 15 seats)
5 commercial (staffed) canteen
6 workshops
7 teaching labs (undergraduate)
8 research labs (staff and post-graduate)
9 IT server rooms
10 other.

Preliminary function electrical energy consumption benchmarks (output) from energy surveys and audits

Based on room by room survey with corrected data, correlated with utility information, the data for the ten categories of room usage is given in Table 3.5.3.

Proposed retrofitting benchmarking rating tools

Further work has involved the development of a proposed Retrofitting Rating Tool (Tables 3.5.4 and 3.5.5). This has involved a number of steps. First, the audit team survey data will be broken down into energy use by:

- building – academic, faculty, or administrative unit; and
- areas of common (category) usage type – proposed ten types being considered.

Table 3.5.3 Preliminary function electrical energy consumption benchmarks (output) from energy surveys and audits

Category	Building A18 kWh/m²/a	Building G04 kWh/m²/a	Building A09 kWh/m²/a	Building G09 kWh/m²/a	Building H69 kWh/m²/a
Office	232	171	77	418	130
Library	76	392			
Computer lab	98	396			415
Lecture room			209		
Commercial			530	1007	
Workshops		145			
Teaching labs	227	227			
Research labs		86			
IT server room		4392			7725
Other				741	

Table 3.5.4 Provisional selection guide for LLO energy consumption rating tool benchmark estimates, campus buildings (for use when room by room audit data or sub-metering not available)

Guide for selection of low/med/high energy consumption estimates for particular areas

	Score
Fabric	
non-insulated roof	5
no external shades on sunlit glass	5
Hours of operation per annum	
approx. 2400	10
approx. 1760	7
approx. 1000	5
Installed equipment (compared to average for this type of usage)	
heavy (around 40 watts per sqm)	10
about average (around 25 watts per sqm)	7
low (around 10 watts per sqm)	5
Lighting	
modern fluro (T8, T5 electric ballasts)	5
older T12 (iron ore ballasts)	10
Air conditioning (central or local)	
area fully air conditioned	15
partially air conditioned	5
Controls (for lights, equipment, air conditioning)	
auto timers, sensors, BMS	5
local switch only	10

Total scores
Around 50 or above: high benchmark
around 25–50: medium benchmark
around 25 or below: low benchmark

Table 3.5.5 Energy Consumption LLO Output Rating Tool Form 2 (Oct. 2010)

Item	Usage/space/area category	Provisional kWh/sqm/a (GFA)		
		Low	**Medium**	**High**
1	office (with associated areas similar to NABERS offices energy requirements	100	130	200
2	library (ref CIBSE Guide F)	50	100	200
3	computer lab (over 15 desktops – ref. EAMP 2009 audits)	300	450	660
4	lecture room (over 15 seats – ref CIBSE Guide F)	110	120	150
5	commercial (staffed) kitchen/café (CIBSE Guide F)	220	300	400
6	workshop (EAMP 2009 audits)	30	40	80
7	teaching labs (undergrad – CIBSE Guide F)	120	140	160
8	research labs (post-grad – CIBSE Guide F)	140	160	180
9	IT server rooms (no. in item 1 – EAMP 2009 audits)	1500	2500	5000
10	other			

The benchmarks for each functional unit will be defined. For example, with offices, consideration of building benchmarks will be carried out by approximations from the NABERS Energy rating requirements. This will include any area occupied and used principally by administration and academic staff, including common areas, stairways, halls and entrances, toilets and amenities, kitchenettes, copy, print and storage areas, IT rooms (serving staff and administration offices only). This description is intended to be consistent (as far as practicable for a campus building) with the NABERS' protocol for an office whole building rating. This will enable use of NABERS (office) energy consumption data benchmarking as appropriate.

This tool will be an essential element where universities wish to connect academic departments directly to the energy they use, for example, utility bill cost recovery. The 2009/2010/2011 audit data is being collated by academic usage to enable these benchmarks to be established. These benchmarks will be refined as more audits are undertaken. Moreover, it will then be practical to consider the extension of the TEFMA benchmarking on Energy Consumption to cover these sub-groups of user. Data to be included is shown in Table 3.5.6.

The aim is to establish energy consumption benchmarks from the selected ten typical area usage types. This information will ultimately be used to increase the energy efficiency of the complete building, and the whole campus.

Table 3.5.6 Energy Consumption LLO Rating Tool Form 3 (Oct. 2010)

Campus location
Building ID
Academic or administration space ID
Area GFA (m²)
Data on input systems and benchmarks Form 1 (see Tables 3.5.4 and 3.5.5)
Select provisional benchmark low/med/high doc. Form (see Table 3.5.5)
Compute the area (m²) x benchmark (kWh/a) = estimated total area kWh/a
    consumption
Optional – if energy costs are available – total kWh/a x $/kWh = estimated
    consumption annual cost

FURTHER RESEARCH

Additional areas for further research involve energy demand control by load shedding and power down non-essential systems before peak KVA is exceeded: load sharing is sharing energy-consuming equipment between academic areas. By setting input benchmarks for these parameters, it is possible to use the tool to better manage the building and other systems pertaining to the building such as the electricity grid.

CONCLUSION

Benchmarking in the context of BEA tools is still a work in progress. While benchmarks resulting from computer simulation have yet to be reconciled with operational performance, the substantive purpose of benchmarking is to encourage 'management by exception' and 'continuous improvement', leading to creative ways of achieving improved performance hence greater sustainability. BEA tools vary in the extent of rigour with which they rely on benchmarking.

ACKNOWLEDGEMENTS

We thank the CIS, The University of Sydney.

REFERENCES

DOECCW (2010) *NABERS*, Department of Environment Climate Change and Water. Available at: www.nabers.gov.au.
Fenn, B. (1991) 'Assessing the efficacy of maintenance management structure: an innovative methodology', MBA thesis, The University of Queensland.

Henriksen, J. and Leifer, D. (2007) 'What is a designer's CO_2-e impact?', in *Proceedings of ANZAScA Conference*, 2007.

Hyde, R.A., Prasad, D., Blair, J., Moore, R., Kavanagh, L., Watt, M. and Schianetz, K. (2007) 'Reviewing benchmarking approaches for Building Environmental Assessment tools (BEA) – rigour versus practicality', in the *Proceedings of 24th International Conference on Passive and Low Energy Architecture (PLEA 2007)*, Singapore, 22–24 November 2007, pp. 58–67.

Leifer, D. (1999) 'A source of error in thermal simulation programs', paper presented at 16th International PLEA Conference, Brisbane, 21–23 Sept. 1999, Vol. 2, pp. 609–11.

Leifer, D. (2003) 'Building ownership and FM', *Facilities*, Research Publication of EuroFM, 21(1 & 2): 38–41.

NBSS (2010) *National Building Energy Standard, SetSCILing, Assessment and Rating Framework*, National Strategy on Energy Efficiency, Canberra: Commonwealth of Australia, March.

Obrart, A. *et al.* (2010) 'Energy benchmarking', unpublished paper TEFMA Conference paper.

Schedler, A. and Udall, R. (2006) *LEED is Broken: Let's Fix It.* Available at: http://grist.org/comments/soapbox/2005/10/26/leed/index1.html (accessed 13 August 2006).

Szokolay, S.V. (1977) *Solar Energy and Buildings*, New York: Halsted Press.

Warren Centre (2009) *LEHR, Low Energy High Rise Building Energy Research Study: Final Research Survey Report*, March.

PERFORMANCE MODELLING TOOLS

What role can computer-based software tools play in informed decisions in predicting?

Lester Partridge

INTRODUCTION

The economic viability of a commercial office retrofit is a function of the return on the capital investment. Often retrofits only consider the aesthetics of a building and in these cases the return is realised by an increase in revenue from reduced vacancy rates. Predicting the return on the investment in these cases is more of an art than a science as aesthetics can be so subjective. In contrast, when a building retrofit involves the upgrade of the façade and services, it is the building performance that is affected. And this performance can have an impact on the occupant comfort as well as the demand and consumption of utilities such as electricity and water.

The increased value of the building will be realised through reduced outgoings (Partridge, 2011). These reduced outgoings include mainly electricity and water charges but there is also an argument to suggest that improved comfort will increase tenant productivity and hence tenant retention rates.

When undertaking a retrofit, building owners require a solid, well-developed business case. Among other things, the business case must provide an accurate estimate of the upgrade works and an accurate estimate of the impact on the building's performance and outgoings (Table 3.6.1). Engineering consultants, cost consultants or trade contractors can determine upgrade work costs once a scope of work is defined. However, an estimate of the building's improved performance requires performance modelling. The relationship between upgrade scope and improved value is well documented in recent research work undertaken in Australia (Arup, 2009). This work was expanded to develop six levels of upgrade scope studied by AECOM and related to building value. The six levels ranged from 'Do nothing' to full 'Knock down and rebuild' (Partridge, 2011) (Figure 3.6.1).

Table 3.6.1 Scope of upgrade works for each of the five levels

Upgrade scope	Level 0	Level 1	Level 2	Level 3	Level 4	Level 5
Do nothing	■					
Painting		■	■	■		
Recommissioning		■	■	■		
Management systems		■	■	■		
Minor repairs		■	■	■		
Upgrading and replacing central plant			■	■		
Replacement of on-floor lighting				■		
Replacement of on-floor air conditioning				■		
Replacement of façade elements				■		
Strip back to structure					■	
Modifications to structure					■	
Replacement of lifts					■	
Replacement of façade					■	
New building services					■	
Demolish and rebuild						■

Source: AECOM.

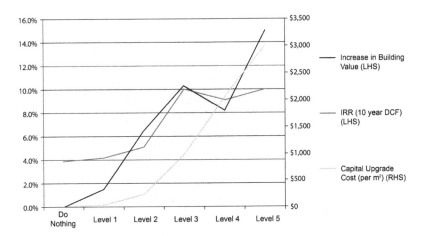

Figure 3.6.1 Relationship between upgrade scope and building value

Source: Partridge (2011).

Performance modelling of buildings is undertaken using Building Environmental and Energy Modelling (BEEM) software (CIBSE, 1998). Although this type of software has been available for a few decades, it is really only since the late 1990s that the development of software and more powerful high-speed personal computers have made its use more prevalent in industry. In addition, the recent introduction of building rating schemes, government incentives and legislation in some countries has also dictated the requirement to use modelling applications to demonstrate performance capability.

BEEM MODELLING

BEEM describes a number of potential applications for modelling a building's performance. Each application is tailored to analyse a certain aspect of the building's performance. For example, there are performance modelling tools for thermal load, energy consumption, daylighting, lift performance, natural ventilation, shadow prediction, duct design, pipe design, evacuation, lighting design and others. There is also Computational Fluid Dynamic (CFD) software, which is often applied for informing the designs with respect to air distribution (CIBSE, 1998).

Over the past ten years a number of commercial entities have acquired the rights to use the codes of some of these software applications to develop an integrated tool. IES (IES, 2012) and TAS (EDSL, 2012) are two common examples used worldwide which provide a common platform for applying various modelling software applications.

Models are typically constructed using a three-dimensional geometry (Figure 3.6.2).with thermal properties applied to solid elements such as walls, roof, floors, glass, doors, etc. Within each space, internal heat loads such as lights and equipment can be applied with hourly profiles for the day,

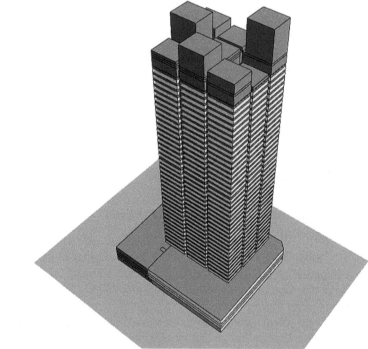

Figure 3.6.2 Typical energy simulation model for tall building

Source: AECOM.

week or year. Neighbouring buildings can also be modelled to incorporate the impact of shadows on a façade. When daylight or lighting analyses are carried out, there is a requirement to include material colour, surface roughness and specularity.

For a large building, the amount of effort and time spent in developing a fully detailed model can be considerable, ranging from two to six weeks for more complex geometries. More recently, the development of Building Information Modelling (BIM) promises to facilitate data entry for software as the new generation of BIM software is designed to encompass all the information of a building's design and construction in the one model (Autodesk, 2012). The model, rather than containing lines to describe walls, doors, plant items, etc., uses components referred to as families. Each of these families contains meta-data which can describe in full what the component is, where it is located, what it is connected to, what it is made of, what colour it is, what thermal properties it has, what it costs, when it will be installed.

Armed with a fully specified BIM model, the engineer has only to apply the software applications to obtain performance output – in theory. However, at the time of writing it is recognised there are still many interoperability issues that prevent full integration of BIM and BEEM software. These issues are not just software capability issues, but also modelling standards and hardware capability.

Often the engineer will need to simplify the architectural geometry to facilitate the thermal simulation model. This is necessary as complex glazed geometry can add an order of magnitude to the run time of the simulation, due to the additional quantity of complex calculations required to be undertaken. It is anticipated that these current issues will be addressed in coming years as more software applications become available and computer hardware becomes more capable. With respect to retrofitting existing buildings, this development will only be realised in the coming decades as the current stock of newly BIM documented commercial developments become ready for their first retrofit.

APPLYING BEEM SOFTWARE TO RETROFITTING

Retrofitting activities are often associated with either aesthetic or performance upgrades or indeed both. When a performance upgrade is required for a building, it is often necessary to determine some key performance indicators (KPI), which it is agreed will be improved. For many upgrades, the KPI is energy-based; however, more recently emphasis has been given to using equivalent carbon dioxide emissions expressed in $kgCO_2$. This move away from the traditional units of energy such as MJ or kWh is a reflection of the current environmental climate change debate. The benefits of measuring $kgCO_2$ are that it is the end product of a building's energy consumption.

It is also a reflection of the primary energy source, which is generated at a remote power plant and incorporates the efficiencies of power generation and distribution. Monitoring MJ/m^2 or kWh/m^2 can be deceptive as is demonstrated by buildings which install gas fired co-generation or tri-generation systems. Because of the inherent low carbon content of natural gas when burnt in comparison to the carbon content of electricity generated by coal burning power stations, there is a significant reduction in CO_2 emitted; however, there is a marked increase in MJ consumption (National Greenhouse Accounts Factors, 2011).

Consideration must also be given to how a building is operated when determining KPIs. Factors, including climatic region, hours of occupancy, level of vacancy and density of a tenant accommodation, can all have an influence on a building's energy or its greenhouse gas performance. Simply comparing raw energy or greenhouse gas emissions with other buildings will provide a flawed comparison. One system developed in Australia in the late 1990s overcomes this comparison issue. Originally known as the SEDA rating scheme, followed by the Australian Building Greenhouse Rating (ABGR) scheme and currently the National Australian Building Environmental Rating Scheme (NABERS), the tool uses an algorithm to normalise a building's carbon emissions based on area, hours of usage, climate and tenancy density (NABERS, 2012). By using this rating tool, buildings of similar class types can be readily compared and a realistic benchmark determined, leading to pragmatic and stretch performance targets.

Baseline calculations

Before embarking on any retrofit to improve the performance of a building, it is necessary to determine how the building is currently performing. Most commercial buildings that are less than 20 years old are fitted with some form of digital Building Management System (BMS). Energy metering can nearly always be relied upon from the building's utility meters for electricity, gas and diesel oil for generators.

However, what is often missing in many older buildings is sub-metering information of electrical and gas energy. Only more recently has the requirement for sub-metering become more valued in trying to develop higher performing buildings. New buildings are routinely being installed with a network of smart meters to collect information on energy usage of major and minor energy consuming items. Armed with this information, energy engineers, designers and building managers are able to quickly identify poorly performing plant and take corrective action through upgrade or control alterations.

Recommended minimum sub-metering requirements include:

- Refrigeration chiller plant
- Cooling towers
- Cooling plant pumps
- Heating plant gas or electricity

- Heating plant pumps
- Tenant condenser water plant
- Air handling plant for air conditioning (per plant room)
- Common area ventilation plant
- Plant room ventilation plant
- Car park ventilation plant
- Lift machinery
- Domestic hot water plant
- Renewable energy plant such as wind turbines, solar hot water pumps or PV arrays
- Lobby air-conditioning plant
- Base building lighting
- Car park lighting
- Base building receptacle power
- Standby generation plant and auxiliaries
- Tenancy lighting
- Tenancy receptacle power
- Switchboard, power factor correction and UPS.

Where existing buildings are not adequately sub-metered, it is necessary to undertake simulation modelling to estimate the building performance. This exercise will allow the energy engineer to estimate where the energy is being consumed in the building. Once this is known, the model can then be used to undertake a number of 'what-if' scenarios to test the effectiveness of performance upgrade strategies.

Modelling methodology

Improving the performance of an existing building requires determining where the energy is being used. This requires the development of a virtual building using one of the available three-dimensional thermal simulation packages that are commercially available. It is recommended that the software used is BESTEST certified (Judkoff and Neymark, 1995).

Simulating the accurate performance of a building is contingent on constructing a model that accurately reflects the attributes of the building. It is therefore recommended that emphasis be given to surveying the building, collecting information on commissioning performance, reviewing as-built drawings and logging any sub-metered data that is available. It is also important to understand how the building tenants have operated it over previous years, as this will have an impact when comparing the modelled performance against the monitored performance.

Infiltration is always difficult to determine or predict. In some of the colder climates in Europe, infiltration levels are legislated to meet certain performance criteria. This criterion is measured in m^3/hr per m^2 of façade area when applied to a 50 Pa pressure differential. New buildings in the UK are readily tested using large fans to literally blow up the building. This method of determining infiltration levels is usually not feasible in existing occupied buildings and so assumptions need to be made as to the 'leakiness' of the façade (CIBSE TM23 2000).

Once the building structure, façade, operating profiles and precinct have been created in the model, the building services must be constructed. The more complex software packages allow the energy modelling engineer to develop a virtual recreation of the building services to serve each zone. It is important that this aspect of the modelling activity is well researched with respect to replicating what is actually on site. Plant capacities, efficiencies, commissioned ratings and zoning must all accurately reflect the actual building.

Once the building services plant has been specified in the model, the most complex and potentially time-consuming aspect of the process occurs in recreating the control functionality. Each plant item will need to be modelled to replicate start and stop profiles, response to high temperature, and low temperature, high-select or low-select, full outdoor air cycle, etc. (Figure 3.6.3). Validating the control strategy by testing one floor and monitoring how it performs on peak days can give the energy modelling engineer an indication of whether the model is mirroring the performance of the actual building space.

When the model is finally completed and operating, it can then be simulated for a full year using hourly weather data. Weather data files include Typical Reference Year (TRY) or International Weather file for Energy Consumption (IWEC) files (Gard Analytics, 2012).

For existing buildings, the energy performance of the model often bears little correlation to the meter readings of the actual building. This is not

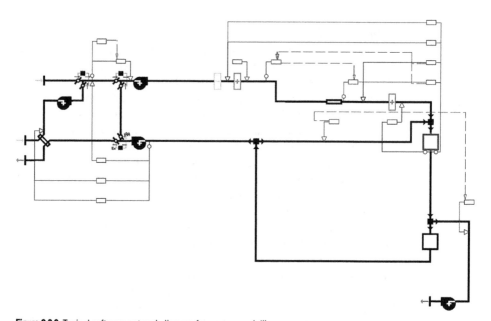

Figure 3.6.3 Typical software network diagram for energy modelling

unusual as buildings rarely operate as well as computer simulations. Existing buildings commonly suffer from problems such as plant breakdowns, faulty dampers or valves, dirty filters or lights being left on after hours, sensors out of calibration, etc. The difference in the energy consumption values between the model and the measured energy can provide an indication of the true potential of the building's performance.

Undertaking a sensitivity analysis of the building model will demonstrate the robustness of a design when in operation. The sensitivity analysis can inform the energy modelling engineer which aspect of the design will have the greatest influence on the energy consumption.

Typical scenarios for undertaking a sensitivity analysis include:

- Outdoor air free cooling economiser cycle is turned off.
- Economiser cycle reduced to 50 per cent only.
- Supply air off coil temperature to one small zone on the perimeter is permanently supplying 12°C.
- One quarter of one level is operational for 5 hours every weekend.
- After-hours tenant equipment loads are left on at 50 per cent capacity.
- Central zone air handling fan variable air volume minimal turndown to 80 per cent from 30 per cent.

There are many other potential scenarios; however, undertaking each of these individually and then in conjunction with each other will provide an indication of how robust the installed system is when operating out-of-calibration. Armed with these results, the energy modelling engineer should be able to validate the model against the actual building performance.

With a validated model, the energy modelling engineer has an opportunity to undertake an upgrade options analysis. The simulation output (Figure 3.6.4) can be interrogated to identify where the energy is being used in systems such as chillers, pumps, air handling unit fans, cooling towers, heating plant, lights, other ventilation fans, etc, and this information can be used to inform the energy modelling engineer what upgrade options are worth considering. For example, if the model demonstrates that the domestic hot water is responsible for contributing 2 per cent of the building's greenhouse gas emissions, while the air handling unit fans are responsible for contributing 35 per cent, it is evident that strategies aimed at reducing fan power will provide a greater benefit than strategies aimed at reducing domestic hot water use.

The scope of any building retrofit will determine the number of opportunities available to the energy modelling engineer. If the building is to be completely gutted back to the slab and core, there are many opportunities to upgrade the façade and plant; however, retrofits often only include limited services upgrades or at the very least aesthetic changes.

Where there is a budget for a building services upgrade, there is an

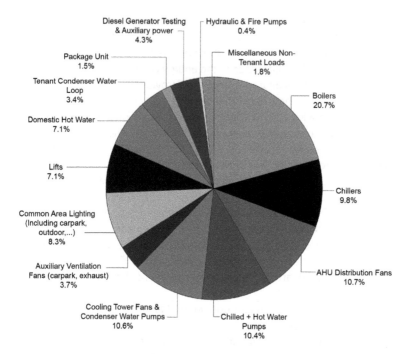

Figure 3.6.4 Typical simulation output demonstrating base building energy consumption

Source: AECOM.

opportunity for the energy modelling engineer to develop a strong business case for each strategy by being able to accurately determine the operational benefit. Modelling the upgrade strategy, calculating the difference in energy, carbon, water and maintenance requirements and providing an upgrade budget construction cost will achieve this. Upgrade strategies, which provide a handsome return on investment, will often be implemented by building owners.

Today with the emphasis on marketing green buildings, it is not uncommon to see building owners implementing upgrade strategies, which have a relatively poor return on investment when accounting for energy savings only. Often their decision is based on a requirement to achieve a carbon emissions target rather than a saving in outgoing energy costs. In cases such as these, the return on the investment is based on reducing vacancy rates by attracting environmentally responsible tenants rather than a direct dollar return on investment by reducing energy costs.

Typical upgrade strategies, which can be considered to improve energy consumption through services upgrade, include:

- replacement of chillers with new higher efficiency models;
- provision of variable speed pumps to serve chilled water circuits;

- replacement of on-floor hot water coil trim reheating systems with direct electric units;
- providing chilled water supply temperature reset to meet the load;
- providing static pressure reset control to variable air volume systems;
- replacing lobby and corridor lighting with low energy compact fluorescent luminaires;
- refitting tenancy floors with low energy single tube 1 x 28w fluorescent;
- installing a low load chiller at 20 per cent of peak load;
- installing low load heating plant at 15 per cent of peak load;
- replacing toilet lighting to achieve below 5 wm^2;
- replacing below deck car park lighting with single tube 36w fittings and painting the ceiling white;
- upgrading the building management system and providing an extensive metering and monitoring system to allow tracking of energy;
- providing variable speed drive to tenant condenser water pump. Ensuring all tenant water-cooled packaged units are provided with solenoid valves so they are isolated when not in use.
- considering using the main system cooling towers to serve the tenant condenser water system;
- providing carbon dioxide sensors in return air ducts to reduce outdoor air when building is not fully occupied;
- providing daylight sensors on perimeter lighting zones to dim lights or turn lights off if daylight is sufficient;
- providing addressable lighting system and control using local occupancy sensors;
- providing photoelectric sensors to activate external lighting to building. Turning off by time
- switch well before dawn or even during early evening;
- providing solar collectors to generate domestic hot water;
- isolating kitchenette boiling water units out-of-hours;
- providing automated blinds on the eastern façade to shade the building on summer mornings;
- providing infrared motion detectors in toilet areas to operate during after-hours periods;
- providing after-hours lights-off to plant rooms by time switch control with warning operation;
- re-configuring lift controls to power down lift and lift controls after hours. Leave one after-hours lift and control system operational for after-hours use.
- turning off lights in lifts if not occupied;
- providing escalators with occupancy sensors to operate only when used;

- providing harmonic filtration;
- providing temperature control shut-off to diesel generator jacket heaters;
- considering gas-fired or waste heating options;
- operating plant room fans on temperature control and providing variable speed drives;
- in meeting rooms where occupancy sensors turn on during after-hours periods, presetting to minimum periods of 1–2 minutes. This prevents regular security guard patrols or cleaning staff bringing on lights which then operate for one hour at a time;
- considering central vacuum system which allows cleaners to work silently during the day thereby preventing after-hours lighting requirements.

Determining which of these or other upgrade strategies provides the greatest benefit is the task of the energy modelling engineer. This can be achieved with a fully validated model of the building as an accurate comparative analysis can be conducted to demonstrate the relative performance improvement for each strategy.

The initiatives listed above should invariably reduce the energy consumption, which in turn will reduce the carbon emissions. However, it is important to simulate each strategy as each strategy can provide a different result for different buildings in different precincts or climatic zones.

As previously mentioned, it is important to be clear on the key performance indicators to target. This is evident, for example, as there are some upgrade strategies that may reduce carbon emission while increasing energy consumption. These include:

- gas-fired engine refrigeration chillers
- gas-fired absorption chillers
- co-generator in conjunction with absorption chiller.

These systems are typically used as a fuel swap strategy by replacing the reliance on mains electricity with that of natural gas. In some regions the carbon emissions associated with mains electricity is significantly higher than that of burning natural gas. It is in these regions that fuel swap strategies can provide significant reductions in carbon emissions.

In order to accurately determine the impact of fuel swap strategies (Figure 3.6.5), modelling is recommended by using hourly cooling load and heating load values from an energy simulation model. The use of hourly data for these analyses ensures that the constraints of plant part load operational capacities can be accommodated in the analysis.

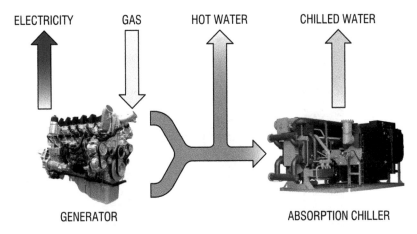

ELECTRICITY   GAS   HOT WATER   CHILLED WATER

GENERATOR   ABSORPTION CHILLER

Figure 3.6.5 Concept of fuel swap using tri-generation

Source: AECOM.

CASE STUDY

The following case study demonstrates the application of energy simulation software for the purpose of reducing greenhouse gas emissions (Partridge and Loughnane, 2008). The building was originally built in the 1980s and in 2006 the owners embarked on a project to reduce the carbon emissions by 40 per cent. Although the 35,000m^2 building had utility meters for electricity and gas, there was little in the way of sub-metering. Extensive energy simulation modelling was carried out to develop a validated model (Figure 3.6.6).

The original model demonstrated that the building was performing well below its potential. An off-axis analysis was undertaken on the model to attempt to validate the energy consumption.

The simulation modelling exercise demonstrated that priority needed to be given to the following strategies:

- Building Management System (BMS) upgrade;
- provision of extensive sub-metering;
- upgrading tenant area lighting;
- upgrading common area (base building) lighting;
- providing intelligent lighting control systems to activate on time of day, occupant sensing and/or perimeter daylight;
- upgrading car park lighting with higher efficiency luminaires and automatic control;
- replacing chillers with new variable speed units;
- providing variable speed drives on all chilled water and heating water pumps;
- providing variable speed drives and controls on all air handling unit fans.

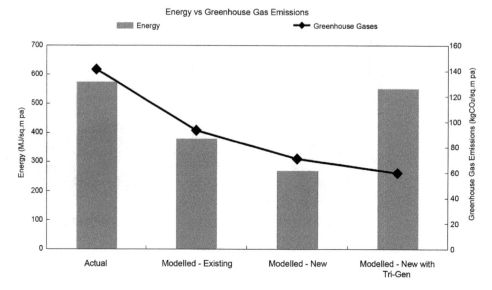

Figure 3.6.6 Energy consumption and GHG emissions chart, demonstrating the comparison between monitored data, existing building operating as per the model and potential savings by implementing upgrade strategies.

Source: AECOM.

A further recommendation was to install a co-generation unit and absorption chiller to provide fuel swap capability. The inclusion of the co-generation and absorption chiller system in conjunction with the energy reduction strategies combined in this case study had the potential to reduce the modelled greenhouse gas emissions by approximately 60 per cent.

CONCLUSION

In the past 10–15 years the development of faster computers and application software with greater capability has provided building engineers with tools to gain greater certainty in energy modelling. These tools have now become a necessary part of the process in demonstrating to clients how their building will perform. In many cases engineers are now being required to validate their predictions 12–18 months after the buildings have been operating. With the drive by tenants and building owners for more energy-efficient buildings, there are sometimes legal considerations associated with not meeting targets. Consequently the role of the energy-modelling engineer has become pivotal to inform the design and demonstrate performance.

REFERENCES

Arup (2009) *Existing Buildings Survival Strategies*, Property Council of Australia, Davis Langdon An AECOM Company, Arup.

Autodesk (2012) *Streamlining Energy Analysis of Existing Buildings with Rapid Energy Modeling*, Autodesk, Inc. Available at: http://usa.autodesk.com (accessed 6 Feb. 2012).

CIBSE (1998) *Building Energy and Environmental Modelling: Application Manual AM11*. Available at: http://www.docstoc.com/docs (accessed 13 Feb. 2012).

CIBSE TM23 (2000) *Testing Buildings for Air Leakage*. Available at: http://www.docstoc.com/docs (accessed 13 Feb. 2012).

EDSL (2012) EDSL TAS software, Available at: www.edsl.net (accessed 6 Feb. 2012).

Gard Analytics (2012) *Building Simulation Weather Data Resources*, Available at: http://gard.com/weather/index.htm (accessed 6 Feb. 2012).

IES (Integrated Environmental Solutions Virtual Environment) (2012). Available at: www.iesve.com (accessed 6 Feb. 2012).

Judkoff, R. and Neymark, J. (1995) *International Energy Agency Building Energy Simulation Test (BESTEST) and Diagnostic Method, NREL, Colorado*. Available at: http://www.iea-shc.org/publications/downloads/task34-BESTEST_Diagnostic_Cases.pdf? (accessed 13 Feb. 2012).

NABERS (2012) National Australian Built Environment Rating System. Available at: www.nabers.com.au (accessed 6 Feb. 2012).

National Greenhouse Accounts Factors (2011) Australian Government Department of Climate Change and Energy Efficiency. Available at: http://www.climatechange.gov.au/~/media/publications/greenhouse-acctg/national-greenhouse-accounts-factors-july-2011.pdf (accessed 13 Feb. 2012).

Partridge, L.E. (2011) 'Strategies for getting the best outcome from retrofitting your building', paper presented at IQPC Retrofitting for Energy Efficiency conference 30 and 31 March 2011, Sydney, AECOM.

Partridge, L.E. and Loughnane, E. (2008) 'Greening your skyscraper: case study in improving the environmental performance of an existing skyscraper', paper presented at CTBUH 8th World Congress, 3–5 March 2008, Dubai, UAE. Available at: www.ctbuh.org (accessed 13 Feb. 2012).

Property Council of Australia (PCA) (1997) *Method of Measurement*, Property Council of Australia. Available at: http://www.propertyoz.com.au/Bookshop/ (accessed 13 Feb. 2012).

3.7
MONITORING BUILDING PERFORMANCE

**How can behaviour and technology changes combine
to improve environmental peformance?**

Craig Roussac

Imagine walking into a supermarket to buy groceries and finding no cash registers, and no price tags on any of the items either, because your bill comes in the mail a few times each year. How would you know if you'd exceeded your budget? Or could you imagine what it would be like to go through life without sensing hunger, thirst, pain or pleasure?

We take feedback signals for granted in our daily lives and arrive at even our most menial decisions, like whether to have a snack . . . or not, by drawing on our bodies' highly evolved monitoring and reporting systems. Everywhere in nature there are mechanisms to monitor the health and stability of systems – ecological, climatic, physiological – and continually make adjustments to restore balance through feedback.

Buildings are not nearly so sophisticated. The modern centrally air-conditioned high-rise building only came to prominence after World War II. And with the arrival of central plant and services, the shape and fabric of conventional office buildings were transformed to such an extent that mechanical systems are now necessary to maintain indoor environmental conditions within an acceptable range (Ackermann, 2002). Yet these systems do not operate like natural systems: they generally are not able to self-regulate. Rather, they rely on competent, committed people to manage them effectively.

The level of competence and commitment displayed by building operators has important implications for building energy use and also the comfort and productivity of occupants. Building operators control the machines which control the buildings. If operators are not reliably informed with timely and actionable feedback, they are unlikely to be able to manage their buildings well. One of the limitations of energy modelling is the fact that these human factors can be very hard to gauge and predict. A motor and fan assembly, for example, may draw 50 kilowatts (kW) while it is operating, and it may be required to operate for 10 hours a day, 5 days a week. But what happens in reality might be quite different. If it is wired or programmed inappropriately, it may run all the time – or it may not run at all. In one example from this author's recent experience, where energy modelling was used to evaluate a range of alternatives for a building services upgrade, the consulting engineers calculated the building's baseline electricity consumption to be 1200 megawatt hours (MWh) per annum when in actual fact it only used 1000 MWh (i.e. 20 per cent less). While it was found that the building's highly competent and committed management team was largely responsible for the difference, the engineers were reluctant to use the building's historical energy consumption as a baseline for predicting the energy use from upgraded plant and equipment. This was on account of the perceived risk of the managers discontinuing their association with the building at some stage in the future and the building's performance declining as a consequence.

Should the engineers have been so cautious? Perhaps not. While it is almost always the case that plant and equipment remain with a building longer than its operations staff, measurement and monitoring systems are able to enhance the effectiveness of any building operator, not just the most talented. Effective feedback systems can be the difference between a building meeting its potential or falling well below it. Fortunately, as we will discuss in the following pages, the steps to building such a system are relatively straightforward.

ACTION REFLECTION APPROACH

This chapter examines an action-reflection approach to retrofitting: collaboration and 'learning through doing' as a means of continuously improving the operation of buildings. The concept of 'action learning' was developed and refined by Professor Reginald Revans in the UK coal industry during the 1940s and 1950s where pit managers were encouraged to meet together in small groups, to share their experiences and ask each other questions. For a summary of the action learning approach, see Revans (1997). The approach was highly successful and Revans went on to refine the method into a formula:

$$L = P + Q$$

Where: L is 'learning', P is 'programming' (or programmed knowledge, i.e. knowledge that is taught or read) and Q is questioning to create insight into what people see, hear or feel (Revans, 1980). Questioning involves the use of closed (e.g. what?), objective (e.g. how many?), open (e.g. why?) and relative (e.g. where?) questions.

The action-reflection approach advocated in this chapter builds on Revans' model ($L = P + Q + R$) where R refers to 'reflection' (see Marquardt, 2004). Through practice we have observed how the effectiveness of measurement and monitoring frameworks hinge on how well information is interpreted, acted upon, and then refined through a structured process of reflection. The following model (Figure 3.7.1) developed by Kolb (1984) illustrates the theoretical basis and shows how the approach works.

Put simply, the action-reflection approach is about solving problems and getting things done. Revans showed that by assembling small groups of 5–8 peers (called an action learning set) and having them meet regularly for a day or half a day over at least 6 months to work collectively on a problem faced in ongoing practice, productivity increased by over 30 per cent (Revans, 1980). An application of this approach is described towards the end of this chapter in the section 'implications for retrofitting'. First, we need to consider the appropriate tools and context.

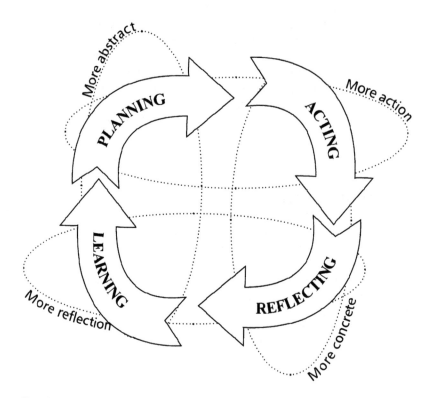

Figure 3.7.1 Learning from experience

Source: Adapted by Serrat (2008) from Kolb (1984).

ESSENTIAL TOOLS

The first step to gain effective control of a large complex building is to ensure that the right monitoring tools are present, starting with metering systems. Almost all buildings have meters to record the main incoming supplies of electricity, gas and water. These meters are generally maintained and read either on-site or remotely by the utility companies for the purpose of preparing monthly or quarterly invoices. Invoices can provide a useful indication of how efficiently the building has operated during the preceding period; however, they provide very little insight into which building components operated effectively and which did not. Figure 3.7.2 illustrates the problem, where each of the boxes represents a significant use of the metered utility and meters are denoted by the letter 'm'. There is simply no way of telling what went where after it passed the meter on the main incoming supply.

Sub-metering can be of great assistance in this regard. Sub-meters can be identical to the main incoming meters in every respect except one: they measure a component of the utility flows and not the whole. As can be seen from Figure 3.7.2, electricity meters at points A, B and C will sum to give the same value as the main incoming meter. The amount of electricity passing through every point can be determined by placing sub-meters at points D, F, G, H, I and J and either summing or subtracting readings from the other sub-meter and main incoming meter readings as illustrated in Figure 3.7.3. The same principle applies for all meter types, creating an effective diagnostic tool to inform management actions. Figure 3.7.3 indicates the bare minimum required to establish distribution patterns though, as a principle, more metering is always better.

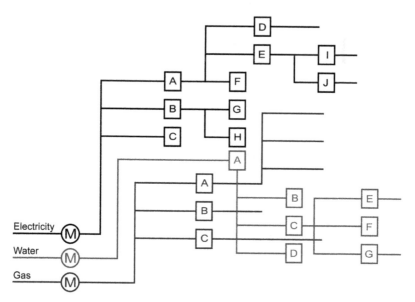

Figure 3.7.2 Building schematic

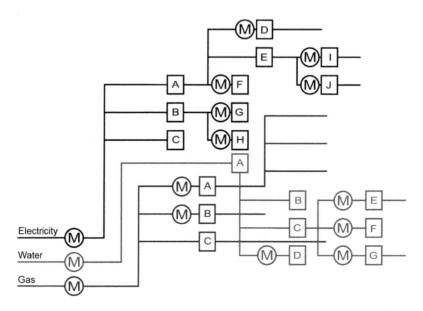

Figure 3.7.3 Building schematic showing a detailed sub-meter arrangement

Sub-meter readings delivered monthly or quarterly will give a useful insight into where resources are being used; however, for the metering data to be useful in an action-reflection approach, it must also be timely. This is because while the location of anomalies may be more apparent, without a sense of what is occurring when, building operators have very little chance of knowing the cause of problems or observing the effects of positive actions – and hence the cycle illustrated in Figure 3.7.1 fails to operate.

A wide range of proprietary monitoring systems are available for purchase. They range from simple 'alarm' systems to sophisticated analytical tools with customised reporting functions and diagnostic capabilities. Often Building Management and Control Systems (BMCS) incorporate a facility to monitor and track meters as well. The key to effective monitoring is to have the relevant information conveyed to the right people so that they can be aware in time to take appropriate action. In that sense, the technological solution matters less than the processes that sit behind it, including allocation of accountability, review frameworks and the general level of engagement that key stakeholders (e.g. building owners, occupants) have with the issues. This aspect will be discussed in the following section.

Investments in metering and monitoring systems produce no directly quantifiable environmental or financial payback. In fact, by themselves they do not produce any benefit whatsoever. Metering and monitoring, which can at one level be described as highly technical, produce results by influencing non-technological factors such as the effectiveness of building operators. Meters merely provide information to inform and motivate action. That is not to say they are anything but crucially important – they are – however, very little research exists to demonstrate that value. Most of the research that does exist is focused on the residential sector where energy savings of 5–15 per cent have been achieved when householders are given direct feedback (via the meter or an associated display) and 0–10 per cent for indirect feedback through informative billing or statements (Darby, 2008). The authors of the chapter on residential and commercial buildings in the Intergovernmental Panel on Climate Change's (IPCC) Fourth Assessment Report (AR4) noted that 'the potential [greenhouse gas] reduction through non-technological options is rarely assessed and the potential leverage of policies over these is poorly understood' (Levine *et al.*, 2007: 409). As a consequence, that influential report did not assess the potential for emission reductions through non-technological options. Yet we know through the experiences of Revans, Darby and others that these factors are highly significant.

POPULAR APPROACHES THAT INCREASE PERFORMANCE

The best commercial buildings exhibit a combination of excellent design and appropriate technology together with highly competent and committed operators. These 'people factors' can be difficult to define and even more difficult to engender in the context of an operational building, with all the competing demands that entails. Hersey *et al.* (2001) define competence as the knowledge and skills an individual brings to a goal or task, whereas commitment is a combination of an individual's motivation and confidence on that goal or task. They found that if either motivation or confidence is considered low or lacking, commitment as a whole will be low (ibid.). Our task in monitoring building performance in order to improve it, therefore, is to influence the commitment and competency of building operators in positive ways.

Benchmarking
One well-established technique for channelling operator interest towards improving the environmental performance of buildings is to make that performance visible to an audience, ideally comprising people whose opinions the operator cares about – for example, the boss! This assumes the audience is interested, of course. Resource flows such as energy and water can have relatively immaterial financial impacts, and, accordingly, may command only a small proportion of senior management time and attention. It should be noted, therefore, that approaches relying on the motivational impact of an engaged audience may be counter-productive without adequate support from an organisation willing to invest in the systems, training and tools to address environmental performance. Furthermore, to a poorly informed audience, more information is unlikely to deliver greater understanding or better behaviour (Janda, 2011). For these reasons, it is crucial that initiatives

be implemented as a suite that combines education with useful information, technology and a forum that facilitates working together.

When the performance of something is being rated, rather than merely 'communicated', you can bet that competitive instincts will take effect and performance will tend to improve. There are now more than 35 recognised 'green' rating schemes around the world that can be applied to commercial buildings. Many of these ratings apply to the design 'attributes' of buildings rather than their performance, but their effectiveness as a means for focusing attention on 'green' issues (as evidenced by the dramatic growth in ratings) is beyond doubt. Ratings such as Leadership in Energy and Environmental Design (LEED) in the USA and the Building Research Establishment's Environmental Assessment Method (BREEAM) in the UK have not only become widely accepted internationally, they now operate as umbrella schemes that include specific ratings for building operational performance along with a variety of other tools designed to service the requirements of stakeholders with an ever-increasing focus on measurement and evaluation.

The distinction between attribute-based tools, which focus on building design, and performance-based tools is important. When we talk about rating the performance of buildings, we are talking about measuring something that requires constant effort and commitment to maintain. And that performance can be independent of the building attributes. Ratings such as the National Australian Built Environment Rating Scheme (NABERS) in Australia and Energy Star in the USA are blind to building attributes, they simply measure performance.

Building performance ratings by their nature make operators feel accountable. In an operating building, or an existing building subject to retrofit, this means that tenants, potential tenants, financiers and a range of other stakeholders can communicate their expectations to those in a position to influence (or control) building performance. Furthermore, the ratings provide a tool for evaluating the extent to which those expectations have been met.

Ratings are highly effective in part because they are easily understandable for a wide audience of interested stakeholders, many of whom might have a relatively basic knowledge of the technical aspects of operating a building. In some instances, a more sophisticated approach is called for. In particular, where portfolios of buildings are concerned, it may be appropriate to set up metrics that can provide detailed insights into operational performance that can be updated daily, weekly or on some other regular basis. For example, if there is an interest in the responsiveness of a building to changes in occupant numbers, a measure of Watt-hours per square metre per week ($Wh/m^2/week$) adjusted for occupancy might be appropriate. Or if it is considered useful to compare energy use with weather data, an adjustment for cooling degree days (CDD) may be appropriate ($kJ/m^2/CDD$). It is important to ensure that the data collected is accurate and meaningful, i.e. that the variables actually make a significant difference, otherwise unnecessary complexity will be introduced for limited benefit.

Engaging

We all know what it feels like to be lacking in motivation: everything seems a little harder than it should. The various systems that exist in buildings need active management from operators who must diligently interact with them. People can be motivated by a variety of things, e.g. by competitiveness, recognition, money, fear, concern for the environment, etc. A motivated operator will be more likely to invest more time and effort in optimising the operation of a building than someone who is not. Motivation is therefore a core concern for efficient building management.

When it comes to operating buildings, some people are more easily motivated than others. Initiatives such as those outlined above that make building environmental performance transparent to a wide audience of interested stakeholders will have varying effects, depending on the confidence, and competence, of the people most directly affected. Those associated with success or 'easy wins' are likely to find disclosure of results motivating, whereas those associated with poor performance may be less enthusiastic. This was the experience of Investa Property Group[1] upon the launch of its 2009 Sustainability Report, the first of its kind to incorporate an interactive data visualisation tool (Investa Property Group, 2010). The 'bare all' approach, which provided insights into detailed monthly performance statistics at an individual building level, was expected to be popular with those staff associated with well performing buildings and less so with those operating the others. Feedback from Investa employees surveyed upon their first exposure to the online tool was enlightening. In response to the question: 'What do you think the consequences of publicly disclosing detailed building-level performance statistics will be for you personally and/or professionally?', 55 per cent of staff (N = 41) from a sample of 74 staff (representing approximately one-third of Investa's workforce) rated the consequences for them as being 'very good' or 'extremely good'. Unsurprisingly, property supervisors working on buildings that had demonstrated significant eco-efficiency improvements were found to respond most favourably to that question, whereas those from poorer buildings were more cautious; though all were more than 'slightly positive'. An overwhelming 70 per cent (N = 52) rated the public disclosure of detailed building-level performance statistics as being 'very good' or 'extremely good' for 'the future performance of Investa-operated buildings'.

Factors beyond the site will also influence a building operator's motivation. In 2007, the British Government's Department for Environment, Food and Rural Affairs (DEFRA) surveyed public attitudes and behaviours towards the environment and found that, for the majority, 'being "green" is seen as the socially acceptable norm' (DEFRA, 2007: 2). They also found that many factors stood in the way of action and, in a subsequent analysis which drew on additional information, produced a segmentation model that 'divides the public into seven clusters, each sharing a distinct set of attitudes and beliefs towards the environment, environmental issues and behaviours' (DEFRA, 2008: 41). The clusters are illustrated in Figure 3.7.4 where they are

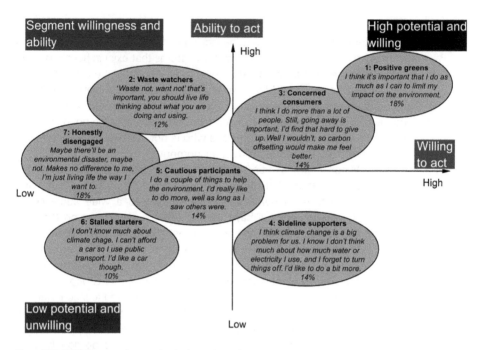

Figure 3.7.4 DEFRA's seven clusters of attitudes to the environment

Source: DEFRA (2008).

plotted according to people's willingness and ability to act on a set of 12 environmental behaviour goals.

DEFRA's model includes detailed profiles of each segment covering, for example, ecological worldview, socio-geo-demographics, lifestyle, attitudes towards behaviours and current behaviours, motivations and barriers, and knowledge and engagement. The clusters were found to vary across population groups and there was evidence that people's categorisation may change over time according to life stage and other individual circumstances (ibid.: 35). The research highlights that building operators will respond differently to approaches aimed at focusing their attention on the environmental performance of buildings and it is necessary to adopt a targeted approach that is cognisant of each individual's mindset.

Mandating

Government-mandated disclosure of environmental performance overcomes the building operator's challenge of finding a commercial rationale to justify focusing on building environmental performance. The flipside is that it can turn the focus from opportunities for value generation to compliance, or 'box checking'. Nonetheless, the perceived benefits of mandating public disclosure of environmental performance are gaining the attention of policymakers around the world. Schemes such as the European Union's Directive on the Energy Performance of Buildings (EPBD), which requires the display of Energy Performance Certificates, and the Australian Government's Building Energy Efficiency Disclosure (BEED) legislation are examples of government intervention to encourage public scrutiny of the performance of buildings. The introduction of mandatory disclosure in Australia (in November, 2011) has led to a significant increase in the number of ratings performed; however, it is still unclear whether this is leading to significant improvements in performance.

Commercial leases between the owners of buildings and tenants are also increasingly being used to set out environmental performance require-ments, often through the use of a 'green lease' schedule. The Australian Government under its Energy Efficiency in Government Operations (EEGO) policy, for example, requires, for the majority of office leases the Australian government enters into, a formal commitment to energy efficiency, including an agreement between landlords and tenants to commit to a minimum ongoing operational building energy performance standard, measured by NABERS. Green leases can be used to create culture change within tenant operations and offices, as well as influencing building managers and building owners by setting out performance and disclosure requirements (Roussac and Bright, 2012).

IMPLICATIONS FOR RETROFITTING

As stated at the beginning of this chapter, we advocate an action-reflection approach to retrofitting commercial buildings – an approach that is about solving problems and getting things done. The question, therefore, is how to put effective tools in the hands of building operators to make it easy for them to identify and address opportunities for improvement. We find that, wherever they exist, such tools can also be useful for engaging stakeholders (e.g. occupants, investors, authorities) by providing a sense of whether performance is improving, or not. This, in turn, generates a focus on per-formance and the motivation to improve as part of a self-reinforcing system. Figure 3.7.5 illustrates how it can work.

The reality for most buildings is the information feedback loop is too slow. Often building operators work in isolation and rely on monthly bills (received some weeks after month end) to give them an appreciation of whether things are getting better or worse. Sub-meter data, where it exists, is usually hidden away and of limited value to all but the most experienced and astute observers. This means actions targeting improvements in environ-mental performance in one area can be unwittingly offset by problems in another. This has led to 'technological' fixes being favoured over 'behavioural' approaches that can often be faster, cheaper and more effective. Technology is crucial for building performance; however, the effectiveness of the people operating the buildings ultimately influences the success of those invest-

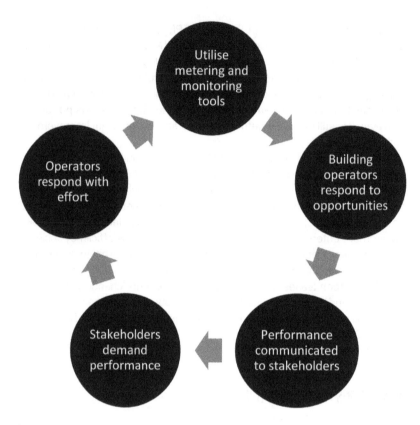

Figure 3.7.5 Monitoring and measurement in a reinforcing loop

ments and, indeed, the likelihood of investments being made in the first place.

Here is a specification for a 'best practice' approach to building performance monitoring. It should build from a base of highly effective metering and monitoring tools that will allow the following:

1   utility usage profiles to be scrutinised in detail;
2   learning opportunities (anomalies between predicted and actual impacts as well as 'cause–effect') to be identified;
3   collaborative team-based problem-solving to be fostered;
4   the success of interventions to be measured;
5   ongoing performance and repeat behaviours to be monitored;
6   a detailed account to be accessible to a wide audience of interested observers.

A sophisticated approach might be one that automatically uploads 15-minute interval energy, operations and weather data to a web-based portal where it is transformed through automated calculations to 'predict' usage based on observed relationships between historical data. Variations between actual and predicted impacts (in particular, energy use) will point to learning opportunities. This is where the 'action learning' comes in. Small multidisciplinary teams can be assembled to evaluate and explain 'better than expected' performance and performance that is 'worse than expected'. Actions can be taken on one day and their effects observed the next. The teams can then consider the measured performance with reference to the desired goals through a process of questioning outlined above. By reflecting, the information can be interpreted and acted upon.

CONCLUSION

It can be safely assumed that most people involved with the operation of commercial buildings would prefer their buildings to impact the environment as little as possible while they set about delivering maximum comfort to occupants. However, there are many demands competing for the time and attention of the people charged with operating buildings. The task of enhancing building performance needs to be made as simple and efficient as possible. It requires timely and actionable data. Furthermore, it requires a framework to analyse that data, deliver meaningful information, facilitate action and then allow reflection to drive improvements.

NOTE

1   Investa is a large Australian-based commercial office building owner and operator.

REFERENCES

Ackermann, M. E. (2002) *Cool Comfort: America's Romance with Air Conditioning*. Washington, DC: Smithsonian Institution Press.

Darby, S. (2008) 'Energy feedback in buildings: improving the infrastructure for demand reduction', *Building Research & Information*, 36(5): 499–508.

DEFRA (2007) *Survey of Public Attitudes and Behaviours toward the Environment: 2007*, London: Department for Environment, Food and Rural Affairs.

DEFRA (2008) *A Framework for Pro-Environmental Behaviours*, London: Department for Environment, Food and Rural Affairs.

Hersey, P., Blanchard, K.H., and Johnson, D.E. (2001) *Management of Organizational Behavior: Leading Human Resources*, 8th edn, Englewood Cliffs, NJ: Prentice Hall.

Investa Property Group (2010) *Sustainability Report 2009*. Available at: http://www.investa.com.au/sustainability/results/ (accessed 27 May, 2011).

Janda, K.B. (2011) 'Buildings don't use energy: people do', *Architectural Science Review*, 54(1): 15–22.

Kolb, D.A. (1984) *Experiential Learning: Experience as the Source of Learning and Development*, Englewood Cliffs, NJ: Prentice Hall.

Levine, M., Ürge-Vorsatz, D., Blok, K., Geng, L., Harvey, D., Lang, S. *et al.* (2007) 'Residential and commercial buildings', in B. Metz, O.R. Davidson, P.R. Bosch, R.

Dave and L.A. Meyer (eds) *Climate Change 2007: Mitigation. Contribution of Working Group III to the Fourth Assessment Report of the Intergovernmental Panel on Climate Change*, Cambridge: Cambridge University Press.

Marquardt, M. J. (2004) *Optimizing the Power of Action Learning: Solving Problems and Building Leaders in Real Time*, Palo Alto, CA: Davies-Black Publishing.

Revans, R.W. (1980) *Action Learning: New Techniques for Management*, London: Blond and Briggs.

Revans, R.W. (1997) 'Action learning: its origins and nature', in M. Pedler (ed.) *Action Learning in Practice*, 3rd edn, Aldershot: Gower, pp. 3–15.

Roussac, A.C. and Bright, S. (2012) 'Improving environmental performance through innovative commercial leasing: An Australian case study', *International Journal of Law in the Built Environment*, 4(1): 6–22.

Serrat, O. (2008) *Action Learning*. Available at: Asian Development Bank: http://www.adb.org/Documents/Information/Knowledge-Solutions/Action-Learning.pdf (accessed 12 June 2011).

3.8

A DIAGNOSTIC TOOLKIT FOR MULTI-DIMENSIONAL TESTING OF BUILT INTERNAL ENVIRONMENTS

How can the physical parameters of existing buildings be measured?

Mark B. Luther

INTRODUCTION

One of the key deficiencies in the entire building procurement process is the lack of feedback to guide future improvements. In bioclimatic architecture it is essential to track weather behaviour as well as interior results as the weather is a significant driver of building performance. Effectively, continuous monitoring, with proactive and reactive responsive building awareness, is needed.

In an effort to better understand the actual performance of buildings the Mobile Architecture and Built Environment Laboratory (MABEL) was conceived and constructed. The methodology is to measure and evaluate on-site indoor environmental building performance in a rigorous manner. In its six years of operation, MABEL has performed the measurement of over 35 buildings including schools, offices, hospitals, sports centres and houses.

MABEL is based on the capability of measurable parameters and their analysis from advanced state-of-the-art equipment. It relates to that which is physically and reasonably possible to be measured on site (non-laboratory) and in compliance with standards or best practices. It is in contrast with controlled environment (closed) experiments.

The greatest challenge in the development of an on-site building environmental performance programme is to recognise parameters that can be measured directly through instrumentation, that can be calculated from results, and that can be combined in recognised relationships to report usable and useful information as well as advancing our knowledge of overall building performance (Luther, 2009). In an attempt to meet some of these challenges of measurement, the MABEL programme has identified several environmental parameters and their deliverables.

The ISO 7730 thermal comfort model applied here is often referred to as the 'Fanger' model after its originator, Dr Ole Fanger. The calculation assesses how comfortable people would be at a given period in time at a specific location in the building. This results in the Predicted Percentage Dissatisfied (PPD) in accordance with the ISO 7730 standard. This comfort model considers the activity level (metabolic rate = MET) and the clothing level (CLO) of the occupant. Comparisons to this model are made with the new 'adaptive comfort' models (de Dear and Brager, 1998) which support the principles of bioclimatic architecture.

The comfort cart instrumentation has been designed according to the ASHRAE Standard (American Society of Heating, Refrigerating and Air-Conditioning Engineers). The carts measure the dry-bulb (DB) temperature, the globe temperature (Tg) and the air velocity at 0.1m, 0.6m and 1.1m heights. An additional relative humidity sensor is also included at a central location on the cart. More recently the carts have been developed into an ADPI (air diffusion performance index) ASHRAE-113 standard for determining draughtiness. Instrumentation includes a 1.7m velocity probe and additional temperature sensor (Table 3.8.1).

A BIOCLIMATIC PROCESSING OF MEASUREMENT RESULTS

Because interior building performance is affected by the external weather conditions, interior results for a particular project are only meaningful in the context of on-site collected weather data. A set-up of the instrumentation is pictured in Figure 3.8.1. Fifteen-minute averages of wind speed, wind direction, solar radiation (direct and diffuse), global illuminance, air temperature, pressure and humidity are sampled and recorded throughout the measurement period.

Such weather data is displayed on a chart on which is also shown a model of neutral temperature (as outlined by Szokolay, 2004) with its 90 per cent (central horizontal shaded) and 80 per cent (external horizontal shaded)

Table 3.8.1 Measured parameters for indoor environmental quality performance.

Parameter	Description and deliverables
Power	
Energy use	Excessive energy use, period of operation, AGBR compliance. Energy monitoring
System defects: equipment efficiency	Diagnostic fault finding in HVAC control systems, equipment scheduling, and operational periods
Flow rates in pipes and ducts	Measurement of flow rates and temperature (energy) in chilled /hot water HVAC systems
Building envelope analysis	Measurement of façade heat transfer and thermal imaging for diagnostic and visual analysis
Lighting	
Background illuminance	Natural and artificial light levels at the workplace
Task lighting illuminance	Workstation light levels from observer to screen and screen to observer: total of six lux measurement points
Correlated colour temperature (CCT)	Investigated light sources and ranges of colour. Light colour variation can have a psychological influence on comfort
Work place brightness/contrast	The glare problem is quantified in accordance with international best practice, regarding glare and discomfort
Daylight autonomy	Daylight factors or rations indicate how much electrical lighting is needed to supplement the natural lighting
Comfort	
Weather, solar and light (external on site data)	Determines the external conditions which influence the internal measured parameters
Thermal comfort levels	A 'comfort cart' is used to measure (PMV/PPD) occupant comfort at the workplace
Surface temperatures and radiant asymmetry	Influence of mean radiant temperature balance within the space
Drafts in air distribution	Draught index at a specific cross-section within a space. Air distribution performance index (ADPI)
Air temperature stratification	The variation of air temperature with height within the space
Ventilation & indoor air quality	
Air change rates	Room ventilation rates (effective air change): HVAC normal operation and HVAC off : air infiltration
Uniformity of supply air distribution	The tracer concentration is measured at a number of locations to determine the balance of the air supply
Indoor air quality (CO_2, VOCS, dust particulate)	Diagnostic testing of air quality levels over time and location within an environment
Fume hood testing	The efficiency of capture of contaminants from laboratories
Air leakage and building envelope analysis	Fan pressurisation testing is used to quantify building envelope leakage
Sound	
Background noise: interior	Background noise levels at each workplace, frequently dependent. Percentages of loudness over time
Background noise: ingress	Speech privacy vs. speech intelligibility
Reverberation time	Acoustic 'liveliness' (sustained sound) in a room
Partition sound transmission	Sound transmission class (Rw) through partitions
Sound intensity; noise sources	Identification of sound leakage areas and sound sources

Figure 3.8.1 Weather station together with solar tracker

comfort bands. Results are produced from measured data and plotted according to the adaptive model of comfort according to de Dear and Bragger (1998), see Figure 3.8.2.

These 'comfort bands', derived from external weather data, set the scene for indoor comfort when occupants clothe themselves appropriately and have the opportunity to adjust and control their environments to some degree (e.g. opening windows, turning on ceiling fans, drawing shading, etc.). As discussed previously, alongside this information are the comfort carts, measuring the internal conditions of the space inclusive of CO_2 levels (see Figure 3.8.3).

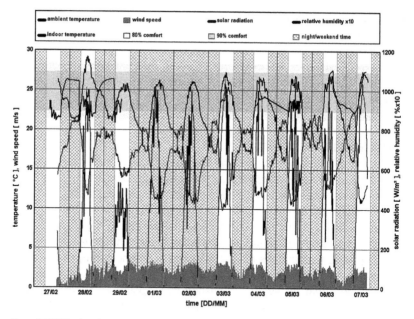

Figure 3.8.2 Weather data and comfort bands in Brisbane, March 2008

Figure 3.8.3 Two thermal comfort carts designed according to ASHRAE by R. de Dear

Measuring energy-efficient strategies

The objective is to illustrate how a measurement facility can contribute to recognising the assets and liabilities of real *in situ* building performance. The idea is to highlight applications where building measurement has provided evidence in improving our construction and design detailing, material use, as well as optimum control.

A variable temperature set-point

In Figure 3.8.2, the exterior air conditions of relative humidity and dry-bulb conditions as well as wind speed and solar radiation are charted. The interior air temperature is plotted alongside these values. Note that a 'comfort band' is provided for the thermal range. For this case, several office buildings are measured locally during the course of the 10-day period. Figure 3.8.2 indicates that the interior temperature level is within the comfort band. However, it is suggested here that the interior temperature set-point may be raised, in several of the measured premises, allowing for greater cooling energy savings. Note that for every 1°C change in set-point, an average energy savings between 8–12 per cent in conditioning can be anticipated (Egan, 2010; Roussac *et al.*, 2010).

Economiser cycle and ventilation control

In reference to Figure 3.8.3, it is recognised that quite often (mostly during non-occupied hours) the external temperature is less than the conditioning set-point or that the external air directly meets the interior air requirements. This indicates that there is an opportunity to condition or pre-condition the interior for the forthcoming day as well as remove excess heat from the building. It can also indicate that there is no reason for the mechanical chilling or heating plant to operate. In other words, the external air provides the conditioning and the only concern is that this air is distributed throughout the building. Figure 3.8.4 shows plots of the quality of the external and internal air for a specific day in an office building at Brisbane Airport, Australia. The air quality, its humidity level and dry-bulb temperature are plotted on a psychrometric chart providing information on the energy treatment required to bring the external air to its desired conditioning. In this case, very little conditioning, if any, would be required since the external air temperature can provide the desired conditions. Conversations with experienced building control engineering companies in Australia state that 80 per cent or more of commercial buildings are not implementing an economiser cycle effectively.

Building air leakage control

In too many observed cases it is witnessed that our buildings are excessively leaky. This is to the point where the successful operation of the mechanical heating and ventilation equipment is severely affected from operating as

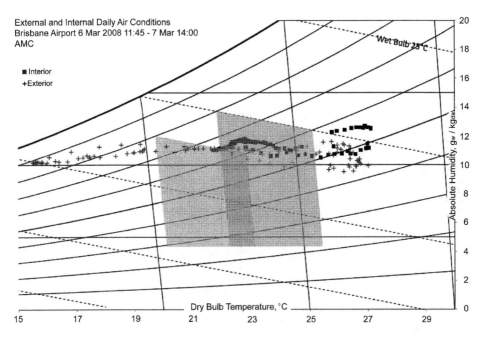

External and Internal Daily Air Conditions
Brisbane Airport 6 Mar 2008 11:45 - 7 Mar 14:00
AMC

■ Interior
+Exterior

Figure 3.8.4 Exterior and interior air conditions plotted on a psychrometric chart

specified. Poor detailing and a lack of attention to construction assembly are to blame. Figure 3.8.5 provides an example of a before and after air leakage testing result according to the CIBSE TM-31 Standard (conducted by Air Barrier Technology, Australia). The results of before and after are shown in Table 3.8.2. Consequently, the building owners were able to save on additional cooling equipment costs. Furthermore, the mechanical conditioning remained in operation during a 40°C+ week heatwave. All of this is due to the reduction of excessive infiltration and exfiltration within the building envelope.

Table 3.8.2 Fan pressurisation before and after tests

Building area	West Wing before	West Wing after	East Wing before	East Wing after
Building volume m³	10394	10394	4269	4269
Test results				
Flow @ 50Pa m³/hr	86604.00	46575	31672	19544
Air exchanges @ 50Pa	8.30	4.5	7.42	4.58
Flow @ 4Pa m³/hr	16770	9019	6133	3784
Air exchanges @4Pa	1.3	0.9	1.45	0.9

Source: Courtesy of ABT Pty. Ltd.

(a)

(b)

Figure 3.8.5 (a) Fan pressurisation testing; and (b) sealed roof testing

Source: Courtesy ABT Pty. Ltd.

Duct leakage and heat loss

Similar to envelope air leakage testing is the leakage experienced by the mechanical ventilation system ducting throughout the building. Unfortunately, this leakage is often underestimated and can account for severe problems in conditioning distribution, contributing to significant energy losses.

An example where this problem was examined and remedied is shown here for an office building in Parramatta, Sydney, Australia (Figure 3.8.6). This prepared the building for the implementation of an optimised HVAC control system. It is therefore important to note that retrofitting often depends upon the synergistic effects of building components and systems that support each other.

Figure 3.8.6 (a) Air duct sealing; and (b) testing within an office building

High efficiency HVAC control systems

One of the best means of saving energy after the building fabric has been properly designed is to optimise the conditioning equipment and its control. The author has experienced (through measurement) that the traditional concept of mechanical conditioning introduces forced turbulent air through sophisticated duct diffusers. A new and empirically proven method is to introduce conditioned air gently and as required through variable speed drive fans and a regulated pressurisation of the space. This can be accomplished through a demand-controlled conditioning concept where the air quality (CO_2) is constantly measured alongside temperature, humidity and pressure differential between the internal and external environment. Such a system was retrofitted on the existing Deakin University Callista Building office as provided by DEOS Australia. See the case study example in Chapter 4.10. This case study demonstrates a change from outdated conventional constant volume control and forced air systems. The measured CO_2 levels for this occupant-packed office have been limited to 650 ppm while providing a constant conditioned external fresh air distribution. All of this has been accomplished with a remarkable 70 per cent energy reduction in mechanical conditioning.

Shading systems

The previous commercial office space experienced an over-heating problem for its east-facing façade, as shown in Figure 3.8.7. Here, external thermal imaging indicates that interior air conditioning for the top floor office space compensates the tremendous heat gain, identified by the lower over-heated levels. Note, in this case, the lower two levels belong to a different tenant and were not in operation on this particular day.

A simple low cost solution, recently measured and tested, is provided by a lightweight micro screen shade (fabricated by Geelong Glass & Aluminium) as shown in Figure 3.8.8. Here between 75–90 per cent of the direct solar gain is blocked out, before the solar beam comes in contact with and is transmitted through the glass. As can be seen here, the shading solution also provides excellent visual benefits and glare reduction.

A measured result, during a peak hour of incident solar gain at 700 W/m², yields an internal energy load of 45 kW for the entire east façade

Figure 3.8.7

Thermal imaging of an external façade

Figure 3.8.8 Images of (a) an external 'test' screen; (b) with an internal view comparison

without the shading. This external load is reduced to 7 kW by the proposed screen. As a result, a tremendous reduction in capital conditioning equipment, energy savings and increased occupant comfort can all be achieved, within a low cost (2–3-year payback) solution. In fact, had the original design strategy implemented this design feature, it would have reduced the initial capital expenditure on the mechanical equipment immediately and paid for itself. For this office building, the focus is on the energy and thermal performance of the façade. These and other strategies all need to be evaluated in the context of a payback period including a life cycle assessment.

Glass façade improvements

Glazing systems can offer tremendous energy-saving opportunities to buildings. Often the problem with existing buildings (as with the previous project) is that they are not glazed properly with double insulated units. Here the glass type, its properties and cavity gas fill can make the difference between energy savings and occupant comfort seated next to the façade. Recently, a manufactured edge-sealing product has allowed the installation of double glazed (IG units) to be installed from the building interior. This is an ideal solution to the single glazed multi-storeyed commercial office buildings throughout the world. For the previous building, two different triple glazed solutions were tested against the existing single pane unit and the previously presented shade screen. Again, the case study in Chapter 4.10 illustrates the outcomes of various glazing and shading types for this project. Note that several levels of incident solar radiation (perpendicular external measurement) are charted. It is interesting to note that the shaded glass still outperforms the best triple glazed unit by over 50 per cent. However, it must be understood that there are other features of performance such as heat loss (insulation) and noise reduction that IG glazing solutions can offer.

Daylight integration and control

Too often one of the largest and often simplest energy saving strategies is overlooked. We forget to introduce the possibility of daylight control and lighting optimisation into our designs. The lighting energy sector alone can account for up to 40 per cent of the building energy total.

We are looking for the optimisation of daylight in our buildings. In doing so, our lighting systems need to be responsive to existing and enhanced daylight integration within the façade as well as skylight systems. The distribution of light, over room surfaces, not necessarily the lux level at the task, is one of the most important factors of perceived lighting. These principles are applied with the direct–indirect lighting distribution curves and fitting in the example of Figure 3.8.9.

In addition to the correct choice of luminaire, for best energy input to light output, the luminaries themselves must be controlled with reference to occupancy, minimum OH&S requirements, and inversely dimmed in reference to the available daylight. This will create the maximum benefit of the lighting output with the least possible consumption of electricity, and also improve the life span of the lamps and associated infrastructure. The presented artificial lighting system here is enhanced by the daylight guiding shading system provided at the clerestory portion of the façade (see Figure 3.8.10).

(a)

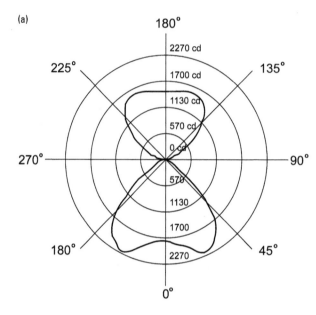

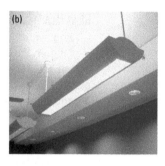

Figure 3.8.9 (a) Light intensity distribution curve; (b) with fixture installed

Figure 3.8.10 (a) Clerestory daylight distribution; (b) with ceiling luminance image

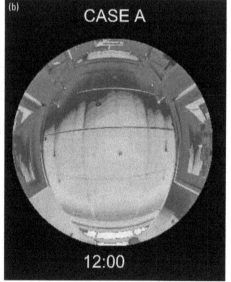

CONCLUSION

In summary, several measurements from a building diagnostic toolkit have been presented. Hopefully, the results speak for themselves, on the importance of measurement and analysis to understanding the significance of bioclimatic performance. In many cases, the findings show that an overheated building can be avoided through improved design and smart choices of materials as well as construction detailing. What has been demonstrated is the importance of measurement and how easily our designs can miss the opportunity of better performance standards. More important, is the proof that measurement provides as demonstrated in several of the previous cases. It is too easy to bypass and appreciate that such *in situ* measurement experience has permitted a reduction in risk taking when bioclimatic sustainable solutions into buildings are introduced.

REFERENCES

ASHRAE (1992) *Thermal Environmental Conditions for Human Occupancy: ASHRAE Standard 55-92a*, Atlanta, GA: ASHRAE.

de Dear, R.J. and Brager, G. (1998) 'Developing an adaptive model of thermal comfort and preference', *ASHRAE Transactions*, 104: 145–67.

Egan, A. (2010) 'The potential energy savings through the use of adaptive comfort cooling setpoints in fully air conditioned Australian office buildings, a simulation study', *Equilibrium Journal*, Australian Institute of Heating and Refrigeration Association, July.

ISO (1994) *International Standard 7730: Moderate Thermal Environments – Determination of the PMV and PPD Indices and Specification of the conditions of Thermal Comfort*, Geneva: International Standards Organisation.

Luther, M.B. (2009) 'Developing an "as performing" building assessment', *Journal of Green Building*, 4(3): 113–20.

Roussac, G.C., Steinfield, J. and de Dear, R. (2010) 'A preliminary evaluation of two strategies for raising indoor air temperature set-points in office buildings', paper presented at ANZAScA (Australian New Zealand Architecture Science Association) annual conference, Auckland, New Zealand.

Szokolay, S. (2004) *Introduction to Architectural Science: The Basis of Sustainable Design*, Cambridge, MA: Architectural Press.

3.9
REDUCING EMBODIED ENERGY THROUGH RETROFIT

How can embodied energy be saved with retrofitting existing buildings?

John Cole and Sattar Sattary

Greenhouse gas emissions associated with the buildings in which we live and work are a major contributor to global warming and climate change. Around the world, buildings are responsible for between 40 and 50 per cent of global energy consumption (Taylor Oppenheim Architects, Lincolne Scott Australia *et al.*, 2005).

In Australia, the commercial sector is the fastest growing greenhouse contributor in the building economy, driving up energy consumption and related CO_2 emissions by about 35 million tonnes a year. Much of this increase is caused by inadequate attention being paid to the principles of sustainable design and the continuing acceptance by many building owners, designers and constructors of unnecessary costs and overheads in the form of energy inefficiency. At some time in the future a carbon price will drive energy efficiency and give impetus to design innovation for reducing carbon impacts. In the meantime, however, meaningful and realistic carbon reduction targets should start with the low-hanging fruit offered by the mitigation potential of the building sector (UNEP SBCI Sustainable Buildings & Climate Initiative, 2010).

It is not just in their operation and occupancy that modern buildings challenge most principles of sustainability. How buildings are constructed also has major impacts on the environment and future operational parameters including efficiency options (Overbay, 1999; Hyde *et al.*, 2009). Until recently it was generally considered that the embodied energy content of a building was small relative to the operating energy over the building's lifetime. Accordingly, the focus on reducing the energy and carbon footprint of the building sector has been mostly on reducing operating energy by improving the energy efficiency of the building envelope (Canadian Architects, 2010; Ecospecifier, 2010). The experience of the past two decades has shown that reflecting the scope of a building's functions and the nature of its tenants' operations, great inroads can be made into achieving operating energy and water efficiencies within a building (Milne and Reardon, 2008).

Irrespective of what technologies or materials are used, however, renovating existing buildings to contemporary environmental and energy performance standards is not easy. Materials are an essential part of construction and building life and the performance demands of buildings are driving innovation in materials science which will deliver a range of composites being integrated into static structures that at once dematerialise, promote durability, and enhance energy efficiency. That is the longer-term prospect. For the next decade overcoming the difficulties with existing buildings, however, may well deliver the best eco-efficiency dividends (Love and Bullen, 2009).

That the opportunity is presented derives very much from the climatic dimensions of many modern buildings being at odds with the needs and good health of living systems. Encouragingly, a growing number of architects and engineers are arguing that the form, fabric and materials of existing commercial buildings can be matched to human and climatic factors through bioclimatic retrofitting (Overbay, 1999; Hyde, 2008).

BIOCLIMATIC DESIGN AND RETROFITTING

Bioclimatic designers seek to conserve resources while optimising the amenity, efficiency and utility of buildings by ensuring that climatic, locational and spatial considerations bearing on the building's operation are closely integrated with its design, alignment and function. Retrofitting to reduce the embodied energy seeks to take this innovation further by preferring materials made using clean renewable energy, promoting dematerialisation where it works, and by achieving higher energy efficiencies by integrating recurrent operational costs with the choice of materials used in construction. Further uptake of this idea has been advocated by the Green Building Council of Australia and adopted in Australian Government programs (Green Building Council of Australia, 2010).

Many older buildings inherently possess design characteristics that make them adaptable to modern environmental retrofits (Steemers, 2003) and materials used in older buildings often possess lifecycle benefits not appreciated at the time of their construction (Love and Bullen, 2009). So rather than demolishing old structures and replacing them with new buildings, it may be more sustainable to aim for better efficiencies in an existing building, albeit one that is also modified with appropriate bioclimatic renovation, refurbishment or maintenance. It all adds up to reducing recurrent energy consumption, embodied energy as well as increasing reuse and recycling opportunities (Sustainability & Innovation, 2005).

As far back as 1963, the Hungarian architect Victor Olgyay determined that climatology along with the disciplines of human physiology and building physics were essential to good building design that took close account of a building's place, space and human functionality (Olgyay, 1963, Hyde, 2008). Olgyay considered bioclimatic design as 'comprising the analysis of site, exposure, climate, orientation, topographical factors, local constraints and the availability of natural resources and ecologically sustainable forms of energy considered in relation to the duration and intensity of their use' (Olgyay, 1963).

Modern leaders of the green architecture movement like American William McDonough argue that building design should be inspired by and mimic nature and aim to minimise environmental impacts (McDonough and Braungart, 2006). Living systems, health and well-being, energy and sustainability are factors now very much front and centre and the thinking about buildings starts with place and function. From there it is not a distant step to the use of bioclimatic design which, when used appropriately, can reduce energy consumption in a building by five to six times as compared to a conventional building (Jones, 1998; Hyde, 2008). Other benefits of such retrofitting include improved health and productivity of workers, and reduction in operating costs for the building (Birkeland, 2002). Australian cities are beginning to act in this area; for example, the Victorian Government and City of Melbourne aim to retrofit more than two-thirds of Melbourne's commercial buildings with a view to improved sustainability and reduced environmental impact (City of Melbourne, 2008).

As the energy efficiency of buildings increases, the relative contribution of embodied energy to total energy impact becomes relatively larger. It follows that if a building is to be retrofitted with a view to reducing its energy impact, bioclimatic retrofitting might provide the initial dividend of reduced environmental impact and climatic efficiency in the buildings' operation, but the real challenge will be downsizing the structure's overall carbon footprint by reducing its embodied energy (Clarke and Pullen, 2008). Linking it all together into a coherent building package constitutes the current innovation edge for retrofitters.

EMBODIED ENERGY AND BUILDINGS

Retrofitting for lower embodied carbon almost could be the subject of a formula. It requires materials manufactured with high energy inputs to be replaced with materials requiring lower production energy. There are several distinct components to the embodied energy of a building, succinctly defined by Milne and Reardon (2008) as 'the energy consumed by all of the processes associated with the production of a building, from the mining and processing of natural resources to manufacturing, transport and product delivery'. CSIRO scientists have shown that embodied energy is very much materials specific, ranging from about two gigajoules per ton for concrete to 180 gigajoules per ton for aluminium (Overbay, 1999). It all adds up to a significant proportion of CO_2 emissions accruing to the building sector deriving from the manufacture of building materials (Table 3.9.1).[1]

Research in Australia and elsewhere has shown that the embodied energy of a building is a significant multiple of the annual operating energy consumed (Overbay, 1999; Technical Guide, 2010). It ranges from around 10 for typical dwellings to over 30 for office buildings (CSIRO, 2000). Making buildings such as dwellings more energy-efficient usually requires more embodied energy, thus increasing the ratio even further. In the commercial sector, however, experience has shown that as the energy efficiency of build-

Table 3.9.1 Embodied energy and CO_2 emission of common building materials in Australia

Material	Embodied energy MJ/kg	Co₂ emission per kg/MJ
Kiln dried sawn softwood	3.4	333
Kiln dried sawn hardwood	2.0	196
Air dried sawn hardwood	0.5	49
Hardboard	24.2	2372
Particleboard	8.0	78
MDF	11.3	1107
Plywood	10.4	1019
Glue-laminated timber	11.0	1078
Laminated veneer lumber	11.0	1078
Plastics – general	90.0	8820
PVC	80.0	7840
Synthetic rubber	110.0	10780
Acrylic paint	61.5	6027
Stabilised earth	0.7	69
Imported dimension granite	13.9	1362
Local dimension granite	5.9	578
Gypsum plaster	2.9	284
Plasterboard	4.4	431
Fiber cement	4.8*	470
Cement	5.6	549
In situ concrete	1.9	186
Precast steam-cured concrete	2.0	196
Precast tilt-up concrete	1.9	186
Clay bricks	2.5	245
Concrete blocks	1.5	147
AAC	3.6	353
Glass	12.7	1245
Aluminum	170	16660
Copper	100	9800
Galvanised steel	38	3724

Source: Lawson (1996).

ings increases, the impacts of embodied energy on the overall building footprint become more conspicuous because the ratio of embodied energy to total energy consumption increases. Not surprisingly, as more and more investors and tenants scan the property markets looking for verified indications of green performance, arguments for reducing the embodied energy in commercial buildings are becoming more compelling.

For some commercial building owners, retrofitting may achieve a significant reduction in their property's embodied energy. In gauging the environmental impact of a building, close attention has to be paid to the embodied energy of the building materials as well as the assembly methods used, the extent of materials reuse and recycling in construction and the actual operation of the building facility (City of Melbourne, 2008). Bioclimatic design characteristics also have to be considered. Such analysis should enable the identification of what can be done to achieve a reasonable reduction in energy use by retrofitting (Liu, 2010; UNEP World Business Council for Sustainable Development 2010).

Assessments of embodied energy levels for common building materials (Tables 3.9.2, 3.9.3) have to also take into account other factors including the energy used in transporting materials from production point to construction site and, as energy savings with recycling can be significant, whether source materials are raw or recycled. Materials with the lowest embodied energy levels such as concrete, bricks and timber, are usually consumed in large quantities, whereas those with higher embodied energy content levels such as stainless steel are often used in much smaller amounts. Construction techniques crucially determine the mix of embodied energy from low or high embodied energy level materials (CSIRO, 2000; Timber Building in Australia, 2010). This means that if embodied energy in buildings is to be assessed, designed and managed, it has to be calculated and accounted for within an 'assembled building element or system'.[2] Besides materials and assembly inputs to embodied energy, there are also the recurring contributions made up by 'the non-renewable energy' used in the maintenance, repair and refurbishment of the building over its entire life (Canadian Architects, 2010).

Retrofitting of some building components such as aluminium window coverings and double-glazing can lead to substantial energy and materials savings (Greener Homes, 2009). Government guides and industry experts point to the re-use of building materials commonly saving about 95 per cent of embodied energy that would otherwise be wasted (Milne and Reardon, 2008; Technical Guide, 2010). The embodied energy savings to be made by recycling of materials will vary according to materials; for example, as can be seen in Table 3.9.4, recycling of aluminium can save up to 95 per cent of energy used in full production, but only 5 per cent of energy can be saved in recycling glass due to the energy used in its reprocessing (CSIRO, 2000; Milne and Reardon, 2008). Table 3.9.5 shows the potential energy savings of some recycled materials.

Table 3.9.2 Embodied energy values based on several international sources (Overbay, 1999): values may vary by location and climatic condition

Material	Embodied energy MJ/kg	Embodied energy MJ/m³
Aggregate	0.10	150
Straw bale	0.24	31
Soil-cement	0.42	819
Particleboard	8.0	4400
Stone (Overbay, 1999)	0.79	2030
Plywood	10.4	5720
Concrete (30 Mpa)	1.3	3180
Lumber	2.5	1380
Cellulose insulation	3.3	112
PVC	70.0	93620
Carpet (synthetic)	148	84900
Paint	93.3	117500
Peters gypsum wallboard	6.1	5890
Aluminum (recycled)	8.1	21870
Steel (recycled)	8.9	37210
Shingles (asphalt)	9.0	4930
Mineral wool insulation	14.6	139
Fiberglass insulation	30.3	970
Zinc	51.0	371280
Brass	62.0	519560
Concrete precast	2.0	2780
Linoleum	116	150930
Brick	2.5	5170
Concrete block	0.94	2350
Polystyrene insulation	117	2350
Glass	15.9	37550
Aluminum	227	515700
Copper	70.6	631164
Steel	32.0	251200

Source: Canadian Architects, 2010, Life-Cycle Energy Use in Office Buildings, www.canadian architect.com. The CO_2 emissions of some common building materials are given in Table 3.9.1, on average 0.098 tonnes of carbon dioxide are produced per gigajoule of embodied energy.

REDUCING EMBODIED ENERGY OF BUILDINGS BY BIOCLIMATIC RETROFITTING

Retrofitting commercial buildings using bioclimatic design principles aims to achieve functional efficiencies by delivering a building product that performs best in the prevailing local climatic conditions (Hyde, 2008). Primary strategies for achieving this include installing passive energy and ventilation systems, securing energy from renewable energy sources, emphasising preference for low embedded energy materials, and sourcing building inputs locally (Clarke and Pullen, 2008).

Sustainable building design emphasises life-cycle thinking, assessment and costing – including taking account of those factors normally dis-

Table 3.9.3 Assembly embodied energy of various construction systems

Assembly	Embodied energy MJ/m³
Single skin AAC block wall	440
Single skin AAC block wall gyprock lining	448
Single skin stabilised (rammed) earth wall (5% cement)	405
Steel frame, compressed fibre cement clad wall	385
Timber frame, reconstituted timber weatherboard wall	377
Timber frame, fibre cement weatherboard wall	169
Cavity clay brick wall	860
Cavity clay brick wall with plasterboard internal lining and acrylic paint finish	906
Cavity concrete block wall	465

Source: Lawson (1996).

Table 3.9.4 Assembly of embodied energy of some building elements

Assembly	Embodied energy MJ/m³
Floors	
Elevated timber floor	293
110 mm concrete slab on ground	645
200 mm precast concrete T beam/infill	644
Roofs	
Timber frame, concrete tiles, plasterboard ceiling	251
Timber frame, terracotta tiles, plasterboard ceiling	271
Timber frame, steel sheets, plasterboard ceiling	330

Source: Lawson (1996).

Table 3.9.5 Potential energy savings of some recycled materials

Material	Energy required to produce from virgin material (million Btu/ton)	Energy saved by using recycled materials (percentage)
Aluminum	250	95
Plastics	98	88
Newsprint	29.8	34
Corrugated cardboard	26.5	24
Glass	15.6	5

Source: Mumma (1995).

counted in traditional approaches. Efficiency ingredients like thermal insulation should only be considered after passive options have been exhausted or discounted. A sustainable bioclimatic approach will also focus on the quality of the indoor environment of the building where the aim will be to promote natural ecological and social attributes within the building that contribute to improved user health and well-being. It can be as straightforward

as providing for natural vegetation or ensuring fresh air by replacing hermetically sealed windows with ones that can open, or replacing elevators with stairs to encourage physical exercise and social interaction in passageways (Burton, 2001; Clarke and Pullen, 2008; Hyde *et al.*, 2009).

Other broad design strategies for bioclimatic retrofitting include reducing demand for heating, cooling, airflow, and artificial lighting (Burton, 2001). Wherever practical, preferences should be given to use of natural light strategies or managed lighting technologies aimed at ensuring the elimination of wasted energy (ibid.) and transport distances for materials should be factored into consideration in calculating the embodied energy reduction goal (Technical Guide, 2010).

In warm climates, reducing the impact of solar radiation is a core feature of bioclimatic design and it applies also to retrofitting. While it might not be possible to re-orient east–west facing buildings to a more efficient north–south axis, low embodied energy shading materials can lessen the solar impact, reduce heat irradiation, and offset many of the climatic negatives of even the most inefficiently oriented building (Givoni, 1998). Reflective light painting and judicious use of landscaping also contribute significantly to reducing heat irradiation (Rellihan, 2003) and play a part in that process wherein 'the form and fabric of the building can be matched to human and climate factors in order to optimise climate response' (Hyde, 2008: 23).

Experience in Australia and other countries yields a range of practical and viable techniques for achieving high performance envelopes (glazing, finishes, insulation and roofing materials) that make possible the achievement of bioclimatic retrofitting in commercial buildings. A high performance building envelope takes into account microclimatic conditions, thermal properties and local environmental responsibility in material selection for envelope. In 2007, the Intergovernmental Panel on Climate Change referred to the savings being achieved through a mixture of bioclimatic design, retrofit and energy management innovation in a number of leading examples including:

- an expected 38 per cent reduction in energy use over pre-retrofit standards at the Empire State Building in New York;
- 40 per cent reduction in heating, cooling and ventilation energy in a Texas office building through conversion of the ventilation system from constant to variable air flow;
- 74 per cent in cooling energy use in a one-storey commercial building in Florida through duct sealing, chiller upgrade and fan;
- 50–70 per cent in heating energy use through retrofits of schools in Europe and Australia.

These results were achieved by selecting materials with low embodied energy in manufacturing and operation and which exhibited the principles of eco-efficiency in their performance, including durability, disassembly, recyclability and reuse. These considerations extended across glazing, lighting, window covering, insulation and ventilation. Bioclimatic designers have also achieved lower embodied energy results by using exterior insulation cladding to create a refreshed appearance and substantially upgraded the insulation value of buildings (Fowler, 2008; Martend, 2010). Work on a commercial building in Italy has demonstrated also the value of linking free cooling ventilation strategies with thermal mass activation to reduce peak cooling loads – particularly by rising the R-value of the concrete hollow core roofs and floors (Corgnati and Kindinisa, 2006).

GUIDELINES FOR REDUCING EMBODIED ENERGY THROUGH BIOCLIMATIC RETROFITTING

The following guidelines can facilitate significant reduction in embodied energy of building materials through bioclimatic retrofitting of commercial buildings in a warm climate (Table 3.9.6).

Table 3.9.6 Guidelines to reduction in embodied energy of building materials through bioclimatic retrofitting of commercial buildings in a warm climate

	Existing conventional building	Bioclimatic retrofitted building
Decision-making	Demolish and construct new building	Decide to retrofit, modify or refurbish instead of demolishing or adding
		Refurbish to accommodate changes in organisation or adapt conditions for a new user;
	Add to existing buildings	Modify or refurbish instead of adding.
		Build no bigger than is needed.
		Consider retrofit of the existing building instead of new construction.
Design	Design standard sizes, and modular	Each design should be optimal for intended use based on climate, transport distances, availability of materials and budget; and balanced against known embodied energy content.
		Ensure materials can be easily separated
		Design for durability (reusable, replaceable or recyclable parts and components).
		Design for long life and 'loose fit' (flexibility of future use).
		Design for adaptability, using durable low maintenance materials.
Operational management	Lack of strategies to improve energy performance	Create management plan for energy efficiency of building
Roof		Select light-coloured roofing materials to help reflect the heat of the sun, reduce the internal temperature of the building
		Cool pavements and whitened asphalt
		Use light colour to reflect solar radiation
		Use roof sheet with higher quality

Table 3.9.6 Continued

	Existing conventional building	Bioclimatic retrofitted building
		Use infrared emittance, *cool roofing materials'*
		Select fly or parasol roof to allow temperature equalisation.
Façade	Glass façades without any purpose	Optimally orientated
		Use high-performance cladding
		Use high-performance cladding
		Use light colour to reflect solar radiation
		Use vegetation to control the temperature rises
		Use reflective surfaces to reduce radiation absorption of fabric surfaces
Construction system	Redundant structure	Avoid redundant structure
		Use construction techniques that allow use of building materials with low embodied energy, and that are recyclable and reusable; small amount can be from higher embodied energy
Insulation		Use well insulated thermal mass to offset energy use
		Use foils that reflect radiant heat
		Use more recyclable and deflector insulation
		Seal ducts, upgrade chiller and check fan controls that can save 12 per cent on its total utility
		Use bulk insulation that air gaps in the fabric material resists conduction and convection through the material
		Embodied energy in well-insulated thermal mass can significantly offset the energy used for heating and cooling
		Use thermal mass to increase the time lag of heat flow through the fabric
		Select reflective foil insulation to reduce downward heat flow
Lighting	Artificial lighting and limited daylight	Use daylight where possible
		Use redirection system to distribute daylight more uniformly and at greater depths
		Reduce demand for artificial lighting, and supply remaining demand by efficient and well controlled electrical means
		Use of daylight strategies can provide large reductions in lighting and energy use, as well as improved amenity, satisfaction, and perhaps work performance.
Cooling and ventilation	Mechanically air conditioned and ventilated	Adapt or replace inefficient building envelope with one designed to reduce or eliminate the need for cooling
	Low performance building envelope and air conditioning	High performance envelope with high efficiency active systems
		Use light colour to reflect solar radiation
	Limited ventilation	Use natural ventilation with optimum flow through
	Lack of fresh air	Use 'fresh' natural air

Table 3.9.6 Continued

	Existing conventional building	Bioclimatic retrofitted building
Heating	Increased energy demands for heating	Design and use efficient building envelope and fittings (energy-efficient building envelope can downsize or eliminate the need for heaters) Supply remaining demand by solar or efficient and well-controlled green technologies Daylight strategies can provide large reduction in need for heating
Material	Weathering of materials	Use low maintenance materials.
	Higher expansion/ contraction of materials	Use appropriate materials and fittings to minimise this
	Quicker fading of materials (e.g. colours)	Ensure materials can be easily separated
	Life benefit of the building materials originally used has not been recognised	Use materials that can be re-used or recycled easily at the end of their lives using existing recycling systems Use materials with lower CO_2 emission in manufacturing process, or processed with renewable energy where possible Use low embodied energy building materials where those materials are consumed in large quantities Use low embodied energy materials with long life and durability suitable to the lifespan of the building. Give preference to materials manufactured using renewable energy sources. Avoid building bigger than is needed (this will save materials) In climates with greater heating and cooling requirements and significant day/night temperature variations, embodied energy in well-insulated thermal mass can significantly offset the energy used for cooling.
Embodied energy		Use low embodied energy materials (which may include materials with a high recycled content), preferably based on supplier-specific data Use low embodied energy materials for the greatest amount of building materials; small amount can be from higher embodied energy Use locally available materials where possible to decrease transportation impact Use embodied energy of assembled building elements or systems to compare with the locally available options, and select most practical option Use long life and durable materials to decrease operational energy during lifetime of building Use materials with low embodied energy and processed partly or fully with renewable energy When retrofitting, ensure materials from demolition of existing buildings and construction wastes are reused or recycled where possible

Table 3.9.6 Continued

	Existing conventional building	Bioclimatic retrofitted building
Reuse and recycle		Ensure off-cuts are recycled
		Use low embodied energy materials that can be reused or recycled
		Where feasible, reuse materials of any buildings to be demolished nearby
		Select materials that can be re-used or recycled easily at the end of their lives using existing recycling systems.
High impact areas		Use long life and durable materials in the high impact areas of a building to decrease operational energy during lifetime of building (e.g. floor finishes) .
Window	Normal window	Use window coverings and retrofitted double-glazing to trap cool indoors, decrease waste
Waste		Decrease waste where possible
		Use replaceable parts, and design for disassembly, to decrease waste.
		Ensure off-cuts are recycled
Supplier	Reliant on fossil fuel energy	Ask suppliers for information about their materials and select low embodied energy where locally available and recyclable
		Use materials salvaged on site to reduce transport
Transport	Increased energy consumption	Decrease energy consumption by selection of locally supplied materials where possible
Energy		Use low embodied energy materials which use clean technologies in their production
		Use low embodied energy materials processed partly or fully with renewable energy
Emissions	Increased greenhouse gas emission levels	Use carbon dioxide levels of common building materials to select low emission materials (e.g. see Table 3.9.1)
Water		Water-efficient taps allow downsizing of water pipes

Source: Hyde, 2000; Burton, 2001; Szokolay, 2004; Sattary, 2007; Clarke and Pullen, 2008; Milne and Reardon, 2008, Hyde *et al.*, 2009, Technical Guide, 2010, Woods Bagot, 2010.

CONCLUSION

In assessing the prospects and utility of bioclimatic design in retrofitting buildings for reduced embodied energy, a useful set of strategic indices was provided nearly fifty years ago by Everett Rogers in his explanation of how innovation works (Rogers, 1962). For innovation to be quickly and widely adopted, Rogers suggested that it had to be: better than existing alternatives to provide relative advantage; not too complex; possible at 'pilot' level so as to reduce risk; observable and measurable so that benefits could be assessed; and ultimately not too radical as to be outside the span of practical comparability within an industry or sector. While bioclimatic design is being adopted in new buildings, other arguments prevail too often in planning retrofits and factoring in reduced embodied energy can almost be one complication too many.

Because in the face of opportunities presented by innovation, change-resistant building owners, investors, and constructors can throw up any number of reasons why retrofitting for bioclimatic efficiency and sustainability outcomes is all too hard. They will justify the status quo by quoting considerations of cost and finance, risk and early adopter costs, market perceptions and undeveloped cost drivers. There is more validity in their claims about the inadequate alignment of building owner and tenant interest, the time-consuming nature of retrofitting and the non-standardised analysis and audit procedures for assessing costs and benefits. That there are system and process inhibitors to change in the building sector is undeniable and indeed the single most commonly cited barrier to retrofitting innovation is the lack of data on retrofit performance, in terms of both energy savings and financial payback (Bloom and Wheelock, 2010). This highlights the need for more pilot projects, collaboration between researchers, building owners, constructors and investors.

In a changing world where climate change and the escalating costs of an increasingly energy intensive lifestyle manifestly challenge old assumptions, the savings and the dividends to be had from building innovation are also increasingly demonstrated. Better data and process are needed to underpin innovation and explain its benefits and its economic incentives and innovative financing to motivate investment. Realistically, not all buildings will offer practical possibilities, even for the innovators. Indeed, by themselves, bioclimatic design, retrofitting and reduced embodied energy are uniquely productive strategies delivering amenity, efficiency and steps along the sustainability pathway. They each merit attention and adoption by the global building sector. In some special places they can be combined to deliver practical experiential change in the structure, character and performance of buildings – taking us closer to the reality of a building that transcends 'green' and attains the mantle of 'sustainable'.

NOTES

1   The volume of CO_2 emissions generated in the manufacture and processing of some common building materials is outlined in Table 3.9.1. Approximately 98 kilograms of CO_2 are produced per gigajoule of embodied energy (Recovery Insulation, 2010).
2   'Assembly of embodied energy' for some building elements and construction systems in the Australian context are given in Table 3.9.2 and Table 3.9.3 respectively (Lawson, 1996).

REFERENCES

ARCHIEXPO (2010) 'Reinforced concrete hollow core slab', The Virtual Architecture Exhibition 2010. Available at: http://www.archiexpo.com (accessed 12 October 2010).

Aye, E., McDaniel, E. et al. (2006) 'High-performance building envelopes', available at: http://www.edcmag.com (accessed 12 October 2010).

Birkeland, J. (2002) Design for Sustainability: A Sourcebook of Integrated, Ecological Solutions, London: Earthscan.

Birkeland, J. (2009) Eco-Retrofitting: From Managerialism to Design, Cleveland, OH: Global Forum.

Bloom, E. and Wheelock, C. (2010) 'Retrofit industry needs assessment study', RetroFit Depot, 3Q, Rocky Mountain Institute, pp. 14–21.

Burton, S. (2001) Energy Efficient Office Refurbishment, London: Earthscan/James & James.

Canadian Architects (2010) 'Measures of sustainability: life-cycle energy use in office buildings', available at: http://www.canadianarchitect.com (accessed 25 August, 2010).

City of Melbourne (2008) 'City of Melbourne unveils cutting edge green building initiative', available at: http://www.melbourne.vic.gov.au (accessed 17 September 2010).

Clarke, B. and Pullen, S. (2008) 'The need for adaptation of existing commercial buildings for climate change', paper presented at Australian Institute of Building Surveyors SA Conference 2008, Adelaide.

Corgnati, S.P. and Kindinisa, A. (2006) 'Thermal mass activation by hollow core slab coupled with night ventilation to reduce summer cooling loads', Energy Efficiency.

CSIRO (2000) 'Embodied and life time energies in the built environment', available at: http://www.tececo.com.au (accessed 26 August 2010).

Ecospecifier (2010) 'Knowledge base: materials impacts in construction', available at: http://www.ecospecifier.org (accessed 26 Aug. 2010).

Fowler, M. (2008) 'High-performance building envelope', Walls & Ceilings, available at: http://www.wconline (accessed 12 Oct. 2010).

Givoni, B. (1998) Climate Considerations in Building and Urban Design, New York: John Wiley & Sons, pp. 114–27.

Green Building Council of Australia (2010) 'Support for energy-efficient public buildings', available at: http://www.gbca.org.au (accessed 26 Aug. 2010).

Hyde, R.A. (2000) Climate Responsive Design: A Study of Buildings in Moderate and Hot Humid Climates, New York: E & FN Spon.

Hyde, R.A. (ed.) (2008) Bioclimatic Housing: Innovative Design for Warm Climates, London: Earthscan.

Hyde, R.A., Yeang, K., Groenhout, N., Barram, F., Webster-Mannison, M., Healey, K. and Halawa, E. (2009) 'Exploring synergies with innovative green technologies for advanced renovation using a bioclimatic approach', Architectural Science Review, 52(3): 229–36.

Intergovernmental Panel on Climate Change (2007) Climate Change 2007: Mitigation of Climate Change: Contribution of Working Group III to the Fourth Assessment Report of the Intergovernmental Panel on Climate Change, Cambridge: Cambridge University Press.

Lawson, B. (1996) Building Materials, Energy and the Environment: Towards Ecologically Sustainable Development, Red Hill, ACT: RAIA.

Love , P. and Bullen, P.A. (2009) 'Toward the sustainable adaptation of existing facilities', Facilities 27(9/10): 357–67.

Martend, R. (2010) 'High performance EIFS', available at: http://www.wbdg.org (accessed 12 Oct. 2010).

McDonough, W. and Braungart, M. (2006) Cradle to Cradle: Remaking the Way We Make Things, New York: North Point Press.

Milne, G. and Reardon, C. (2008) 'Your home technical guide', Embodied Energy, available at: http://www.yourhome.gov.au (accessed 25 Aug. 2010).

Mumma (1995) Energy-Efficient Window Retrofits, Home Energy Magazine Online January/February 1995, www.homeenergy.org.

Olgyay, V. (1963) Design with Climate: Bioclimatic Approach to Architectural Regionalism, Princeton, NJ: Princeton University Press.

Overbay, D.W. (1999) *Windows on the Past: The Cultural Heritage of Vardy, Hancock County, Tennessee*, Macon, GA: Mercer University Press.

Rellihan, S.S. (2003) 'Design with climate: a retreat for Viegues, Puerto Rico', Master of Architecture thesis, University of Maryland.

RetroFit DEPOT (2011) 'Empire State Building', available at: http://retrofitdepot.org (accessed 10 Feb. 2011).

Rogers, E. (1962) *Diffusion of Innovations*, New York: Free Press, reprinted 1995.

Sattary, S. (2007) 'Construction processes: low impact construction', PhD thesis, Department of Architecture, University of Southern Queensland, Toowoomba.

Szokolay, S.V. (2004) *Introduction to Architectural Science: The Basis of Sustainable Design*, London: UK Architectural Press.

Taylor Oppenheim Architects, Lincolne Scott Australia, *et al.* (2005) 'Building energy brief for commercial buildings', *Environmental Design Guide*, DES 34 (May 2000): 1–2.

Technical Guide (2010) 'Embodied energy: technical manual', *Your Home: Design for Lifestyle and the Future*. Available at: http://www.yourhome.gov.au (accessed 16 Sept. 2010).

Timber Building In Australia (2010) 'Environmental properties of timber', in *The Range of Building Materials*, available at: http://oak.arch.utas.edu.au (accessed 16 Sept. 2010).

Woods Bagot (2010) 'Woods Bagot and Buro Happold unveil "ZERO-E" to lead carbon economy drive: Joint pilot project advances construction industry's contribution to achieving a zero carbon economy by 2050', *Zero carbon economy by 2050*, http://www.woodsbagot.com/en/Pages/zero_original.aspx (accessed 14 June 2012).

3.10
A CHECKLIST FOR PEAK ENERGY REDUCTION STRATEGIES IN BUILDINGS

How can retrofitting strategies be conceptualised into simple principles?

Mark B. Luther

INTRODUCTION

The method for improving our building stock primarily requires changes to the building procurement and design processes. A bioclimatic approach to this process can benefit greatly the carbon reductions and excessive energy use. One of the first steps is to consider the strategies for reducing the peak load and operational energy of a building. This can be accomplished through recognition of the various sectors of energy use and careful strategic bioclimatic planning. An example of the energy use in a typical Australian office building is illustrated in Figure 3.10.1.

From Figure 3.10.1 it can be noted that approximately 80 per cent of the energy is related to building conditioning and lighting. This implies a huge potential for investigating energy savings within office buildings in regard to bioclimatic influences. This chapter examines first some of the strategies for reducing peak energy demand. This is important on a global and community scale, since power plants are becoming deficient and incapable of supplying the ever-increasing peak demand. As industry and the load of conditioning buildings increase, our infrastructure to support this energy

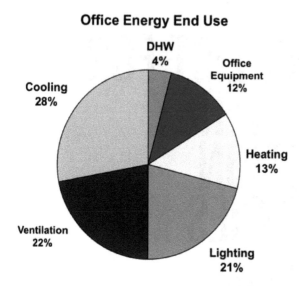

Office Energy End Use

Figure 3.10.1
Sectors of energy use
for a typical commercial
office building

demand is failing. Therefore, it is most economically viable and essential to reduce our overall peak demands within our building stock. In fact, retrofitting towards a reduced peak demand among our existing buildings holds the key to approaching a sustainable power supply infrastructure.

REDUCING THE PEAK ENERGY DEMAND (PED)

In line with energy reduction is recognising the period and demand of energy use. If the peak energy loading of a particular project is dramatically reduced, the size of mechanical equipment and period of operation are reduced. The key point here is that tremendous capital expenditure in mechanical equipment, ongoing maintenance and equipment replacement costs can be minimised.

A bioclimatic approach involves using the microclimate, the form and fabric in a building design to reduce demand for energy in the building as an important strategy. This offers capital savings, reduced operational and maintenance costs as well as providing an effective interior climatic-controlled building environment. These are synonymous with a life-cycle benefit analysis approach.

Perhaps the business case argument for implementing bioclimatic architecture principles within a project is to produce a *balanced peak demand load* for a building. The following discussion will provide several reasons as to why this balanced peak load is desired for our buildings.

Figure 3.10.2 charts the peak demand load (kWh) per month for a particular office building (black bars) against the same building with implemented energy strategies (white bars). There are several important messages, aside from the basic difference in reduction that are important to the

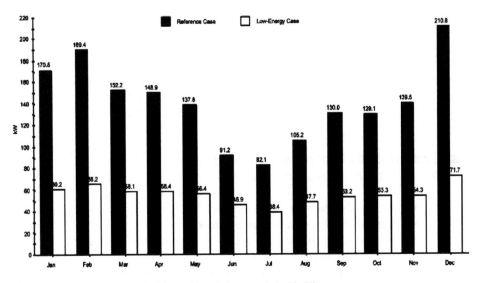

Figure 3.10. 2 Monthly peak electrical demand for existing vs. optimised building

building owners and operators. Much of the peak load is attributed to additional environmental loads brought about by summer overheating. So, for example, the energy flux external is doubled or tripled in summer months, reducing the effectiveness of the mechanical systems and increasing the risk of breaches to peaks in energy use.

A monthly maximum peak demand load reduction in kWh

A significant reduction in the peak energy demand is seen in the retrofitted case (white bars) of Figure 3.10.2. If the cost of electricity is charged based on a peak demand rate cost for the entire year, then there is incentive to reduce this previously large peak load amount. This is called by utility companies 'a 12-month rolling peak demand charge'. This means that a lower demand charge rate can be applied to a building's energy costs by the utilities company; however, there are penalties of exceeding this hourly rated peak demand load agreed with the company. Hence for retrofitting, benefits can be achieved though reductions in the peak load demand

A reduced variance in monthly peak demand loads

Under a 'rolling peak demand charge' scheme the user pays a 'peak demand' privilege for the entire year. In the retrofitted example, only a 30kW monthly expected difference between base load and peak load is shown. However, the owner or occupier is primarily paying for the cost of electricity based on the actual rather than the expected peak use.

The significance of this problem can be seen if there is a variance in the actual operation of over 160 per cent in relation to 88 per cent. However, with the proposed peak loading, the owner will need to pay a penalty for the energy used over the 30kW difference among monthly peaks. It is therefore argued that a major technical and management strategy is to negate variance, which in turn leads to improved and more predictable control expectancies in the building operation. The entire building will operate with a smoother energy cycle. Discrepancies can be more easily detected because the variance is not as great and any excessive loading will be immediately recognised.

Capital cost savings under the reduced peak demand load

A further benefit from this approach is as follows:

- the transformer size can be reduced;
- the switchboard capacity can be reduced;
- smaller equipment implies improved maintenance and lower capital cost for replacement;
- increased potential for renewable energy and its effectiveness.

Furthermore, a reduced peak load variance can now be managed effectively through solar PV energy. In this case, it is achieved by reducing the risk of exceeding the peak demand as well as reducing it overall.

In some business cases, the reduction of a peak demand load could be the sole economic reason for implementing a methodology that can provide such a result. It is argued that a bioclimatic approach to design offers a way of addressing peak loads while reducing the need for excessive mechanical equipment in buildings. In a sense, the antithesis of bioclimatic architecture is to limit the capacity of equipment while reducing the overall energy consumption. In an effort to achieve reduced and balanced peak demand loading, a bioclimatic approach for the building must be realised. However, it should be recognised that migrating the use of other loads such as plug loads and other heat sources is complementary to if not ancillary to the problem.

A STAGED APPROACH TO REDUCING ENERGY PEAK LOAD

What is presented here is a step-by-step methodology to integrating bioclimatic principles together with innovative conditioning strategies, resulting in a high performance low-energy design. The following presents a staged bioclimatic approach towards conditioning a building, yielding a positive result in energy reduction:

1   *Consider the assets of the site and its potential manipulation of the landscape*: These principles rely on optimising the site terrain and its vegetation to create a local micro-climate, providing wind protection (when needed), guiding for natural

ventilation, blocking the sun and considering water collection (Landscape Planning for Energy Conservation, 1982).

2   *Introduce fundamental passive solar design*: This includes solar orientation, proper window sizes, internal thermal mass sizing and effective building envelope shading.

3   *Ensure an insulated and airtight building envelope*: Our buildings need to operate under the slogan 'build tight and ventilate right'. In today's technology, there can no longer be excuses for unwanted air leakage. An insulated building envelope with thermal operable windows is inclusive of this strategy.

4   *Consider passive pre-conditioning methods through the construction fabric*: An example may be to use shaded sub-floors areas, structural labyrinths or gabion walls. One of the most stable bioclimatic conditioning resources is the ground or a subfloor foundation system. Figure 3.10.3 is an example of the sub-floor preconditioning process. Here hot external air enters the 'basement' where a gabion wall sprayed with water provides an initial cooling of this air passing through it. Additional cool water storage tanks (thermal mass) add to stabilising this temperature. Finally, the external air temperature is exchanged with the internal exhausted air through an air-to-air energy recovery system. After this initial treatment, if the set-point is not obtained, additional mechanical conditioning is then provided.

5   *Consider a total non-conditioning (mechanical) bypass for fresh air intake*: When conditions are appropriate, we would want to allow external air to enter directly into the building, conditioning or recharging its thermal mass.

6   *Use an energy recovery system for exhausted air*: These low energy systems take account of pre-conditioning incoming fresh air while exhausting the unwanted used air. In cases where the air temperature difference between internal or external is large and where voluminous amounts of fresh ventilation air are required, these systems make good sense.

7   *Consider other active solar thermal mechanical processes*: After providing the pre-conditioning prior to this stage it is reasonable to consider that additional conditioning may be required to obtain a desired adaptive model set-point temperature. If additional conditioning is required, it is possible to consider several low energy conditioning systems such as evaporative (direct or indirect) cooling, solar hot water heating, convective solar air heating systems, etc. before resorting to conventional mechanical equipment (Smith, 2003).

8   *Finally, an advanced energy-efficient conventional mechanical conditioning system may be necessary*. By this stage, we anticipate about a 50–80 per cent reduction from conventional

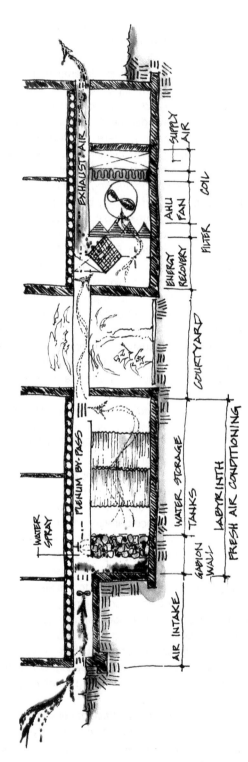

Figure 3.10.3 A pre-conditioning system of external air for the ventilation and conditioning system

equipment sizing. The fact is that our conventional conditioning peak load requirement has been substantially reduced and may even be eliminated.

9 *An optimised control strategy for all the above stages.* One of the most critical factors is often the lack of an optimised or effective building operational strategy. It is almost common practice that complex buildings waste energy and often operate out of schedule. When the capacity of active systems' mechanical equipment is reduced and passive bioclimatic systems take over, the buildings operate under a more simplified structure. Set-point temperatures fluctuate more. Users are more in control. However, there are several intelligent mechanical conditioning control systems presently on the market which can contribute to enhancing performance and overall energy saving.

In summary, the conditioning of buildings needs to begin with an engineering approach that considers the site, building form and orientation, as well as its applied materials as an asset to conditioning. Too often the engineering process begins at the mechanical end after calculating a peak energy load.

A CHECKLIST OF ENERGY-EFFICIENT STRATEGIES

The problems with rating tools are that they do not provide information which could assist in producing the required building performance results. In many cases, building rating tools do not require the degree of sophisticated detail as an energy performance programme. Although there are many software packages available, it has been difficult to quantify and rationally justify the implementation of staged energy-efficient strategies and systems into a building project. This has partially been attributed to the output weaknesses of energy programmes. Rating schemes are in need of a component-by-component justification to building performance improvement (Luther and Cheung, 2003).

Several energy-efficient strategies, which can be ranked in terms of their contribution to performance, for a particular building, are listed in Table 3.10.1. A brief explanation of each in regards to the energy-efficient strategy and the software assumptions follows.

A variable temperature conditioning set-point

Unfortunately, this is not directly accounted for in most energy simulation programs. A 'moving set-point' offered by adaptive comfort models, and in line with bioclimatic principles, often needs to be programmed. Setting a wider 'comfort band' within the set-point assumptions, provides a simplified estimate of this result. It is an established rule of thumb that for every degree

Table 3.10.1 Energy-efficient strategies used in a ranking system in terms of their contribution to performance

Energy-efficient strategies
A variable temperature conditioning set-point
Building air leakage (weatherisation) control
Ducting air and temperature leakages
High efficiency HVAC control systems
Economiser-cycle ventilation control
Energy-efficient lights
Daylighting
Glass façade improvements
Shading systems
Insulation
Thermal mass and storage
Passive solar conditioning

(°C) in a relaxed set-point, between 8–12 per cent of energy conditioning can be saved (Egan, 2010).

Building air leakage control

One of the most effective conditioning savings for Australian commercial buildings is to reduce their air leakage. In general, building construction in warmer climates tends to be more relaxed in its detailing, allowing for excessive infiltration and exfiltration. A tighter, more controlled building envelope is desired even in warmer humid climates so that internal temperatures can be maintained and effectively controlled for longer periods (Luther, 2009).

Ducting leakages

Similar to that of building envelope leakage is the fact that the delivery system (i.e. the ducting) can often be leaky and uninsulated. This leads to energy losses and an ineffective distribution of air to where it is needed.

High efficiency HVAC control systems

This category can range from an efficient coefficient in performance (COP) of the equipment to an effective control and conditioning strategy. For example, a demand-controlled ventilation strategy monitors the interior air quality better than conventional systems and provides conditioning based upon a series of criteria (not just set-point temperature). As a result, excessive ventilation conditioning is reduced (see Chapter 4.10 Case study).

Economiser-cycle ventilation control

Perhaps going hand-in-hand with a high efficiency HVAC system is the specific capability of an economiser ventilation control. The ability of a mechanical system to avoid the needless conditioning of air is offered via the economiser cycle where external air conditions can directly provide the ventilation conditioning needs to the building. In this case, an 'assisted' natural ventilation system can take place through the running of fans and dampers in the HVAC system.

Energy-efficient lights

This category is conveniently related to the next one on daylighting. However, the basic overall efficiency of the selected luminaire is important. This begins with the electric light source (its efficacy; lumens/watts) and the lamp fixture's effectiveness to distribute this light source. The design layout with its separate perimeter row switching capability and potential dimming control are all important aspects here.

Daylighting

This category is linked to 'energy-efficient lighting' as well as that of the glass selection and type in the façade or clerestory of the building. Daylight is only effective in energy savings if and when the internal lighting can be dimmed or turned off when available to those specific areas of the building. It is also essential in warmer climates to exclude the direct solar radiation and to diffuse and filter the light through shading and glass selection of the wanted visible light spectrum to enter the building. Note that this category relies heavily on design knowledge to be effective. Too often daylight penetration is excessive and uncontrolled in bioclimatic designed buildings. This leads to shading systems completely drawn shut and full dependence on electrical lighting.

Glass façade improvements

Linked with the above two categories, the selection of a proper glazing size and its spectral filtering potential are important characteristics to introduce into the building for an effective daylight-linking (daylight with electrical lighting control) strategy. Furthermore, the insulated benefits of the glazing system also need to be considered.

Shading systems

As realised with several of the energy-efficient strategy categories, many are linked to each other. Again, the shading of a building is directly related to the size of the glazing aperture of the façade and how effective this system is in blocking out the unwanted solar gains. However, clever shading systems also need to consider how the shading device itself can provide visibility and diffuse light to the occupant space. Basically, shading systems keep the heat off the building and should be considered for any component receiving excessive solar radiation.

Insulation

It is generally common practice in warmer climates to under-estimate the value of good insulating practices. Insulation alone may not be the answer if leaky construction detailing practices are in effect. Too often we mistake more insulation with better air tightness. Good construction practices should consider insulation to be synonymous with building air tightness, but this is often not the case.

Thermal mass and storage

This category is one of the founding principles of passive solar architecture. However, it is often mistaken for buildings, which need to be heavy in construction. Thermal mass can in fact be separated from the building structure to some degree. Lightweight, airtight and well-insulated components need not be excluded from a building that can provide thermal mass in storage.

Passive solar conditioning

Solar heating of a building can occur in several forms. Glazed solar facing apertures can allow the collection of heat gain directly into the building fabric (i.e. floor, walls). Indirect systems of solar gain linked to storage should also be considered. This is possible, for example, if we are to consider solar hot water collection, which is stored in an insulated vessel as an aspect of solar conditioning.

PRO-ACTIVE VS. REACTIVE BUILDINGS

A dozen of the most influential components a building can have on energy performance have just been discussed. A useful analysis is to rank the importance of each of these strategies for a particular building type, its design and climate region. An example of such a performance ranking, using the ENERGY-10 software package, is provided for a hypothetical office building in Melbourne, Australia in Figure 3.10.4. Note that the same strategies could be ranked differently depending on their characterisation of energy efficiency. For example, Figure 3.10.4(a) indicates those strategies which are best for annual energy consumption. However, if the reduction in the peak load is desired, Figure 3.10.4(b) identifies those strategies that best achieve this.

(a)

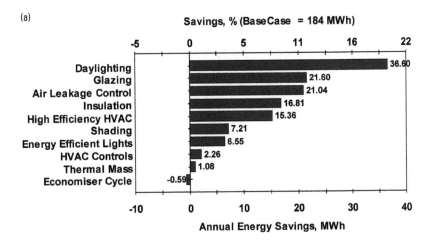

(b)

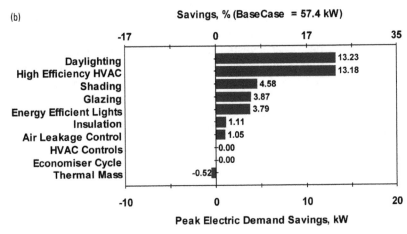

Figure 3.10.4 A detailed ranking of several energy-efficient strategies: (a) annual savings ranking; and (b) peak demand ranking

The new bioclimatic approach is to intelligently consider building conditioning beyond the single day in which it is needed. What we require is proactive conditioning through implementing several of the energy-efficient strategies presented here. Planning for thermal storage conditioning is only a part of the solution. Our buildings still need to 'be tight and ventilate right'. We need to prevent unnecessary solar overheating and reduce our overall energy demands. The need to reduce the 'peak energy demand load' is a fundamental and required achievement for our commercial buildings. Adhering to the bioclimatic architectural principles, as presented here, can assist in providing this requirement. The incentive for this comes from the carbon incentives for retrofitting.

CARBON TAX INCENTIVE FOR RETROFITTING

The building industry and owners in Australia and other parts of the world are being confronted with mandates on energy targets and the legislation of a carbon tax. In Australia commercial office buildings are subjected to a ranking of their CO_2 emissions and energy consumption. Recently, tenants' fit-outs in office spaces have been subjected to a lighting energy assessment through the Commercial Building Disclosure (CBD) programme in Australia (Gardner, 2011).

In Australia, there is an imperative urgency to provide an immediate response to the National Research Priorities and their 'associated priority goals' for developing an Environmentally Sustainable Australia (Australia's National Research Priorities, 2011). A carbon tax and energy mandates have left building owners in immediate need of solutions requiring a pre-retrofitting assessment for decision-making. The Garnault Report, as well as the international Intergovernmental Panel on Climate Change, have identified that building performance retrofitting offers the largest potential for greenhouse gas reduction per dollar invested (see Figure 3.10.5) (Garnault Climate Change Review, 2008; Intergovernmental Panel of Climate Change, 2007).

In a 2002 report by the Australian Greenhouse Office, over \$4.3 billion was spent annually on operating commercial and industrial buildings (GBCA Garnault Review, 2002). Since this study, an increase of over 50 per cent in energy costs have occurred. Therefore an estimated \$8 billion would be spent on energy alone. The need to identify and implement the behaviour of simple, clear-cut, energy solutions for all buildings will help in 'transforming

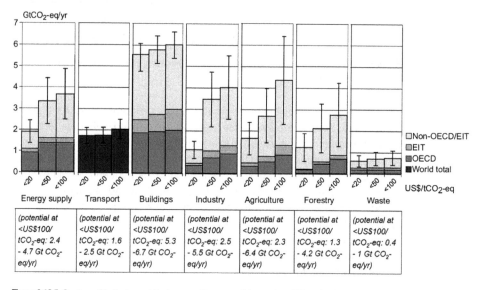

Figure 3.10.5 Sectors of industry and their respective potential to reduce CO_2 emissions per dollar investment
Source: IPCC (2007).

the way we utilize our . . . energy resources through a better understanding of human and environmental systems and the use of new technologies' (Department of Climate Change, 2011).

The key issue for Australia in the coming decades is the achievement of greenhouse gas emission reduction targets. If Australia takes no action by 2020, our carbon pollution could be 20 per cent higher than in 2000. The Government's targets are equivalent to a reduction in every Australian's carbon footprint of nearly one third to one half (ibid.). In order to achieve the 2020 target of 5 per cent below 2000 greenhouse gas emissions, there is a need to undertake aggressive energy efficiency measures, in which the retrofitting of buildings can play a major part.

What is missing is a standard for the entire retrofitting procedures, stages and decision-making processes for buildings. Standards on Energy Audits exist and these are well suited to be part of the proposed standard. However, at present there is a gap between the energy assessment rating and a pre-decision-making tool providing a methodology in ranking the solutions. The Energy Audit Standard does not take into account life-cycle-costing benefits such as comfort, spatial (floor area) gains, and operational time reduction in achieving HVAC set-point temperatures, all of which affect the productivity levels of the users.

Furthermore, there is an incentive for building owners to retrofit and turn their liabilities into assets. A carbon tax or an emissions trading scheme would imply that a high performing building will accumulate credits. These credits could evolve into becoming a revenue asset for owners of efficient energy producing buildings. Of importance is the carbon abatement curve.

The global 'carbon abatement cost curve' developed jointly by McKinsey and the Vattenfall Institute of Economic Research provides a map of the world's abatement opportunities ranked from least-cost to highest-cost options (Figure 3.10.6). This cost curve shows the full range of actions that we can take with technologies that either are available today or look very likely to become available in the near future. The width of the bars indicates the amount of CO_2e that we could abate while the height shows the cost per ton abated. The lowest-cost opportunities appear at the left of the graph and the highest-cost to the right.

The building sector, of all industries, has the greatest potential to reduce carbon emissions. Energy mandates and a carbon tax are creating anxiety within this industry. Pathway solutions towards effective resource management often remain blue-sky concepts, yet do exist, and require implementation. Conventional procurement methods will be critically reviewed and benchmarked against a state-of-the-art integrated systems approach, using a genuine case study. It will be discovered which methods provide the most significant reduction in carbon footprint across the building lifecycle.

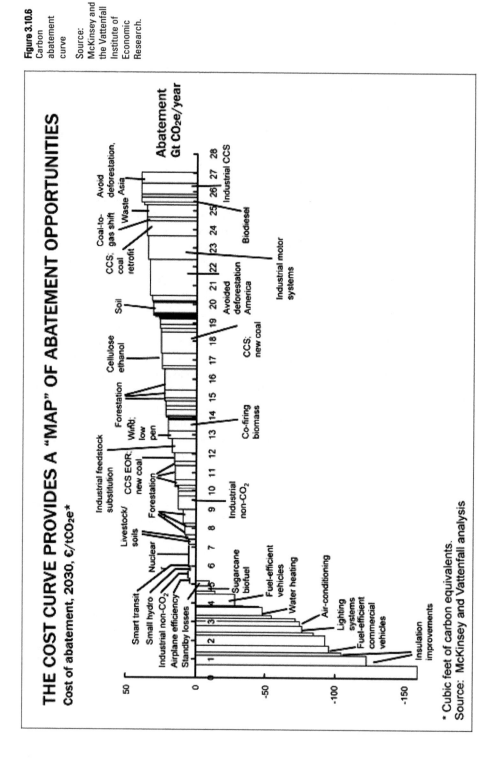

THE COST CURVE PROVIDES A "MAP" OF ABATEMENT OPPORTUNITIES

Cost of abatement, 2030, €/tCO₂e*

* Cubic feet of carbon equivalents.
Source: McKinsey and Vattenfall analysis

Figure 3.10.6
Carbon
abatement
curve

Source:
McKinsey and
the Vattenfall
Institute of
Economic
Research.

CONCLUSION

Finally, in order for us to act upon bioclimatic conditioning and control, an awareness of our environmental conditions needs to occur. In fact, our building control systems as well as we ourselves need to be informed. It has often been said that we cannot act upon what we cannot measure. Today's information and electronic technology has excelled to the point where such information is possible and at a low cost. We can access information in real time via the internet on our choice of device, including our phone.

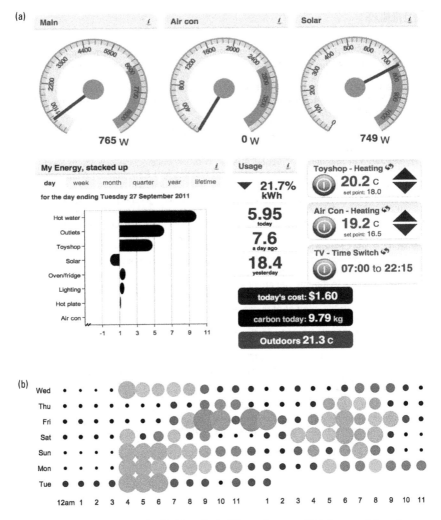

Figure 3.10.7 (a) An informative 'dashboard' to building performance; (b) daily energy information

Source: Courtesy SmartEnergyGroups.com.

Advances in microcontroller technology, especially over the past 5–8 years, are enabling the 'Internet of Things' to become a reality (Ashton, 2009; Gershenfeld *et al.*, 2004). This latest wave of internet expansion is transforming the internet from the people-driven technology we have today, to an internet where 'things' are directly and actively connected to the online world. This is significant on two accounts: it provides flexible access to the accurate, real-time information we need to understand what our environments are doing, and it provides the ability to respond immediately to act on that information (see Figure 3.10.7).

REFERENCES

Ashton, K. (2009) 'That "Internet of Things" thing: In the real world, things matter more than ideas', *RFID Journal*, available at: http://www.rfidjournal.com/article/view/4986 (accessed 22 June 2009).

Australia's National Research Priorities (2011) available at: http://www.dest.gov.au/NR/rdonlyres/AF4621AA-9F10-4752-A26F-580EDFC644F2/2846/goals.pdf (accessed Nov. 2011).

Department of Climate Change (2011) available at: http://www.climatechange.gov.au/government/reduce.aspx (accessed 26 Oct. 2011).

Egan, A. (2010) 'The potential energy savings through the use of adaptive comfort cooling setpoints in fully air conditioned Australian office buildings, a simulation Study,' *Equilibrium Journal*, Australian Institute of Heating and Refrigeration Association, July 2010, 9(6): 32–5.

ENERGY-10, National Renewable Energy Laboratory-Buildings-Energy-10, available at: http://www.nrel.gov/buildings/enrgy10.html (accessed 11 September 2009).

Gardner, I. (2011) *Commercial Building Disclosure*, Melbourne: Facility Management Association.

Garnault Climate Change Review (2008) *Interim Report to the Commonwealth, State and Territory Governments of Australia*, Canberra: Australian Government Printer.

Gershenfeld, N., Krikorian, R. and Cohen, D. (2004) 'The Internet of Things: the principles that gave rise to the Internet are now leading to a new kind of network of everyday devices', *Scientific American*, available at: http://www.scientificamerican.com/article.cfm?id=the-internet-of-things (accessed 27 Sept. 2004).

Intergovernmental Panel of Climate Change (IPCC) (2007) *Working Group III Contribution to the IPCC Fourth Assessment Report*, New York: IPCC.

Luther, M.B. (2009) 'Ventilation research on Australian residential construction', *Architecture Science Review*, 52(2): 89–98.

Luther, M. and Cheung. J. (2003) 'High performance low-energy buildings', paper presented at Australia and New Zealand Solar Energy Conference 2003, Melbourne, 26–29 Nov. 2003.

Robinette, G.O. and McClenon, C. (eds) (1983) *Landscape Planning for Energy Conservation*, Newton, MA: Environmental Design Press.

Smith, P. (2003) *Sustainability at the Cutting Edge: Emerging Technologies for Low Energy Buildings*, New York: Elsevier.

3.11
PENALTY–REWARD –PINCH (PRP) DESIGN FOR IMPROVING THE SUSTAINABILITY OF EXISTING COMMERCIAL BUILDINGS

What alternative design approaches can be used for retrofitting?

Edward Halawa

INTRODUCTION

When the term 'renovation' is applied to an existing building, it implies a fundamental assumption that the building is worthy of existence and that the main task – by definition – is to 'restore' it to a good condition. In sustainability terms, building renovation mean transforming an existing building into one that is capable of existing much longer in a healthy condition which brings to its occupants the necessary comfort they demand, to its owner a peace of mind regarding profitability and marketability in the highly competitive environment, and to the planet a set of criteria (conditions) which translates into wiser and more efficient use of various resources.

None of these factors should be left out when renovation takes place. However, it is often the case that at the end of the renovation process, some of the factors could not be addressed. For example, the thermal comfort or energy performance target may eventually be compromised; or the cost involved in realising these targets becomes prohibitive. It may also happen that improvement in the energy performance is realised at the expense of another important element of renovation, the reinvigoration of the building itself (Martinaitis *et al.*, 2004). This is not necessarily the fault of parties involved in the process; it may be due to various factors beyond their control.

This book presents sustainable retrofitting based on a 'bioclimatic design' approach, when dealing with building renovation, and gives an alternative to a 'conventional' design approach to this process. It is more likely that the majority of existing buildings which are the targets of renovation did exist before the bioclimatic design approach emerged and therefore were constructed based on a conventional 'active systems' design approach. This will pose various constraints on the design options and paths not encountered in the design of new buildings. In such a case, insistence on relying on a passive design approach alone cannot be justified. In most situations, the ideal combination of passive and active design approaches is likely to bring about the optimum performance better than a single (unbalanced) passive or active approach.

This chapter introduces a new concept, termed the penalty–reward–pinch (PRP) design approach, in which passive and active approaches are optimally synergised to bring about sustainability in the construction of new buildings and renovation of existing buildings. While some have now abandoned the so-called sequential design approach and switched to the more integrated approach, lack of detailed principles governing the latter makes it difficult to implement as this process still has to find its way to make a significant impact as a building design concept for sustainability. The proposed penalty–reward–pinch (PRP) design concept is built upon the premise that in many situations building sustainability can be achieved through the optimum matching of passive and active systems.

CURRENT ISSUES

Before discussing the current issues, some terms need to be defined. In this chapter: active design approach refers to design relying heavily on conventional mechanical refrigerative systems to provide comfort in buildings. They normally require high electrical power and consume a large amount of electrical energy. On the other hand, there are emerging 'mechanical' systems generally known as thermally driven heating and/or cooling systems which are 'active' systems in that they need mechanical devices such as pumps or fans to circulate or deliver the working fluid(s). However, in this chapter, they are classified as part of passive systems as they generally have a high electrical COP (coefficient of performance) and therefore require low electrical power and consume a small amount of electrical energy to operate. These systems can be powered by waste heat and solar energy and therefore can be classified as 'green' technologies. They are also typically expensive as they consist of various additional components to 'tap' the free or cheap energy sources.

Proponents of passive and active designs of buildings claim the credit for achieving sustainability but neither party has been able to prove their claims convincingly. Currently, there is no clear or undisputed method of showing the minimum energy consumption achievable at a 'sustainable' level by employing each approach. Bioclimatic design principles form an appealing concept, which has attracted enormous attention in recent years. Bioclimatic design relies on passive solar systems and environmental resources to provide heating, cooling and lighting of buildings (CRES, 2010). Passive systems can reduce the energy requirements of a building. There are many examples: good insulation or double glazing can reduce heat loss or heat gain; natural ventilation can reduce the need for energy from a mechanical system. The passive concept also exploits the building's envelope for energy generation from 'green' technologies. The combination of various passive actions can substantially reduce the energy requirements for heating and cooling. The main issue with the implementation of this approach lies in the selection of the most appropriate technologies, which satisfy the following criteria: (1) environmental benefits; (2) human comfort improvement; and (3) economic feasibility. Various green technologies are already available and others are in the various stages of development. The technologies already entering the market can easily satisfy the first and the second criteria, however, the majority of these emerging technologies have yet to prove to be economically feasible. This is the main reason why such an appealing approach is still finding its way to becoming a 'standard' design approach.

On the other hand, the conventional approach relies heavily on keeping the building isolated from its environment and providing comfort by employing 'active' mechanical systems. Active systems can easily provide energy required for heating/cooling and in some situations installation of active systems cannot be avoided due to their high reliability in providing the services 'at short notice'. However, active systems have proven to be a burden to the environment through their massive energy consumption with consequent greenhouse gas emissions.

For many situations, the combined passive and active design approach rather than a single unbalanced approach will deliver the optimum results in terms of energy consumption level, thermal comfort and cost. If the main target is to minimise the energy consumption, then the 'architectural design first rather than the engineering aspect' or sequential approach as suggested, for example, by Karyono (2000) can be justified. However, if the simultaneous targets mentioned above are to be achieved, a new design approach needs to be developed.

'REWARD' AND 'PENALTY' ASSESSMENT OF SUSTAINABILITY ACTION(S)

The main problem related to the adoption of new emerging green technologies in building so far is the cost involved; or in economic terms, the long period of investment payback. The proponents of these green technologies seem to ignore this reality very often for the 'sake of the environment'.

The Reward and Penalty concept *describes*, *records* and *assesses* the consequence(s) of an action or a bulk of actions for sustainability. An action can be clearly explained by a number of examples: adding a layer of insulation to the ceiling of a building, increasing the tightness of doors or windows to reduce infiltration, setting the building orientation (for a new building), etc.

An action for sustainability always comes at a cost, termed *penalty* (P) and promising positive outcomes, termed *reward* (R). Figure 3.11.1 shows a schematic of the penalty and reward (PR) concept. As shown, the main penalty of an action is the capital cost of investment. However, it may also come in other forms such as reduced comfort or reduced (loss of) productivity. A naturally ventilated space may result in compromised comfort of the occupants which in turn affects the occupants' productivity.

A positive outcome of an action is the reward, which in sustainability terms is expressed as the reduction of energy used. Every action has its own reward and penalty, no matter how small they are.

The difference between the reward and penalty determines the *worth* of an action. A *zero-worth* action means such an action contributes

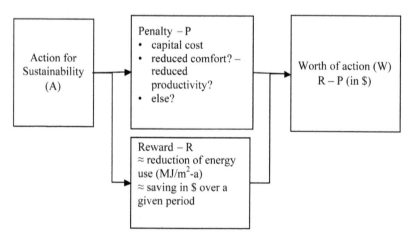

Figure 3.11.1 Description, record and assessment of the worth of a sustainability action

nothing to sustainability and should not be adopted. Choosing a very high value of insulation to reduce heat gain from a building surface may come into this zero-worth action category when such an action results in the highly increased cost of building. It is because of the uncertainty in the worth of an action that many appealing options for sustainability have not been 'marketable'. An innovative passive solar system may never come on the market due to the uncertainty of its worth in terms of sustainability.

One may argue that such a simplified definition of worth of an action in monetary terms may be misleading as the reward, such as a benefit to the environment is 'hidden'. The answer to this is: unless a mechanism has been devised, formulated and formalised (e.g. through government legislation), then such a 'hidden' reward remains irrelevant in terms of the penalty and reward concept.

Penalty and reward must have the same units; otherwise they cannot be summed up. They also have to be measurable, in energy terms (MJ, kWh, etc.) or in financial terms ($). If they have different units, there must be a way of converting one to another: they must be convertible. The decision as to whether an action can go ahead (i.e. be approved) will depend on the total worth of action (TWA) defined as the difference between the reward and the penalty. A positive TWA means that a particular action is worth considering.

INTRODUCING THE PENALTY AND REWARD (PR) CONCEPT IN PRINCIPLES, SOFTWARE AND DESIGN

In the majority of websites advocating sustainability, we can see various prescriptive advice such as 'use high thermal mass to reduce temperature swings', 'open windows during mild weather conditions to exploit maximum natural ventilation /cooling'. For residential buildings, the worth of such action can be easily assessed, and therefore – without the need to use complicated procedures – can be easily adopted.

Such a simple approach, however, cannot be applied to commercial buildings without additional design, construction and maintenance. Thus the introduction of the PR concept in this circumstance helps avoid controversy and uncertainty.

The current approach to passive design without proper evaluation of penalty and reward can be misleading. For example, the 'increase thermal mass will minimise temperature swing' may not be acceptable for the following reasons. This statement may in fact lead to further unnecessarily high construction cost (penalty) where no guarantee that the claimed reward will prevail. In this case computer validation is imperative. Without critical assessment of any *action* we may in fact be led into creating another problem from a supposed solution.

The PR concept is a must; it stands at the very heart of the problem – the sustainability problem. This concept eliminates or minimises the controversies surrounding the ways of creating sustainability through various options or actions.

The PR concept should be introduced into principles, design and software. This will help the designer, builder and the owner of the building to discuss various options more clearly in economic and sustainability terms. The introduction of the PR concept in principles enables all the parties involved to easily understand the nature of certain actions. In the design stage, the PR concept will help the designer to easily sort out reasonable or potential actions and filter out non-potential actions. Quantification of this concept will help produce modelling and simulation software packages which are able to select the most appropriate actions for a given project.

A BUILDING AND ITS ENVIRONMENT: A THERMODYNAMIC VIEW

A building and its environment exchange heat every second at a different rate depending on the external conditions (environment), internal conditions and the building itself. Between the building envelope (surfaces) and ambient environment, there is heat exchange through radiation, convection and conduction. Building surfaces exchange heat with ambient air through radiation (on a warm day) and convection (most notably during windy days). Heat absorbed by the surface is conducted to the internal surface of the walls and can take hours before it reaches the inner surface. (When the input varies and a building has high thermal mass, this heat may never reach the inner surface of the building envelope.) On the other hand radiation can instantaneously reach the building interior through the transparency of the glass in the windows or any other openings. The same is true for heat conveyed through infiltration. Thus, the building envelope acts as a 'thermal' boundary between the B(uilding) and the E(nvironment). It also acts like a heat exchanger between the two.

The concept of temperature difference, ΔT, is used to quantify the extent to which heat exchange occurs between the building and environment. ΔT is created by the building envelope. Without the building envelope, ΔT would not exist. Once a building envelope is set, ΔT will depend on: ambient conditions, envelope properties and the building's internal conditions.

Depending on the ambient conditions (or weather), B and E can act as a heat sink or a heat source. During the summer (high irradiance, high ambient temperature), B is forced to act like a heat sink by E (cooling load). In this circumstance, possible passive actions (PA) may include: shading (artificial or natural) to block some of the heat entering the building. When PA cannot totally eliminate the cooling load, then active action (AA) is sought: an air-conditioning system.

PINCH TEMPERATURE CONCEPT

From this discussion of the thermodynamic interaction between a building and its environment through the ΔT concept, it is worth mentioning a 'pinch temperature' concept familiar to most chemical/mechanical engineers. The 'pinch temperature' concept (Linnhoff et al., 1982) was introduced in designing Heat Exchanger Networks (HEN). In order to maximise the heat exchange between the available hot and cold streams in HEN, the 'pinch temperature' is introduced. The 'pinch temperature' is a point, no more than a minimum temperature difference (ΔT_{min}) between hot and cold streams. The 'pinch temperature' is regarded as the 'bottleneck' in HEN. The pinch point temperature will determine the minimum external utility that must be given to the system to achieve the target temperatures of hot and cold streams.

The effect of lowering the 'pinch temperature' is to reduce the external heating or cooling required. However, a lower 'pinch temperature' means a relatively large heat exchanger area compared to that needed with a higher 'pinch temperature' and therefore high initial cost.

The 'pinch temperature' concept stems from very fundamental thermodynamic principles: the first and the second laws, and proves to be a very valuable tool in assessing the actions taken in terms of rewards and penalties they bring about. The pinch concept can be useful and possibly adopted in developing the design approach or principles for commercial building. The complexity of commercial building structures is comparable with that of a HEN. However, the adoption is admittedly not straight-forward. The characteristics of the two systems (HEN and commercial buildings) are very different. There is no direct or clear parallel between 'components' of the two systems. However, they share the very concept of temperature difference (ΔT) from which an analogy can be traced and developed.

Figure 3.11.2 is a hypothetical graph showing the effect of adopting passive action which is borrowed from the 'pinch temperature' concept (Linnhoff et al., 1982). Originally, the quantity in the horizontal axis is the minimum temperature difference, ΔT_{min}, across the HEN. As shown, there is an optimum ΔT_{min} which gives the minimum total cost of the HEN.

In adopting such curves to the building design approach, the quantity in the horizontal axis is taken as the fraction of energy reduction achieved using passive actions. It can be expected – although may not be always the case – that relying heavily on the passive design approach (as shown by the arrow pointing to the left) will result in a higher capital cost of the building but lower energy cost. Also as shown in Figure 3.11.2, there will be an optimum point for adopting the combined approach of passive and active systems, a point where a set of passive actions results in lower energy requirement at an acceptable comfort level and is economically affordable.

It should be noted that not all passive options available are feasible; certain options may only improve the building performance marginally, or the

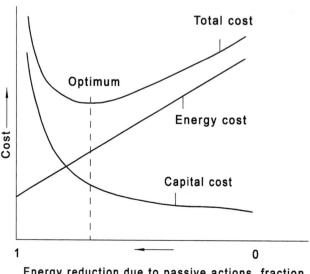

Figure 3.11.2
Hypothetical graph showing the effect of adopting passive actions

Source: Linnhoff *et al.* (1982).

cost of its adoption may be too prohibitive. With the help of the penalty–reward concept discussed previously, these options can be easily evaluated. However, more research is required to establish the appropriate quantity that should lie on the horizontal axis. The invention of the ΔT_{min} in the HEN design has led to a standard design procedure in the design of HEN (Linnhoff *et al.*, 1982). It is postulated that an equivalent of ΔT_{min} exists in establishing the optimum match between passive and active systems in building design.

POTENTIAL AND CHALLENGES

The combined penalty–reward–pinch (PRP) concept introduced here can be a very powerful tool to assess various options available to bring about building sustainability. It should help to easily identify various options worth considering in the design of either new buildings or the renovation of existing buildings. In addition, the proposed concept should also be able to optimally match the feasible passive options with that of an active system.

The concept, however, relies heavily on the availability of a much more powerful computer package than is available today. In addition to the existing capability of current computer modelling packages, the new modelling package should be able to perform co-simulation (Trcka *et al.*, 2007) to enable the analysis of coupled building fabric and form (envelop) and energy systems serving the building. Such a package should also be equipped with a database for building materials and various (green and conventional) technologies used in buildings which consists of technical specifications required in the modelling such as found in Saman *et al.* (2009) and the DCCEE energy rating web site (DCCEE, 2010). And, finally, such a package should have built-in economic analysis tools to enable economic assessment of various options.

CONCLUSION

This chapter has discussed a new approach, termed the penalty–reward–pinch (PRP) design approach with the aim of identifying the feasible passive systems and optimally matching them with the active system to bring about building energy sustainability. The concept relies on assessing the *penalty* and the *reward* of every action taken during the design of new buildings or design for renovation. The pinch temperature concept from heat exchanger network design has been discussed and adopted to help establish the optimum match between passive options and the active system for building energy sustainability. Research work is being directed to identify the equivalent [DEL]T_{min} in the design of building energy systems.

REFERENCES

CRES (2010) *Bioclimatic Design and Passive Solar Systems*. Available at: http://www.cres.gr/kape/energeia_politis/energeia_politis_bioclimatic_eng.htm> (accessed 19 November 2010).

DCCEE (Department of Climate Change and Energy Efficiency) (2010) *Energy Rating*. Available at: www.energyrating.gov.au (accessed 23 November 2010).

Karyono, T.H. (2000) 'Report on thermal comfort and building energy studies in Jakarta, Indonesia', *Building and Environment*, 35: 77–90.

Linnhoff, B., Townsend, D.W., Boland, D., Hewitt, G.F., Thomas, B.E.A, Guy, A.R. and Marsland, R.H. (1982) *User Guide on Process Integration for the Efficient Use of Energy*, Rugby: IChemE.

Martinaitis, V., Rogoža, A. and Bikmanienè, I. (2004) 'Criterion to evaluate "twofold benefit" of the renovation of buildings and their elements', *Energy and Buildings*, 36: 3–8.

Saman, W., Mudge, L., Halawa, E., Cheng, T.Y. and Bruno, F. (2009) *ANZHERS: Space Cooling Rating Tool*, Final Report, Canberra: Department of the Environment, Water, Heritage and the Arts.

Trcka, M., Wetter, M. and Hensen, J. (2007) 'Comparison of co-simulation approaches for building and HVAC/R system simulation', in *Proceedings of the 10th IBPSA Building Simulation Conference*, 3–5 September, Tsinghua University, Beijing, pp. 1418–25.

3.12

DRIVERS FOR RENOVATION OF COMMERCIAL BUILDINGS

David Leifer

INTRODUCTION

The life-cycle of buildings can include renewal, retrofitting, refurbishment or recycling. Since new buildings add only 3 per cent per annum to the sum stock of buildings in Australia, it is clear it would take a generation in order to bring half of our buildings up to present-day standards. Consequently more can be achieved by improving the sustainable and energy performance of the existing stock through renovation.

Improved environmental performance of this building stock can be 'pushed' through legislation or 'pulled' by economic inducement. While the Building Code of Australia 'pushes' energy standards through minimum levels of thermal transmission and maximum limits to solar gain, water heating in residential buildings (BCA, 2007) or electric lighting in all other classes (BCA, 2008), governments have been reluctant to fundamentally 'pull' the real estate market's activities through taxation or carbon trading until very recently.

However, as the costs of energy increase and a longer-term view of building life-cycles emerges, the economic impetus for improved per-formance becomes stronger. This chapter looks at the economic 'pull' issues surrounding decisions for the renovation and energy improvement of commercial office buildings.

BACKGROUND

The existing building stock can be improved by renovation when upgrading shell and/or service systems to promote better energy performance.

Refurbishment can be minor, medium or major as defined below:

- Minor: the minimum work necessary, little more than a superficial treatment of the building including making good faults and enabling it to operate without obvious deficiencies, would include: cleaning and checking water-tightness of the façade; remodelling the main entry and lobby; upgrading the finishes in the lettable space; refurbishing the toilets; minor improvements and repairs to HVAC and electrical services; and refurbishing of lifts.
- Medium: as above, moreover: *partial upgrading of the façade*; new tiling and fixtures in the toilets; upgrading of staircases and core components; upgrading of HVAC and electrical services *ensuring that they are energy efficient.*
- Major: as above, moreover: *new façade and visible elevations*; total upgrade of lettable space; remodelling and upgrading of the core and components;
- modernising or new lifts; and full upgrading to state-of-the-art HVAC and energy efficiency systems (Rawlinsons, 2010: 706).

Evident from the above (in the authors' italics) are the interventions for directly improving the energy performance through these renovations. Building shells and envelopes have a physical life of more than 60 years and are often thought of as permanent, whereas major elements of the building services systems have economic lifetimes in the region of 25–30 years (BLP, 2001). The ratio of construction cost of the entire building compared to the building services systems given in Table 3.12.1 is roughly one to three (i.e. Structure & Envelope vs. Building Services ratios of 32:35 per cent, 33:34 per cent, and 34:35 per cent for the three building specifications in Table 3.12.1). It is therefore evident that major renovation decisions are likely to arise when these building services systems approach the end of their service lives due to wear and tear.

As these building services systems involve pumps, compressors and motors that wear out and require regular inspection and maintenance, they are also increasingly controlled by sophisticated digital electronic systems that are prone to technological redundancy, meaning that their operating efficiencies are comparatively degrading with respect to state-of-the-art equipment.

The requirement to replace these expensive building elements and the disruption caused to tenants during the replacement process together will often trigger the decision to renovate the entire building. Similarly, changes to legislation (for example, more stringent access for the disabled

Table 3.12.1 Construction costs of Sydney CBD office buildings: costs of fabric and services components

Number of storeys	Prestige		36–50 storeys		21–35 storeys	
	($)	(%)	$/sqm	(%)	$/sqm	(%)
Preliminaries	1,178.50	24.8	1,076.00	24.9	885.00	22.7
Substructure	36.00	0.8	36.25	0.8	40.00	1.0
Superstructure						
Columns	102.00	2.1	99.50	2.3	87.25	2.2
Upper floors	318.50	6.7	311.25	7.2	308.25	7.9
Staircases	29.00	0.6	25.25	0.6	25.00	0.6
Roof	23.75	0.5	21.00	0.5	25.00	0.6
External walls & windows	857.75	18.1	777.25	18.0	721.00	18.5
External doors	5.00	0.1	4.75	0.1	4.25	0.1
Internal walls	142.75	3.0	137.25	3.2	131.25	3.4
Internal screens	14.50	0.3	13.75	0.3	11.75	0.3
Internal doors	30.00	0.6	28.50	0.7	27.50	0.7
Sub-total	1,523.25	32	1,418.50	33	1,341.25	34
Finishes						
Wall	59.25	1.2	45.25	1.0	44.25	1.1
Floor	50.25	1.1	42.50	1.0	39.00	1.0
Ceiling	100.50	2.1	75.75	1.8	72.50	1.9
Fittings						
Fitments	29.75	0.6	26.25	0.6	24.50	0.6
Services						
Plumbing	151.50	3.2	131.75	3.0	125.50	3.2
Mechanical	716.25	15.1	660.50	15.3	585.75	15.0
Fire	115.50	2.4	119.75	2.8	114.25	2.9
Electrical	313.50	6.6	254.50	5.9	238.25	6.1
Transportation	342.25	7.2	320.00	7.4	282.50	7.3
Special	9.75	0.2				
Sub-total	1,648.75	35	1,486.50	34	1,346.25	0.35
External services	5.00	0.1	5.00	0.1	5.00	0.1
Contingency	118.75	2.5	108.00	2.5	97.50	2.5
TOTAL	4,750.00	100.0	4,320.00	100.0	3,895.00	100.0

Source: Rawlinsons (2010).

requirements) can mean that properties may be downgraded if they no longer comply with current statutes and this can have a dire impact on the rental revenue. In some cases the intending relocation of a major tenant may cause the owner to initiate renovation (as did the owner of the building housing 14 floors of the Queensland Attorney General's Department in the early 1990s).

At these 'crises of dilapidation', or end of a building's economic life, the owners are left with the decision to sell, replace or renovate. The renovation decision will depend upon a number of factors, the major issue being the relationship between the property and the owning/using organisation (Leifer, 2003). The possible relationships are:

- the owner being a property developer who will renovate the building and then sell it on;
- the owner renting the entire building to a single tenant;
- the owner renting the building to multiple tenants; or
- the owner occupying the entire building themselves.

DRIVER FOR DIFFERENT TYPES OF BUILDING OWNER

The property developer

The building developer's concern is to sell the building at a higher cost than is used to build it; the greater the difference, the greater the profit. As such, the developer's time-horizon ceases post-sale. Their major concern will be to neither over- nor under-capitalise their investment in order to optimise the land use, i.e. to attain the maximum amount of building under the planning rules and controls. The developer will worry almost exclusively about CAPEX (Capital Expenditure) and exceeding the minimum regulatory standards or standard demanded by the market.

Table 3.12.1 shows that the average construction cost per square metre for Grade A, B and C high-rise office towers (in 2009) in Sydney CBD was $4,750, $4,320, and $3,895 respectively. Given the enormous sums that will be required for a 10,000 sqm building, it is clear that the capital will be raised through lending institutions, and therefore the longer the building remains unsold, the greater the interest repayment that will be called for.

The owner of a single-tenanted building

If the owner rents the building to a single tenant, the lease terms will usually involve a full repair lease, and all of the operating and maintenance costs will be placed upon the tenant. Here the operational costs on the tenant arise through (1) the repair clauses in the lease with the owner merely responsible for the main building structure, rates and land taxes; (2) the utility costs; and (3) other operating costs. Again the owner will be concerned with whatever CAPEX is committed in the ownership of the building, and also the sinking fund (where an annuity paid by the building owner is accumulated over time to meet at maturity the predicted costs falling due in the future) necessary to maintain the major items of building fabric and plant for which he/she has the responsibility, e.g. electrical switchboards, lifts, HVAC etc. A single tenant will usually sign a mid-term lease (say, five years with an option on a further five), and renovation might become an issue for the owner as the lease approaches expiry.

Owners will be most interested in the service lifetime of the build-ing components for which they are responsible. Some examples of economic lifetimes (average service lives where the cumulative costs of maintenance exceed the costs of renewal) of typical components are shown in Table 3.12.2. Where a significant number of expensive components face renewal within a similar time period, renovation becomes an alternative to individual component replacement.

The owner of a multi-tenanted building

Where the owners rent the building to multiple tenants, they are financially responsible not only for the maintenance (as in the case above) but also for the operation and provision of services of the base building. Often the owner will employ an agent to run the building. As in the preceding case the owner must consider the schedule for replacing dilapidating items and countering fair wear and tear. The owners will also retain the responsibility for meeting statutory responsibilities including the operation of the fire and emergency warning systems, inspection and maintenance of lifts, treatment of cooling towers, etc. They will also be responsible for day-to-day operations such as

Table 3.12.2 Economic lives of selected building fabrics and building services components

Component	Economic life-years
Brickwork	40–75
Curtain-wall systems	35–40
Sealed double-glazed units	10–15
Oil-based mastics	10–15
Silicon-based sealants	20–25
Compressed fibre sheeting	20–25
Precast concrete	10–70
Profiled steel cladding	
Plain	15–40
Pre-coated	15–20
Services equipment	
Refrigeration plant (medium and large compression and absorption)	15–25
Distribution systems	20–25
Terminal units	15–25
Cooling coils	15–20
Heating coils	15–20
Fans	15–20
Pumps	20–25
Pipework and valves	20–25
Cooling towers (depending on materials)	10–25
Control equipment	15–20
Packaged air handling units < 35kW	8–15

Source: NPWC (1988, Schedules 5 & 6).

management of the loading docks, garbage collection and disposal, maintenance of power supply, and the management of the common areas such as
stairwells, toilets, and lift lobbies.

The owners or their agents will generally sub-contract other operations such as cleaning to the base building and encourage their use by the
tenants. The owners will recoup their costs from the tenants through a service
charge, where the tenants pay proportionately according to the area of space
they occupy. Therefore, the owner is concerned not only with the CAPEX and
sinking fund, but also OPEX (Operating Expenditure). (OPEX can be deducted
from the owner's income to reduce income tax liability.) Table 3.12.4 on p. 302
gives the 2009 operating costs of base buildings in Sydney for Grade A, B and
C buildings, being $128.01, $130.77, and $110.86 per sqm respectively,
meaning that in around forty years as much will be expended on operating the
buildings as was paid for their construction.

Generally, construction or purchase cost of the building will be
compensated over time by the income generated. This is illustrated in Figure
3.12.1 that shows the cumulative cost of ownership (per sqm) and income
over a 20-year timeline for a Sydney Grade A office building (in 2009 dollars).
It will be seen that a 'break-even' point occurs after nine years. However,
the situation becomes slightly more complex if the owner intends to hold
onto the building as an investment for significantly longer. As will have been
noted in Table 3.12.2, a significant number of major services items will have
reached their service lives between their 25th and 30th years, hence
necessitating replacement, renewal, or disposal.

It has historically been the case that property appreciates in value
over time despite intermittent and temporary setbacks. The benefit from the

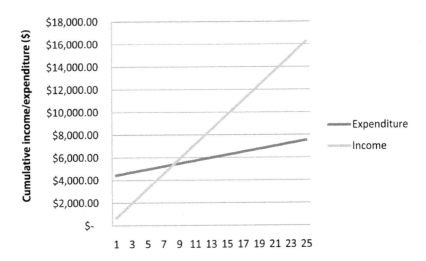

Figure 3.12.1 Cumulative cost of ownership (per square metre) and income over a 25-year timeline
for a Sydney Grade A office building (in 2009 dollars)

Note: Based on Excel spreadsheet using data from Tables 3.12.1 and 3.12.4.

sale option must be compared to the benefit of the replacement and renewal options

The owner-user

The owner-user is concerned with both the capital value of the property and also its operating costs. These are to be compared with those premises' costs that the organisation would otherwise incur were they not property users, i.e. the market costs they would incur were they tenants. Where the value of the property represents a significant proportion of the organisation's assets, there will be a strong pressure to ensure that the asset value is maintained and improved; more so where the property is used as security against borrowings.

Summary of ownership

It goes without saying that owners will also be concerned with the value of their properties in regard to insurance coverage for replacement should something untoward happen to them.

It is also worth adding that tenants will have little financial interest or investment in the buildings they occupy other than the fit-out and furnishings that they have supplied. However, they will be very concerned with the rental and services charges that they pay, and that the premises assist them to carry on their core business effectively.

The perceptions and commitment of the various parties identified above can be summarised in Table 3.12.3. It is only when the longer-term owners are responsible for the maintenance and replacement of the major building infrastructure that they will take a serious interest in life-cycle costs, and this in turn will influence their decisions in regard to renovation.

FINANCIAL ANALYSIS

Table 3.12.4 shows both the income and operating costs of Sydney office buildings. It will be seen that the loss of revenue from allowing an A or B grade building to degrade to C is $293 and $240 per sqm down from $650 and $596, a drop of 45 per cent and 40 per cent respectively.

Table 3.12.3 Interests of owners and building users

Ownership and/or use	CAPEX	OPEX	Life-cycle costs
Developer	X		
Owner and single tenant	X		
Owner and multiple tenants	X	X	X
Owner and occupant	X	X	X
Tenant		X	

Table 3.12.4 Average income and operating costs for Grade A, B and C Sydney office towers, 2009

Income /cost item	Premium & Grade A		Grade B		Grade C	
	$/sqm	(%)	$/sqm	(%)	$/sqm	(%)
Income						
Rental income	518.68		502.34		322.60	
Car park income	36.07		30.39		12.92	
Naming/signage income	4.98		3.46		n.a.	
Cleaning recoveries	13.39		10.79		n.a.	
Total income	649.88		596.48		356.31	
Statutory charges						
Municipal/council rates	13.97	10.9	17.08	13.1	14.96	13.5
Water & sewerage rates	2.17	1.7	1.23	0.9	2.76	2.5
Land tax	15.81	12.4	20.80	15.9	16.54	14.9
Other statutory	1.98	1.5	2.30	1.8	n.a.	
Total statutory charges	31.33	24.5	39.37	30.1	34.13	30.8
Operating expenses						
Electricity	13.40	10.5	14.66	11.2	12.75	11.5
Administration/management fees	12.28	9.6	11.60	8.9	9.73	8.8
Security/access control	9.69	7.6	3.72	2.8	3.79	3.4
Common area cleaning	8.90	7.0	10.24	7.8	10.38	9.4
Repairs and maintenance	8.54	6.7	7.78	6.0	11.71	10.6
Building supervision	7.04	5.5	5.76	4.4	7.45	6.7
Lifts and escalators	5.73	4.5	5.35	4.1	6.76	6.1
Air conditioning and ventilation	5.02	3.9	5.90	4.5	5.23	4.7
Insurance premiums	3.53	2.8	3.30	2.5	3.03	2.7
Fire protection/public address	2.42	1.9	3.14	2.4	3.79	3.4
Energy management/building automation	2.41	1.9	2.25	1.7	2.69	2.4
Miscellaneous costs	1.45	1.1	1.26	1.0	2.45	2.2
Gas and oil	0.79	0.6	1.04	0.8	1.43	1.3
Emergency generators	0.61	0.5	0.67	0.5	0.73	0.7
Gardening/landscaping	0.27	0.2	0.42	0.3	0.20	0.2
Pest control	0.11	0.1	0.13	0.1	0.32	0.3
Car parking	0.11	0.1	1.75	1.3	1.12	1.0
Total operating expenses	85.24	66.6	82.95	63.4	78.01	70.4
Total expenditure	128.01		130.77		110.86	
Total profit	521.87		465.72		245.45	

Source: Property Council of Australia.

It will also be seen that of the total operating costs of $128.01, $130.77 and $110.86 per sqm per annum, the energy costs (electricity and fuel oil) account for $14.19 (11 per cent), $15.70 (12 per cent), and $14.38 (12 per cent) respectively, which is a very small proportion (around 1/40th) of the construction costs. Unless serious efficiencies could be made, or energy costs rise dramatically, there is little financial justification for CAPEX investment in energy consuming systems *per se*. However, if achieved during an impending renovation, the investment can be amortised over a wider cost base.

The cost of refurbishment

It must be appreciated that, in a warm climate such as Australia enjoys, most commercial buildings and offices, in particular, containing a substantial amount of ICT equipment, overheat and require artificial cooling. This means that whenever the external temperature is lower than internal comfort, there is an advantage in heat transfer into the building. Table 3.12.5 gives the cost of major and medium refurbishment. While average costs of minor refurbishment are available, these costs are remedial and do not cover the improvement of systems and improvement of energy efficiency.

The envelope of a building acts as a passive climate modifier. The materials have thermal resistance and thermal capacity which both affect the rate of thermal conductivity, i.e. the rate of heat transfer, their external surfaces affect the sol-air temperature – that is the temperature differential driving heat transmission between outside and inside – during insolated periods, and the amount and transmission of glazing affect solar penetration. Given the variability of climate and its modification by the envelope as a system, energy-consuming mechanical climate adjustment is needed to ensure user comfort during those times when the passive modification is inadequate. Therefore the better the passive performance of the envelope, the less mechanical modification is required.

Energy improvements through renovation can be effected by decreasing the unintended infiltration of air through the envelope, i.e. the airtightness of windows and doors. The other major opportunity will arise when the building services, in particular lighting, HVAC, control systems, and other energy-consuming plant are replaced and upgraded.

It will be seen from Table 3.12.5 that in a major refurbishment 93 per cent of the original cost of the external windows and walls is invested. We may assume that in this case glass with decreased solar heat transmission characteristics will be installed.

Treatment of the opaque parts of the external walls can be made to increase its thermal mass, thereby delaying the entry of daytime heat until the cool of the night. Along with night-purge by ventilation the amount of daytime cooling can be reduced. Thermal mass – dense building structure – when it is charged with excess heat or coldness cannot be controlled and will impact upon the internal environment whether wanted or not, driven solely by the temperature differences between its core, its surfaces and that of the air.

The reduction in the need for cooling – assuming that no extra energy is required for artificial lighting – can mean smaller HVAC equipment is required, hence the lower its power rating and operating energy, subsequently the smaller the cabling and switchgear needed and less space required for the equipment. If this space can be recouped as lettable area,

Table 3.12.5 Refurbishment costs for building fabric and services

21–35 storeys	Major		Minor		Recycle	
	$/sqm	(%)	$/sqm	(%)	$/sqm	(%)
Preliminaries	959.50	25.4	286.25	19.5	1,336.50	28.3
Substructure	6.00	0.2	5.25	0.4	6.25	0.1
Superstructure						
Columns	5.25	0.1	–	0.0	10.50	0.2
Upper floors	22.75	0.6	–	0.0	44.25	0.9
Staircases	11.50	0.3	10.50	0.7	29.25	0.6
Roof	20.00	0.5	1.25	0.1	33.75	0.7
External walls & windows	798.00	21.1	113.75	7.8	867.25	18.4
External doors	11.50	0.3	12.25	0.8	11.75	0.2
Internal walls	67.00	1.8	23.50	1.6	157.50	3.3
Internal screens	9.25	0.2	9.00	0.6	9.25	0.2
Internal doors	31.25	0.8	15.25	1.0	31.25	0.7
Finishes						
Wall	80.00	2.1	37.75	2.6	86.25	1.8
Floor	111.25	2.9	104.25	7.1	110.00	2.3
Ceiling	92.50	2.5	84.25	5.8	96.75	2.1
Fittings						
Fitments	31.00	0.8	27.00	1.8	32.75	0.7
Services						
Plumbing	139.50	3.7	72.50	4.9	155.75	3.3
Mechanical	566.00	15.0	280.75	19.2	657.25	13.9
Fire	125.00	3.3	46.75	3.2	135.25	2.9
Electrical	268.00	7.1	213.00	14.5	323.50	6.9
Transportation	200.25	5.3	42.00	2.9	291.75	6.2
Special	–	0.0	–	0.0	–	0.0
Demolitions	78.25	2.1	2.50	0.2	111.50	2.4
External services	–	0.0	–	0.0	4.00	0.1
External works	6.00	0.2	2.50	0.2	7.75	0.2
Contingency	132.25	3.5	73.25	5.0	165.00	3.5
Total	3,775.00	100.0	1,465.00	100.0	4,715.00	100.0

Source: Rawlinsons (2009).
Note: Also shown, costs as a proportion of Grade A 'new build' per square metre.

then the rental income will be increased. Conversely renovations must not reduce a building's net lettable area as each square metre is a loss of $650 and $596 per annum (in perpetuity) for Grade A and B buildings from the rental income.

It is proposed that the first step in reducing electrical energy in a building is to ensure that equipment is switched off when not in use, this being a low or no-cost measure.

The second intervention is to adjust the maximum and minimum set-point temperatures on the HVAC equipment. Adding 1°C to the deadband can potentially give a 5 per cent reduction in heating and cooling energy required. This is also a no-cost measure (Roussac *et al.*, 2011).

A third intervention is to have the load power factor monitored. If it is less than 0.90, investment in power factor correction equipment will probably be warranted, and most certainly cost-effective energy savings will be achieved with power factors below 85 per cent. The cost of this equipment is almost certainly within the budget for electrical refurbishments indicated in Table 3.12.4. Water heating should be powered by other means than grid delivered electricity: gas, or preferably solar should be employed.

Improvements in the energy efficiency of lighting has a capital cost, but not only can more energy-efficient luminaires be installed, it will also have a flow-on to the power requirement for cooling. This is because most of the energy that is put into the lighting system is experienced as heat that needs to be mechanically cooled to maintain comfort conditions.

Upgrading electronic control systems on all plant and equipment can make a very big difference, being able to heuristically optimise operations by learning how the building behaves. The equipment does, however, require periodic recalibration and re-commissioning to ensure that it all works as it ought. Replacement of major items of old equipment such as motors and pumps will probably now be more efficient due to technological improvements. They are likely to deliver the same output power for less input energy, and variable speed drives – although expensive – can be installed on items with fluctuating load patterns. There is some speculation that co-generation (heating and power) and tri-generation (heating, power and cooling) will become increasingly economic as energy process development proceeds. Tri-generation involves absorption chillers, but a discussion of the pros and cons of equipment is beyond the scope of this chapter. A further direction to be discussed, although also beyond the scope of this chapter, is the adoption of renewable energy systems.

FINANCIAL REASONING

One of the major impediments working against the investment in energy saving technologies is the use of Discounted Cash Flow and Net Present Value (NPV) calculations in investment decision-making. This is illustrated by Figure 3.12.2 where the value of an initial investment of $10,000 decreases in value over time depending on the 'discount rate' (being the annual difference between the interest rate and the inflation rate).

Clearly a saving of $10,000/year for 25 years would amount to $250,000 in present-day dollars (that is, assuming that inflation is zero). However, the argument behind discounted cash flow (DCF) is that a present investment of a lesser amount will, at compound interest, accumulate to the desired sum in the future. Figure 3.12.3 demonstrates how the future value of an initial investment of $75,000 in year 25 accumulates to $254,000 at

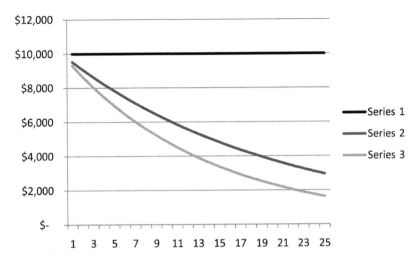

Figure 3.12.2 Discounted cash flow value of savings of $10,000/yr at 0 per cent (Series 1), 5 per cent (Series 2) and 7.5 per cent per annum (Series 3) discount rates over 25 years

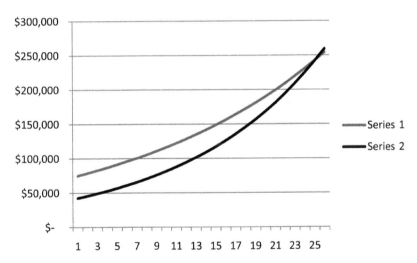

Figure 3.12.3 Comparison of Future Value of $75,000 and $42,500 in 25 years at compound interest rates of 5 per cent (Series 1) and 7.5 per cent (Series 2) respectively

5 per cent interest; or an investment of $42,500 accumulates to $260,000 in the same period at 7.5 per cent interest.

On the same basis, the Net Present Value of the saving of $250,000 in 25 years' time is only worth $75,000 (at 5 per cent), or even only $42,500 (at 7.5 per cent) *today*. However, this holds true *if and only if* the investor actually invests the principle sum at that enduring rate of interest. This is rarely done in actual fact, nor is the prevailing interest rate that predictable (except for long-term bonds) meaning that decisions are made on false economics.

What happens to future energy savings? This can be illustrated in Figure 3.12.4 showing the negative effect NPV analysis has on energy savings. The cumulative effect of $10,000/yr over ten years at 5 per cent depreciation – indicated by the area under the line – only appears to be a saving of $6,000 (or even $2,750) today.

CONCLUSION

This chapter has discussed the economic 'pull' drivers that influence the different situations of building owners. Long-term consideration involves a life-cycle view of the economics including income, CAPEX, OPEX and renovation costs. The economic lives of the capital expensive plant are much less than that of the building structure and envelope. Rising energy costs make retrofitting of more energy-efficient technologies desirable. However, the analytical accounting practice of reducing costs and benefits to Net Present Values diminishes the value of energy savings in the long-term future making them appear less viable.

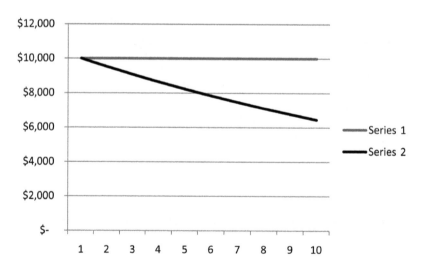

Figure 3.12.4 Net Present Value of energy savings of $10,000/yr over 25 years at 5 per cent depreciation rate. Actual savings in 2009$ (series 1) and Net Present Value of future investments at a 5 per cent depreciation rate (Series 2).

Source: Excel spreadsheet.

REFERENCES

BCA, Building Code of Australia (2007) *Standards Australia*, Volume 1, Canberra: Australian Government.

BCA, Building Code of Australia (2008) *Standards Australia*, Volume 2, Canberra: Australian Government.

BLP (2001) *Building Services Component Life Manual*, Building Life Plans & Building Performance Group, UK.

ISO (2000) *Buildings and Constructed Assets – Service Life Planning – Part 1 General Principles*, Geneva: International Standards Organisation.

Leifer, D. (2003) 'Building ownership and FM', *Facilities*, 21(1/2): 38–41.

NPWC (1988) 'Life cycle costing', paper presented at National Public Works Conference, Australian Government, Canberra, Oct. 1988.

PCA (2010) *2009 Benchmarks: Survey of Operating Costs – Office Buildings*, Canberra: Property Council of Australia.

Rawlinsons (2010) *Australian Construction Cost Guide 2009*, WA: Rawlinsons Publishing.

Roussac, C., Steinfeld, J. and de Dear, R. (2011) 'A preliminary evaluation of two strategies for raising indoor air temperature setpoints for office buildings', *Architectural Science Journal*, 54(2): 148–56.

A BIOCLIMATIC DESIGN APPROACH TO RETROFITTING COMMERCIAL OFFICE BUILDINGS

What are the 'rule of thumb' passive strategies for retrofitting?

Marci Webster-Mannison, Brett Beeson and Kat Healey

INTRODUCTION

The urgency and rationale for designers to create low energy commercial office buildings that respond to climatic conditions and indoor comfort in an environmentally sensitive manner are well established. Energy consumption from existing buildings, and the inherent embodied energy already spent, represent the majority of the commercial building sector's energy use well into the future. Much of this stock is 20–30 years old and due for major refurbishment. Therefore, most significant opportunities for reducing the energy use of commercial buildings arise in retrofitting rather than new-build.

Early in the design process there is a need to create credible retrofitting options prior to the development of building simulation models. Designers draw on a repertoire of understandings, memories and imaginings to help view new design situations in different ways; this chapter outlines some 'rules of thumb' that may be useful in the early design decision-making process. These 'rules of thumb' do not aim to provide a formula for the analysis of the complexities of an ecological approach, simply a starting point that may be useful in providing a framework for conceptualising environmental and technical issues, and defining some boundaries to the design problem.

Essentially, there is no one 'right' answer; however, the 'rules of thumb' enable the designer to make sensible decisions about some key aspects critical to the pursuit of a bioclimatic solution: the need to re-conceptualise thermal comfort, enhance daylighting, reduce air-conditioning loads, and augment natural ventilation.

RE-CONCEPTUALISING COMFORT

The warm humid climate region of Australia that includes south-east Queensland is relatively benign and highly suitable for natural ventilation of buildings. Based on the reference mean year (RMY) data for Brisbane, outdoor temperatures fall within the 90 per cent acceptability limits of the adaptive comfort standard (ASHRAE, 2004) for just over half of all office hours throughout the year. Heating is required for only a small part of the year, mainly in the winter mornings. Passive heating opportunities are enhanced by high solar irradiation during winter. Nevertheless natural ventilation is rarely considered as a thermal comfort strategy – either alone or in a mixed-mode configuration – for office buildings in south-east Queensland which are, almost without exception, fully air-conditioned. The replacement of air conditioning with natural ventilation and the resultant operational energy savings remain a missed opportunity.

Warm humid/subtropical climate defined

The warm climate region of the east coast of Australia consists of a warm humid, or subtropical, climate in south-east Queensland, ranging to a warm temperate zone further south towards Sydney where it is cool temperate. The climate is more moderate closer to the coast, and greater diurnal ranges and less humid conditions are found further inland.

High-rise commercial buildings in this region are restricted to the near-coastal urban centres including Brisbane, the Gold Coast, Newcastle and Sydney, although the westernmost parts of Sydney sprawl a considerable distance from the coast. Across this region, the transitional seasons of spring and autumn are generally good for human comfort. In the case of commercial offices, occupancy is limited to daytime hours so managing overnight temperatures is not crucial. In this context, winter in the more northern coastal parts of the region is also generally comfortable. High temperatures and/or humidity during summer are the greatest challenges to passive design across the region.

In the warm humid part of the region, maximum temperatures in summer are not excessive, with an average maximum temperature of 29.1°C in Brisbane during the hottest month of February; but humidity is reasonably high which exacerbates summer discomfort. Further south in Sydney the summer maximum temperatures are lower on average (26.4°C for the hottest month of January), but far more variable; Sydney averages 5.0 days per year with temperatures in excess of 35°C compared to 1.2 days for Brisbane. In western Sydney, twice as many days per year are hotter than 35°C than in Sydney's central business district, despite being part of the same greater metropolitan area.

Re-conceptualising comfort

When re-conceptualising thermal comfort in a warm climate, it is conventional to quantify thermal comfort using the predicted mean vote (PMV) scale. The associated Predicted Percentage Dissatisfied (PPD) estimates the proportion of dissatisfied occupants, depending on their attire and activity level and the physical conditions within the building. In doing so, the appropriate temperature set-point is obtained. This system is applicable to sealed, fully air-conditioned buildings, but not for naturally ventilated buildings where occupants have some level of control, which could be as simple as being able to open a window. An alternative system for establishing the comfort zone, the adaptive comfort model, is preferred, as it relates the comfort zone to outdoor ambient conditions.

The adaptive comfort model highlights the importance of occupant control to comfort in naturally ventilated buildings. While passive design features improve the thermal performance of the overall building, occupant controls allow people to both improve their own comfort and reduce frustration associated with centralised control systems, which allow no occupant intervention.

ENHANCE DAYLIGHT

Typical commercial high-rise buildings often have poor daylight, yet occupants indicate a high preference for good natural daylight. The challenge in retrofitting is to increase daylight penetration without significantly increasing heat load, which would necessitate upgrading of ductwork and diffusers. The following two strategies are the basic methods for increasing occupant daylight: (1) either bring the window and occupant closer together, or (2) increase the daylight penetration through the windows.

Reduce occupant distance to glazing

Good daylight can be defined as at least 200lux. This is sufficient for general tasks, but not for detailed work. For Brisbane, the design sky is 8500lux so we need about 2.5 per cent of this, which may be expressed as a Daylight Factor of 2.5.

As a 'rule of thumb', we expect good daylight to penetrate two and half times the distance from the floor to top of the window. For a window 1000mm high with a sill of 900mm, we would use a figure of 1900mm times two and a half. Hence we predict good daylight up to 4800mm from the façade. However, the quality of daylight and the associated heat gain are dependent on many factors such as the sky condition, the type of glass used in the façade, the visible light transmission of the glass (VLT), and the façade orientation. Hence, for heavy shading or darker glass, this estimate should be reduced.

Applying this 'rule of thumb' to a typical commercial office floor (58m x 15m x 2.7m), daylight penetration may be expected to be about 5.4m from the windows. If the façade is fully glazed, daylight penetration of about 6.8m is expected. Therefore, this typical 15m deep floor plate will not receive good daylight throughout and requires lighting by permanent artificial lighting (PAL) usually provided by electric lighting. The floor plate geometry, therefore, has an impact on the area of PAL and hence energy consumption. Narrower floor plates with a width of around 12m facilitate good daylighting and natural ventilation.

Applying this rule of thumb to the typical office building using the base case façade with 70 per cent VLT and minimal shading, 75 per cent is well lit. If the floor plate was square, only 34 per cent would be well lit.

To reiterate, daylight is associated with solar heat gain and trade-offs need to made between the amount of heat and light that will be transmitted through the façade.

Trade-offs for daylighting and solar heat transmission

The range of glazing available has expanded over the past decade. Older buildings may have poor daylighting or high heat gain due to excessively dark or clear glass. This glazing can be replaced with more suitable glazing which enables better tradeoffs between daylight and heat gain (see section on 'Reduce Heat Load' for details on the fundamentals of glazing). A typical selection of glass types is shown in Table 3.13.1. For illustration purposes, a similar VLT (55–70 per cent) has been chosen.

Light shelves

Light shelves are often proposed, and rarely well designed. The theory is that light is bounced onto the ceiling by internal, or external, horizontal blades which may have the dual purpose of shading. In practice, significant scattering occurs in diffuse conditions. During bright, sunny conditions light shelves can be effective, but are not often needed in such conditions. They must be kept clean for good performance. It is recommended that careful consideration and specialist advice are sought on light shelf design.

Table 3.13.1 Rule of thumb: trade-off between daylighting (VLT) and solar heat transmission (SHGC)

Type	VLT (%)	SHGC	Insulation	Costs	Notes
Tinted monolithic	58	0.69	Low	Low	For shaded or southern façades
Tinted monolithic	67	0.52	Low	Low	
Laminated low E	53	0.56	Medium	Higher	Moderate performance
Laminated low E (S4)	68	0.62	Medium	Higher	Moderate performance
DGU tinted	59	0.45	High	Moderate	Moderate performance
DGU low E	62	0.35	Very high	Higher	High performance
DGU low E	63	0.38	Very high	Higher	High performance

Augment daylight with minimal artificial lighting

Once ambient lighting drops below about 200lux, there is a high probability people will want to turn on lights to complete their tasks (Hunt, 1979). This is dependent on the task: low background lighting may be quite suitable for screen-based work. People further from windows will have a greater need for artificial light so the common practice of centrally controlled lighting on a single switch will result in unnecessary lighting for the perimeter occupants.

Some strategies for the augmentation of daylight with minimal artificial lighting include the following:

- Task-lighting where each occupant has a low-wattage task light with a manual switch.
- Perimeter lights on a separate circuit to internal lights, so they can be turned off separately.
- Light sensors which can monitor and automatically dim perimeter lights as required.

Of these, task-lighting provides the occupant with the best control, and potentially results in the lowest energy use, as individuals may choose to have lights off, or at lower levels than an automated system may determine. Automated dimming of lights on the perimeter is feasible, but must be handled with care as it can be frustrating for occupants.

Control glare via adjustable blinds

To control glare, which is particularly relevant when higher VLT glazing is used, all occupied façades should have blinds, except where very heavily shaded. Glare is subjective, difficult to model and even harder to accurately predict. Depending on the task, time and person, the need for blinds varies. Blinds that allow individual control are recommended to avoid occupant complaints. The control system for blinds can be automated, or manual. Manual control works well in private offices, or where there is a high level of user commitment. In some cases the occupants will not adjust the blinds, and they stay closed even when not required. Yet frequently operating blinds, such as is the case with many automated systems, are an annoyance.

The following strategies for controlling glare are recommended:

- Use manual blinds where possible.
- Consider how natural ventilation elements and blinds will function together.
- Limit use of automated blinds to more public and larger spaces.
- Provide easy user override of all automated blinds.
- Minimise automated blind movements.

REDUCE AIR-CONDITIONING USAGE

The reduction of air-conditioning usage will be discussed in terms of the following two broad approaches:

1 the 'Esky' approach;
2 the 'Cool Room' approach.

The Esky approach

The Esky approach assumes a sealed building – typical of high-rise commercial buildings in Brisbane. It assumes it is hotter outside than inside (given the average temperature in Brisbane of 23.8°C, this may seem an invalid assumption; however, when the surface temperature increase from solar gain is considered, it is sound), and that solar gain is usually bad. Hence, the building envelope should be highly insulated and the glazing area should be minimised. Traditional air conditioning is used to handle the internal heat gains of people, equipment and lights. Outside air is assumed to be undesirably hot and is limited, or its heat is exchanged with outgoing air.

The Esky approach can be very energy-efficient, particularly in combination with a fresh air economy cycle. It is common and therefore easy to design for mechanical engineers and architects. It may seem an extreme design for a benign climate, and it is. However, less extreme versions are common in Brisbane – these are commonly either glazed with highly-reflective, dark glass (commercial buildings) or heavily shaded, minimally glazed large boxes (government offices).

There are two important downsides to the Esky approach, which mean it is not recommended. First, it can lead to sick building syndrome, where the low levels of outside air, recirculation of contaminants, and poor external connection create an unhealthy environment. Second, and more subtly, it sets a limit to the energy efficiency of the building. All of the internal heat load is handled by mechanical plant which has efficiency limitations (following thermodynamic laws). Therefore, air-conditioning energy consumption cannot be reduced below a certain limit.

The Cool Room approach

The Cool Room approach recognises that the subtropical climate provides an opportunity to open the building to expel heat and take advantage of the benign ambient conditions. Commercial offices need cooling year-round in this climate, with some minimal winter morning warm-up. Instead of trapping the heat inside, as in the Esky approach, the building can be opened up and the heat expelled without the need for air conditioning. Solar gains still need to be minimised, for example, by using shading, smaller windows and/or high-performance glass, and the building still needs to be well insulated. It is important to note that opening the building serves at least two independent functions. One is to expel heat, as mentioned above. The second is to provide air movement to cool people which can be achieved in other ways such as fans (see the section 'Increase air movement').

How much air movement is required to expel a typical office's heat? For a well-shaded, typically dimensioned floor plate in Brisbane, a minimum of 5 air changes per hour (ACH) are required. Beyond 10 ACH, there is little improvement in performance (based on thermal modelling for a building in Brisbane).

Increase air movement

In cooler climates mechanical engineers and building managers have traditionally feared draughts, believing they lead to occupant complaints. This same thinking does not apply to all situations in warmer climates, despite it being ingrained in common practice. When people are neutral to warm, they actually prefer some gentle movement (Arens et al., 2009).

There are a variety of ways to provide air movement. In this section we consider air movement solely for occupant thermal comfort (i.e. breeze on the skin) which can be decoupled from air movement for heat removal.

Fans

Without fans, people start to feel excessively hot (PMV 1.0, PPD 25 per cent) at 27.5 °C; with fans people can be comfortable up to 29.3°C. Fans can reduce the perceived temperature by up to 2.7°C (Arens et al., 2009). Ceiling or wall fans are among the cheapest, easiest and most reliable techniques to provide air movement. Yet fans are rarely seen in offices – why? Conventional design practice is for commercial offices to have ceiling tiles, fluorescent lighting and carpet; anything else is a challenge for most clients and designers to accept. Adding to the complexity is the base-building–tenant divide. These challenges are surmountable.

As a rule of thumb, initial sizing of ceiling fans can be estimated using medium-sized (1200mm diameter) fans at 5000mm centres, which gives one fan per 25m². Smaller fans need higher speeds to move a given volume of air, so larger fans are preferred. The extreme end of this spectrum is large (over 3000mm diameter), high-volume, low-velocity fans. Wall fans (perhaps interconnected with wall louvres), that deliver air more directly to the occupant, are another possibility. Running fans in reverse helps to de-stratify the air, pushing the hot air down for winter operation. Outside areas can also benefit from fans.

Natural ventilation

In sub-tropical houses, natural ventilation is the norm. Yet in commercial buildings, it is a rarity. Natural ventilation assumes that by opening windows (or any external opening) the heat load within a building will be expelled. This section considers natural ventilation for heat removal, as opposed to natural ventilation to create an internal breeze. Something needs to drive the heat removal process. The simplest method is to use prevailing winds to create a differential pressure on openings. This relies on wind availability. For a typical commercial building in Brisbane to expel heat, the following is needed:

- Air changes per hour: 5 ACH
- Wall to operable window ratio: 15 per cent
- Wind velocity: 0.3m/s

About 0.3m/s of wind is needed to drive the natural ventilation. Given the need for flyscreens, and the variance of wind direction, a minimum of 0.5m/s is a reasonable guide. It is essential to know if there is enough wind when it is required (i.e., at comfortable temperatures during operating hours). In Brisbane, only 10 per cent of the time is without such suitable wind (Figure 3.13.1). Note this is in unsheltered conditions: typically surrounding buildings

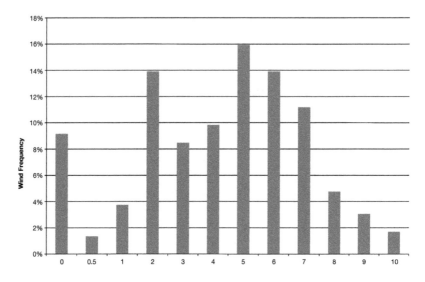

Figure 3.13.1 Wind frequency in Brisbane (8–18h, 18–26°C)

Source: Brett Beeson.

will reduce wind velocities. There will also potentially be wind tunnel effects. For buildings up to five storeys, no special consideration of height effects is required. For taller buildings higher wind velocities can be expected, so further consideration is needed by the designers.

The final factor to consider is the depth of the floor plate. For cross-flow ventilation, widths greater than 12–14m will struggle to achieve good ventilation. For single-sided ventilation, 6–8m is the maximum feasible depth. These rules of thumb are summarised in Table 3.13.2.

With retrofitting, it may be necessary to increase ventilation. Addressing the requirements for natural ventilation can be a complex task; a summary of methods is given in Table 3.13.3.

AUGMENTING NATURAL VENTILATION

We've seen that wind-driven ventilation has a number of challenges such as noise and low wind speeds. The following strategies can be used to improve natural ventilation, particularly when wind speeds are low.

The stack effect

Hot air rises. The air in a building is heated by people, lights and equipment. The hot internal air rises through the building, drawing in more air from cooler areas below. An atrium is typically used to provide an exit air path for the hot air. See the section on 'Atria' for details. The following factors improve the stack effect:

- Height of room to allow for stratification of hot air layer above occupants' heads

Table 3.13.2 Rule of thumb: sizing metric for cross-ventilation

Sizing metric	Value	Example: typical office floor (58m x 15m x 2.7m)
Window:Floor Ratio	10%+	For a 870m^2 floor plate, 87m^2 of open area is needed. Open area is the clear passage through a window. For a 1m^2 sash window, 10% will be frame and 50% will be openable, so the maximum open area is $(1 - 0.1) \times 0.50 = 0.45\text{m}^2$.
Window:Wall Ratio	20%+	The 870m^2 area has 400m^2 of wall so 80m^2 of open area is needed. It agrees roughly with the above.
Maximum floor plate depth (Single-sided openings)	2.5 x ceiling height	With a ceiling height of 2.7m, 7m maximum depth.
Maximum floor plate depth (Cross-ventilation)	5.0 x ceiling height	With a ceiling height of 2.7m, 14m maximum depth.

Table 3.13.3 Natural ventilation strategies and methods

Strategies	Methods	Notes
Maximising wind effectiveness	Orientate building openings to 'catch' prevailing breezes.	Vents may supplement windows, be careful in selecting leak-proof baffle for modulating/closing air intake. Flyscreens may block significant breeze in the order of 50%, and more if not kept clean. Outlet side may need to be baffled.
	Use fins to direct angled breezes inside.	Long fins can be extremely effective for channelling breezes.
	Use wind catchers.	Baffle uni-directional wind catcher.
	Floor-by-floor 'cut-outs' from the façade will effectively reduce building depth.	Additional daylighting benefit.
Controlling airflow	Use high-level louvres to avoid the occupied zone.	High level glazing may further benefit daylighting.
	Reduce opening sizes as the building height increases.	Measures from the Bureau of Meteorology are usually at 10m, so scale appropriately above this.
	Ensure windows or louvres can be modulated by BMS.	Some actuators cannot be half-open. Variable opening control is needed.
	Height of stack above building is critical.	Necessary to avoid overheating of upper levels.
Managing noise	Use acoustic screens.	Avoid line of sight noise transmission.
	Use awning windows.	
	Minimise perimeter openings, prefer atrium openings.	See 'Atria'.
	Provide individual control.	Higher occupant satisfaction with individual control for when aural privacy is needed.
Natural ventilation	For cross-flow ventilation, a minimum of 15 per cent of wall area should be open.	The larger the openings, the more effective the building will be at low wind speeds.

- Adequately sized lower external openings (see the section on 'Natural ventilation')
- Adequately sized openings to the stack or atrium. These can be considered the second side of a cross-ventilation strategy.
- Adequately sized openings at the top. Detailed analysis is recommended (using computer simulation) as many factors are involved. A rule of thumb is to size openings at the top with an area of half the total openings into the stack/atrium.
- If the strategy is to draw air from the stack/atrium, the outlet openings on the external face of the building will need baffles.
- Consider interaction with wind, which can help or hinder operation. Wind catchers and baffles can be used to manage wind.

- Higher stacks/atria are more effective. The build-up of hot air at the top of the stack needs to be prevented from moving back into building; usually at least a full storey or two is needed.
- Higher internal heat loads. Usually this is undesirable in occupied areas, but for atria consider tolerating higher solar loads and offsetting with fans to maintain comfort. This also increases daylight.

The interaction of wind and stack effect is important to consider. Results from the computer simulation of a building with one-sided ventilation are shown in Figure 3.13.2. Note that at low wind speeds, the stack effect works well (6000 L/s of flow). However, when the wind is coming from the 'wrong' direction, the effects of wind and stack cancel out (seen where the lines cross the horizontal axis). When using the stack effect, design to optimise operation with wind effects by placing inlet openings on the prevailing windward side, and outlet openings (at the top) on the lee.

In retrofitting, the opportunities for introducing the stack effect include the following design strategies:

- Use existing lift wells (or fire stairwells), often well located centrally to the floor, if being replaced. New lifts may be added to a suitable façade, perhaps in combination with a naturally ventilated foyer and staff amenities and meeting rooms, or to buffer an unwanted orientation.

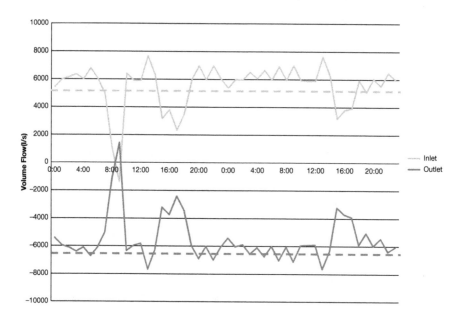

Figure 3.13.2 Wind and stack interactions over two days for single-sided openings

Source: Brett Beeson.

- Add new stacks to the external face which may also provide additional insulation to the façade and mitigate the need for façade replacement.
- Stacks may be integrated into a new façade system.
- Use service shafts, and find alternative space for replacement of services, such as external to the building.

Add stairs to enhance floor-to-floor connection and combine with stack function.

Atria

Atria can solve a number of problems simultaneously. We prefer narrow floor-plates for natural ventilation, but existing commercial office floor plates tend to be deep, and sometimes square. The introduction of an atrium can effectively cut the building in two which creates narrow floor plates suitable for natural ventilation, and provides daylight without excessive direct sunlight that can add to the air-conditioning load. Atria also provide logical locations for meeting spaces, shared facilities and stairwells. Multiple floor atria usually require smoke solutions via fire engineering. The introduction of an atrium to an existing building will add cost and complexity to a building, and potentially reduce the usable floor area, unless space trade-offs can be identified, for example, reduced plant space, or the addition of new lifts or services or amenities external to the original floor plate. Therefore, atria should be integrated early into the design and serve multiple functions. Daylighting is usually the driving factor in determining the sizing of atria (See 'Natural ventiliation' for sizing for ventilation). Atria are often undersized, not extended high enough above the building, and disappointing in performance. A clear minimum size is therefore necessary. The 'well index' (Calcagni, 2004) is a simple rule of thumb to determine the depth of an atrium, and is highly correlated to daylighting effectiveness.

$$\text{Well Index} = \frac{\text{height} \times (\text{width} + \text{height})}{2 \times \text{length} \times \text{width}}$$

A number of examples are given in Table 3.13.4.

Table 3.13.4 Rule of thumb: atrium sizing and well index

Width m	Length m	Storeys	Well index	Daylight factor (%)	Notes
2	35	2	1.4	7	Just big enough
4	35	2	0.8	16	Good minimum
8	35	4	0.8	15	Higher means greater width required
12	35	6	0.9	14	Deep atria must be wide

We aim for a daylight factor of 10–15 per cent at the base of the atrium, to allow plant growth and provide a pseudo-outdoor experience, which activates people's circadian rhythms. This requires a well index of less than 1.0. For higher atria, the width can be problematic. To minimise atria size while maintaining performance:

- Focus on lower levels, as they get less daylight. Use larger, clearer windows lower down.
- Use a V-shaped atrium which is also advantageous in terms of the stack effect as air from all levels is combined, increasing the required throat dimensions of the atria as the number of levels increase.
- Use white or light-coloured walls to increase light reflection.
- Use more wall area and fewer windows on high levels.
- Minimise plants on the wall or in higher-level planter boxes.
- Use plants and acoustic baffles (light-coloured) to manage noise.
- Maximise incoming light by using the highest VLT glazing possible. This will increase heat load. An operable shading system is ideal to minimise gain in high summer.
- Use blinds on top levels as high daylight levels will produce glare here.

The question of atrium orientation is difficult – do we prefer east–west or north–south? Usually this is a moot point as the building form defines the orientation. For sub-tropic climates we are focused on reducing solar gain. Back to the Brisbane example, winter sun penetration is better in a north–south orientation; however, due to high daylight and heat levels throughout the year, overall, an east–west orientation may be preferred.

Thermal chimneys

By retrofitting a tall 'chimney' on a building, the air in the chimney is heated and rises. This, in turn, pulls cooler air from the bottom. A thermal chimney can be connected to an atrium, or to the 'rear' of rooms to provide exhaust ventilation.

Thermal chimneys work most effectively when the following factors are taken into account in the design:

- The taller the better – extended higher than the roofline by several metres at least.
- Sufficiently sized for the volume of air to be moved.
- Paint a dark colour, or with a selective coating, or use a solar accelerator, to increase temperatures.
- Multiple narrow chimneys are used instead of one big one of the same cross-sectional area.
- The sun is shining – cloudy condition has poor performance.

The hotter the sun, the more heat exhaust may be required, so thermal chimneys should have a good profile match. In sub-tropical climates, however, it is often hot and overcast with high humidity. In these conditions, performance is poor. In Brisbane there are good operating conditions for about half of operating hours. More precisely, from October to March, 8:00 to 18:00, when the temperature is above 25°C, cloud cover is below half the sky for 47 per cent of the time. Since tall chimneys are necessary for good performance, attach them to the north or west of the building, starting from the first floor and rising above the roof. If reliable operation is essential, install inline axial fans to augment flow in poor conditions. Furthermore, the addition of fans presents opportunities to run the chimney in reverse to provide winter heating. Incorporate wind baffles to prevent unwanted air movement.

Mechanically assisted natural ventilation

When natural ventilation and its augmentations are not sufficient, fans and ducts can be used to expel heat. This is appropriate in deep-plan offices with single-sided ventilation. By exhausting air from the back of the room, the room is kept more comfortable. Typically a mechanical designer will be enlisted to design the system, but it is worth bearing in mind a few considerations. The location of specific elements in the system is vital, so the system may not work if not directed to the hotspots in the space. If an air-based system is already used for air conditioning, the cooling coils are simply turned off (or better – bypassed to reduce pressure drop) and the fans are run. This is similar to a traditional economy cycle. Oversized ducts compared to traditional air conditioning reduce fan energy. For large air volumes, services' space is needed to allow for an installation without tight bends. Hence additional ceiling space or bulkheads may be required. Install variable speed drive fans, which the BMS should modulate based on traditional thermostats to provide just enough mechanical ventilation.

Activate adaptive comfort

As we have seen in the section on 'Comfort', there are a variety of ways to predict comfort. There is also a lot of debate about the 'best' method. For our purposes we can say that Predicted Mean Vote (PMV) is suitable for air-conditioned, or stable temperature spaces. In particular, it is suitable when people have limited exposure to the outside temperature changes. This is the case in very cold, or very hot climates (when people often stay inside) and in sealed, air-conditioned offices (where people stay inside for long hours).

Adaptive Comfort (AdC) is suitable in predicting comfort when people have significant exposure to outside temperatures. This is the case in mild climates and where buildings are naturally ventilated for some of the year. If we compare adaptive comfort to PMV or a simple temperature metric, we predict that people will be less hot in the same conditions, see Figure 3.13.3.

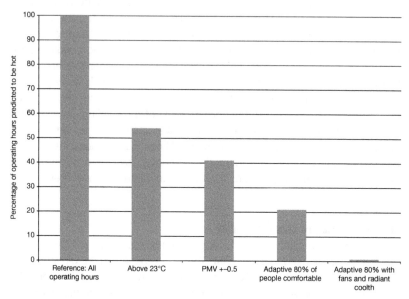

Figure 3.13.3 Predicted percentage of time when occupants are hot

Source: Brett Beeson.

The average temperature in Brisbane during operating hours is 23.9°C. Typical set-points in Brisbane are 22.5°C plus or minus 1.5°C giving a range of 21°C to 24°C. Trials have shown that wider bands are possible without occupant dissatisfaction. See Figure 3.13.4 where the suggested set-points are on the extreme end of the spectrum and need careful client consideration.

Successful set-point operation relies on occupant 'buy-in' as they must expect higher temperatures in summer and dress accordingly. Satisfaction with comfort in an office is higher when the level of individual control increases (e.g. individual control of openable windows and air movement, sun-shading, daylighting and task lighting, heating and cooling). For mixed-mode buildings the controls' settings, or occupant training, must strongly encourage natural ventilation. This sets up a positive feedback loop, where people get used to higher temperatures in summer, and so use less air conditioning which further adapts them. Positive or negative feedback is seen in mixed-mode buildings where people are divided into two groups – the 'air-conditioners' and the 'natural ventilators'. Encouraging variance in occupants' dress habits to suit the climate, the season and the nature of work, and deliberate interventions designed to connect people with the outside environment – by providing external seating, rest areas, water fountains, meeting rooms and the like – will activate their natural adaptive tendencies.

Toilets, kitchens and low-occupancy spaces can be naturally ventilated. Situating toilets and kitchen on the perimeter will facilitate this; an east

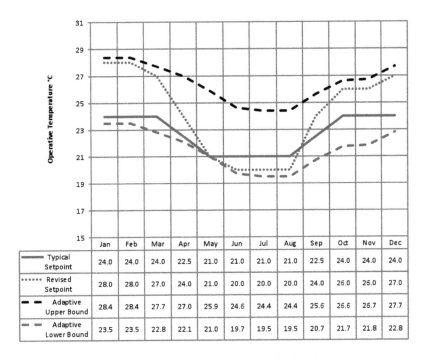

	Jan	Feb	Mar	Apr	May	Jun	Jul	Aug	Sep	Oct	Nov	Dec
▬▬ Typical Setpoint	24.0	24.0	24.0	22.5	21.0	21.0	21.0	21.0	22.5	24.0	24.0	24.0
∙∙∙∙∙∙ Revised Setpoint	28.0	28.0	27.0	24.0	21.0	20.0	20.0	20.0	24.0	26.0	26.0	27.0
▬ ▬ Adaptive Upper Bound	28.4	28.4	27.7	27.0	25.9	24.6	24.4	24.4	25.6	26.6	26.7	27.7
▬ ▬ Adaptive Lower Bound	23.5	23.5	22.8	22.1	21.0	19.7	19.5	19.5	20.7	21.7	21.8	22.8

Figure 3.13.4 Adaptive comfort set-points showing extreme set-points bounds

Source: Brett Beeson.

side, or highly shaded west side may also buffer some unwanted heat gains. Air quality also has an effect on thermal comfort as poor air quality is often associated with static air (perceived as stale) and high CO_2 concentration (leading to sluggishness).

Reduce heat load

When the building is in air-conditioning mode, we move to strategies used in the 'Esky approach'. In essence, this means reducing the heat entering the occupied zone as early as possible. It is a useful way of thinking for all heat gains. For example, the addition of external insulation to the façade will stop the heat earlier, but also may allow the exposure of internal thermal mass if the original façade is thermally massive. The creation of an air gap can be an effective way to contribute to the insulation of an existing external façade, and vegetation can be a cost-effective way to create this 'double-skin'.

Shade glazing

For warm climates, the summer sun rises in the south-east (in the southern hemisphere) and will strike the southern façade. These latitudes also have very high summer sun positions. For example, in Brisbane (latitude 27° south), the summer sun at noon is almost directly overhead. Rules of thumb for shading in Brisbane are given in Table 3.13.5.

Minimise glazing extent

The cost-effectiveness of curtain wall construction for office buildings, and the perception that more glazing is better, have led to floor-to-ceiling glazing becoming the norm in Brisbane. Yet research tells us that people are happy with glazing areas of just 30 per cent of wall area. Indeed, satisfaction goes down slightly when windows are larger than 50 per cent of wall area (Keighley, 1973).

Glazing size is the major driver of solar gain – which leads to increased discomfort and air conditioning. Typical walls have an insulation value of R2.0 while even double-glazed windows have a value of R0.3. In other words, seven times more heat is gained from windows per unit area than compared to walls.

Table 3.13.5 Rule of thumb: shading requirements

Method	Effectiveness	Notes
Northern façades		
Fixed horizontal overhangs	High	To shade October to April, 8:00 to 17:00, a VSA of 65° is required. For a 1000mm high window, a 500mm overhang is required. Fixed shading is appropriate.
Eastern façades		
Minimise glazing	High	Position toilets, core, stairs, storage and the like on the east or west.
Fixed shading horizontal	Medium	To shade October to April, 8:00 to 17:00, a VSA of 42° is required. For a 1000mm high window, a 1100mm overhang is required. This can be difficult to achieve in practice. A smaller overhang is often combined with fins or blinds.
Operable shading	High	Modulate to close in the morning and open automatically when sun if off the façade. A PE sensor can be used. Blinds are still required for glare control.
Fixed shading fins	Medium	Fins can assist horizontal shading and double as breeze catchers. However, perpendicular fins will sometimes allow direct sunlight, while angled fins block view.
Internal blinds for heat control	Medium	Combine with some fixed shading. Early morning heat is less of a problem, so although blinds are less effective than external shading, they are suitable for this façade. Use a VLT < 10% and a reflective backing.

Note: Vertical Shading Angle (VSA) is the angle from horizon to shade (0–90°). Visible Light Transmittance (VLT) is the percentage of light passing through a material (lower means darker).

Hence the strategy is to minimise the extent of glazing and to locate windows on the south and north orientations. The west and east can lead to unavoidable solar penetration from the early morning and late afternoon sun. Also consider the use of high sills (at desk height) because a window to the floor provides no daylight benefit, is costly and energy-ineffective. Higher windows provide more even daylighting and harness the brighter zenith sky. (Note, glare can be a problem with large high-level windows.) Some further strategies are:

- Use task lighting to supplement daylighting since it may be difficult to achieve enough daylighting for all workers using windows.
- Place vents under the windows to increase open area.
- Simple measures such as light-coloured materials can improve performance for a given area of window.
- Consider light shelves and light tubes in the walls or ceilings to throw light deeper into the floor plate, or to harvest light from a better orientation.
- Light shelves' effectiveness is often disappointing due to dirt but can improve daylighting slightly for a given window size.

High performance glazing

There are three performance attributes to consider when selecting glazing:

- SHGC (solar heat gain co-efficient): shaded glazing can have higher values (~0.60) while unshaded glazing, particularly on exposed western or northern façades, needs lower values (< ~0.30).
- U-Value: single glazing has a U-Value of ~7.0 $W/m^2/K$. For cold mornings and hot afternoons, this increases air conditioning and decreases comfort for nearby occupants. Double glazing and low-e glazing can achieve ~3.0$W/m^2/K$.
- VLT: clear glazing has a VLT of 80 per cent. Darker commercial glass can be as low as 20 per cent.

Glazing is a complex topic, but some key metrics are:

- Solar heat gain co-efficient: the fraction of (the sun's) energy passing through the glazing.
- U-Value: is a measure of the insulation provided by a material. Higher means less insulation.
- Low-E: low emissivity which means less energy is radiated from a surface, indirectly improving insulative performance.
- VLT: Visible Light Transmittance: the fraction of light passing through a material.

In general terms, in a cooling-dominated sub-tropical climate we aim for a low SHGC and a low U-Value. The more shading a window has, the less important the SHGC is, and the more important the U-value is. People consider a VLT of 35–60 per cent to be most acceptable (Boyce, 2003). In hotter climates it is likely that these are on the high side of acceptable. However, for moderately sized, well-shaded windows (which we should be using), these values are suitable.

Exposed thermal mass

For a given air temperature, the mean radiant temperature (the average of the surfaces' temperatures surrounding us) can have a significant effect on our thermal comfort. For example, when we walk into a massive stone building (think of an old church), the air temperature inside is the same as outside but it feels much cooler: the surrounding stone is cool. In an office environment, as a generalisation, we are comfortable at about 24°C. If we can expose lots of thermal mass, and reduce the mean radiant temperature by five degrees, we will be just as comfortable at 29°C. The following methods can be used to activate thermal mass:

- Cool the mass overnight by night purge, for example, by automatically openable windows to allow cooler night air to enter. Fans can be used to augment this and ceiling fans will further improve the cooling of concrete soffits. As seen in Figure 3.13.5, Brisbane's diurnal range is about 11°C in summer, which is adequate, although not excellent for night purging due to the relatively high overnight minimums.

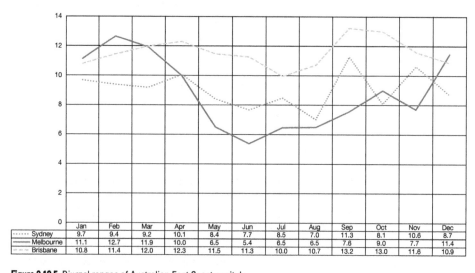

	Jan	Feb	Mar	Apr	May	Jun	Jul	Aug	Sep	Oct	Nov	Dec
Sydney	9.7	9.4	9.2	10.1	8.4	7.7	8.5	7.0	11.3	8.1	10.6	8.7
Melbourne	11.1	12.7	11.9	10.0	6.5	5.4	6.5	6.5	7.6	9.0	7.7	11.4
Brisbane	10.8	11.4	12.0	12.3	11.5	11.3	10.0	10.7	13.2	13.0	11.6	10.9

Figure 3.13.5 Diurnal ranges of Australian East Coast capitals

Source: Brett Beeson.

- Expose mass by removing ceilings and coverings such as carpet. Occupants must be exposed to (i.e. be able to see) the thermal mass to benefit. Hence in offices, the ceiling is most effective, followed by the floor. Ceiling tiles and carpet negate any thermal mass benefit and need to be removed or minimised in this method.
- Shade mass by making use of mass in ceilings or away from windows. Any solar gain on thermal mass – except in the coldest part of winter – is detrimental.

When hard, thermally massive surfaces are exposed, acoustic problems (such as reverberation) are very likely and will need to be managed. Vertical baffles hanging from the ceiling are an effective way to minimise acoustic problems and maintain exposure to the thermal mass.

Insulate roof and walls

An obvious way to reduce the heat load is to insulate the external faces of the building. In sub-tropical climates, the temperature is moderate (consider Brisbane's design condition of 32°C) while the sun is very hot (i.e. high W/m^2 of solar radiation). This leads to the following priorities (highest first):

- Insulate roofs. A reflective barrier below a lightweight roof is effective. BCA specified values should be used as per their intent: they are absolute minimum values and typically should be exceeded.
- Insulate walls exposed to the sun. Note the BCA comment above.
- Insulate walls not exposed to the sun. Note the BCA comment above.
- Insulate exposed floors. Generally the BCA does not require insulation here, but again, this is the minimum legal requirement, not best practice.

Tables 3.13.5 and 3.13.6 show climate-appropriate designs for commercial office buildings in a warm temperate climate, with emphasis on improving daylight and air-conditioning load.

CONCLUSION: A SPECIAL RESPONSIBILITY

This chapter makes some observations about the activity of designing , and provides 'rule of thumb' guidance on the bioclimatic design strategies encountered by the architect, and others, relevant to commercial office retrofit in a subtropical climate. A summary of passive strategies for retrofitting appropriate to a subtropical climate is given in Table 3.13.6.

Table 3.13.6 Passive design strategies for retrofitting commercial office buildings in the subtropics

Strategy	Techniques	Notes
Improve interior daylight		
Reduce occupant distance to glazing	Narrower floor plates	Increases environmental heat loads and glare. Daylight design requires the design of external shading devices appropriate to orientation.
Modify floor plate	Internal atria, 'cut-outs', etc.	Cleaning is needed.
Modify floor-to-floor ceiling height to increase luminance.	Remove suspended ceilings.	Supplement with task lighting.
	Add light shelves.	Use of solar glazing to moderate solar heat gain.
Modify façade to attenuate glare and enhance luminance.	Create higher amounts of vertical light transmission through the glazing. Add light shelves. Control glare via adjustable blinds.	Select appropriate glazing ratio (transparent to opaque) for plan depths.
Reduced air-conditioning usage		
Increase air movement	Install ceiling fans or desk fans.	Ceiling height should be modified to provide headroom. Electrical load from fans is a fraction of the energy of air conditioning.
	Natural ventilation	Comfort, exposure to wind variations, external noise may be issues.
	Mechanical ventilation	Duct sizing, fan energy and services integration may be issues.
Activate adaptive comfort	Modulate set-points seasonally	Set-point scheduling requires seasonally based fine tuning.
	Zone building for natural ventilation	Naturally ventilate toilets, kitchen, etc.
	Natural ventilation	Retrofitting operable windows. Mixed-mode may be appropriate.
Reduce heat load	Minimise glazing extent Shade glazing	Avoid over-glazing.
	High-performance glazing Add insulation to roof and walls/Reduce infiltration	Cost may be prohibitive.
Localised air conditioning	Workstation air supply	Ductwork needs integration.
Use thermal storage	Exposed thermal mass	Acoustics need to managed. Removal of suspended ceilings will require more extensive services integration.

Note: The typical office is assumed to be 58m x 15m x 2.7m. Highlight focus on improving daylight.

Current building design, construction and management methods in Australia are not sensitive to contemporary environmental understandings and commitments. In fact, current building regulations push designers to use air conditioning and are a disincentive to the natural ventilation of buildings.

Environmental design is often treated as an 'added extra'; however, the benefits to industry are real and designers may need to position the strategy in the industry context by identifying these benefits as the primary rationale for 'doing the right thing'. A comprehensive bioclimatic design strategy:

- provides a legible and detailed account of the environmental status and performance of the development;
- develops a clear direction for ongoing improvement;
- reduces operational costs of energy and water;
- enhances flexibility for future uses;
- contributes market credibility;
- improves indoor air quality;
- contributes to occupants' productivity and self-esteem;
- benchmarks the building in terms of environmental performance;
- helps protect building owners and property managers against environmental liability (potentially future environmental insurance liability claims); and
- highlights specific areas for potential environmental legislation.

There are many limitations of our understanding of building physics and environmental analysis. There is no book or computer model that will tell us how to design; however, this chapter provides a 'rule of thumb' approach useful to the early decision-making process for the retrofit of commercial office buildings. The design process is at risk of becoming merely an academic exercise, undertaken not to explore potential solutions, but to build a case for a predetermined view. All the factors that would allow a defence of a particular solution will never be known. However 'scientific', information is shaped by social assumptions and values, which necessarily guide decision making. Designers have a special responsibility for the values that guide the decisions that shape our work.

REFERENCES

Arens, E., Turner, S., and Zhang, H. (2009) 'Moving air for comfort', *ASHRAE Transactions*, 18–30.

ASHRAE (2004) *Comfort Standard*, Atlanta, GA: ASHRAE.

Boyce, P.R. (2003) *Human Factors in Lighting*, Chennai: Lighting Research Centre.

Calcagni, B.M.P. (2004) 'Daylight factor prediction in atria building designs', *Solar Energy*, 76(6): 669–82.

CIBSE (2001) 'Natural ventilation', in CIBSE, *CIBSE Guide B2 Ventilation and Air-Conditioning*, London: DTI.

Hunt, D. (1979) 'The use of artificial lighting in relation to daylight levels', *Building & Environment*, 14(1): 21–33.

Keighley, E. (1973) 'Visual requirements and reduced fenestration in offices: a study of multiple apertures and window area', *Building Science*, 8(4): 321–31.

3.14
SUMMARY
Richard Hyde

Part III has examined the issues surrounding how to achieve better performance through retrofitting of buildings through utilising new technologies
and better integration with human systems. The notion of how this influences behavioural factors is based on the premise that synergies between
technological and human systems are necessary for the whole to operate at
its optimum performance.

It is tempting to talk about the findings from these chapters under
the separate issues of technology and behaviour regarding technology but
that artificially partitions technology from its human masters since that is the
purpose of technology, which is often forgotten. It is interesting to note
which research and practice opportunities arise from this frame of view.

The work on typology clearly identifies some of the major building
types that have been built in recent times yet for each one it became clear
that there was a matching retrofitting response which would be reasonably
consistent for the type (Chapter 3.2). More work is needed to define these
types more consistently and to identify the strategies and solution sets.
Furthermore, the adaptive reuse of existing buildings is bringing into the
frame heritage issues concerning the existing building. Hence performance
standards for retrofitted buildings are likely to vary with those that can be
achieved through new building. In this sense, when examining what the
potential is for retrofitting in existing commercial buildings, more emphasis
should be placed on the typological characteristics that distinguish types of
office building. Also further exploration of the historical origins, development
of the building typology and the current issues driving typological trends in
commercial architecture with a view to retrofitting should be considered.

Many of the issues that arise out of the need for retrofitting come
from whether or not improved levels of occupant comfort are needed in
existing buildings (Chapter 3.3). This often occurs when naturally ventilated
buildings are perceived to have performance problems and are converted
into air-conditioned buildings. When designed carefully, naturally ventilated
indoor environments need not compromise occupants' comfort, well-being
or productivity. Indeed, some argue it is quite the opposite, that naturally
ventilated buildings provide indoor environments far more stimulating and
pleasurable compared to the static indoor climate achieved by centralised air

conditioning. When retrofitting for comfort and indoor environmental quality, occupants must be at the centre of the design concept. Occupants' expectations and attitudes, perceived control and availability of behavioural thermoregulatory options are essential when providing not only comfort but also satisfaction.

It is important to place the behavioural dimensions in a broader context. The conclusion works on the assumption that most people involved with the operation of commercial buildings would prefer their buildings to impact the environment as little as possible while they deliver maximum comfort to occupants. However, there are many demands competing for the time and attention of the people charged with operating buildings. The task of enhancing building performance needs to be made as simple and efficient as possible. It requires timely and actionable data. Furthermore, it requires a framework to analyse that data, deliver meaningful information, facilitate action and then allow reflection to drive improvements.

How to link both design and operations with regard to building performance is a further issue that has been investigated to aid the sustainable retrofitting process. Benchmarking is seen as a useful process in this endeavour (Chapter 3.4). While benchmarks resulting from computer simulation have yet to be reconciled with operational performance, the substantive purpose of benchmarking is to encourage 'management by exception' and 'continuous improvement', leading to creative ways of achieving improved performance, hence greater sustainability.

Benchmarking as a methodology has been integrated into Building Environmental Assessment (BEA) systems that have become a new pathway for green design. Yet within BEA there are two main approaches to assessment that have yet to be fully integrated in terms of the benchmarking methodology to be useful: design phase assessment and operational phase assessment. Typically design phase assessment is carried out prior to construction. Operational phase assessment is carried out after construction.

Both approaches use benchmarking to establish performance levels of buildings (Chapters 3.5 and 3.6). Recent research has questioned whether these systems are achieving this objective. There are two main criticisms of the current suite of rating tools. First, the assessment process does not support the design process but is reducing its effectiveness. Some designers are questioning the compatibility of the building assessment systems and the design process.

Second, the benchmarking methodology provides evidence of performance for the assessment process (Chapter 3.7). Sources of data such as that from computer simulation are used to calculate the data for benchmarking; however, this evidence may be at variance with the operational performance of the project. This may lead to misleading information. Hence, it is concluded that BEA tools vary in the extent of the rigour with which they apply benchmarking (Chapter 3.8). Theoretically, highly rigorous tools comprehensively match the principles and/or policy frameworks with the indicators; indicators are assessed with information, which is easy to collect, but once it is collected, it yields valid data; and finally 'best practice standards' are not biased or based on impractical levels of performance. Very often highly rigorous tools are data-hungry and expensive to service, making them impractical to use in the schema of some organisations. Lack of rigour can make the tools useless and hence here lies the conundrum for this type of approach.

As rigour is often traded against practicality of use, many tools usually include a checklist for pre-assessment, which allows potential users to quickly check to what extent their project engages with the environmental criteria (Chapter 3.9), but the process becomes an end in itself, a 'snapshot' in time. To avoid these problems, further research is needed to validate benchmarks through larger-scale studies, which look at larger populations of buildings, using a range of types of benchmarking. It is important that benchmarking is more strategic, that is, based on a number of sources of information drawn from both design and operation conditions. A more strategic focus to benchmarking might also engage building users to assist with facilitating change, which is the central aim of benchmarking.

For retrofitting, the inherent limitations with BEA can be resolved and both the design phase and operational phase are integrated round the existing building. A tool for rating the retrofitting process has yet to be devised. While there are operational standards in Australia which determine the protocol assessment of operational energy usage, these suffer from many of the limitations of existing systems.

Some work developing the energy rating tool for retrofitting has been carried out. Sets include looking at larger cohorts of buildings within a type and carrying out functional benchmarking of particular functional room sub-types within a particular building type. So for university buildings we start benchmarking lecture theatres, office space and other types of building. With this approach it is possible to reverse engineer these rooms through deeper benchmark analysis using sub-system and product benchmarking.

This kind of approach not only requires redefinition of design for retrofitting but also the roles of building managers and engineers. The conclusions suggest that using sources of evidence from both building monitoring and energy modelling is necessary. Formulated into tools they have now become a necessary part of the process in demonstrating to clients how their building will perform. In many cases engineers are now being required to validate their predictions 12–18 months after the buildings have been operating. With the drive by tenants and building owners for more energy-efficient buildings there are sometimes legal considerations associated with not meeting targets. Consequently the role of the energy modelling engineer has become pivotal to inform the design and demonstrate performance.

Complementing modelling is the use of monitoring of building performance though measurement of the condition of the physical environment. This can be more effective if a range of measurements of the physical condition are brought together in a comprehensive suite of instrumentation to create a building diagnostic toolkit. The physical measurement of the building performance complements other data from the building such as user feedback and satisfaction. This helps to validate the opinions of users about the performance of the building and also provides an important basis for retrofitting.

Finally, the last four chapters (Chapters 3.10–3.13) provide further conclusions concerning the drivers and challenges for retrofitting.

- While sustainability is being adopted in new buildings, additional arguments prevail too often in planning retrofits and factoring in reduced embodied energy also 'can almost be one complication too many'. The current inertia in the system prevails against innovation and creates a slow process of change. 'Better data and process are needed to underpin innovation and explain its benefits and its economic incentives, and innovative financing to motivate investment.'
- To this end a new approach to design, termed the Penalty–Reward–Pinch (PRP) design approach with the aim of identifying the feasible passive systems and optimally matching them with the active system to bring about building energy sustainability is advocated
- This approach can be coupled with the economic 'pull' drivers that influence the different situations of building owners. The obsolescence of the mechanical systems in buildings is less than that of the building structure and envelope. Rising energy costs make retrofitting of more energy-efficient technologies desirable. 'However, the analytical accounting practice of reducing costs and benefits to Net Present Values diminishes the value of energy savings in the long-term future making them appear less viable.' Therefore, other arguments need to prevail.
- This argument comes from the final chapter, which summarises the 'bioclimatic design strategies encountered by the architect while trying to steer an ethically and ecologically informed path through the world of contemporary commercial office design'. This is termed a special responsibility, that we have to go beyond the economic boundaries for sustainable retrofitting.

RETROFITTING EXEMPLARS

4.1
INTRODUCTION
Francis Barram and
Nathan Groenhout

Part IV provides a range of case studies, which can be examined as exemplar projects that demonstrate sustainable and more specific bioclimatic approaches to retrofit. The aim is to identify lessons learned from a parishioner's approach to renovation and retrofitting.

Part IV starts with a study, which examines an adaptive reuse retrofitting approach. The PMM project examines adaptive reuse of a building located in the central business district of Brisbane, Queensland Australia (Chapter 4.2). Sustainable retrofitting is achieved through introducing new retail functions to enhance the economic viability of the building. Recycling, functional and technical upgrading enhancements and the addition of bioclimatic elements reduce environmental impacts. It is interesting to see how this has been achieved.

The case of 55 St Andrews Place, Melbourne, Victoria, Australia, follows a different approach (Chapter 4.3). It identifies what the practice strategies are that can be used for retrofitting. This involves reverse engineering a building to meet environmental performance targets. Built in a similar era to PMM, it follows a more conventional approach to retrofitting but uses a number of innovative practices to achieve performance standards that have become common in the industry at present.

The next case study involves a critical examination of bioclimatic retrofitting applied to retrofitting university buildings (Chapter 4.4). The chapter examines what issues arise in the precinct scale in retrofitting of large cohorts of buildings and provides a view of which challenges and opportunities exist for improvement at the policy and technical level.

The next two cases are more active systems oriented with a focus on retrofitting technologies that utilise energy sources other than the grid. The case study of 503 Collins Street, Melbourne, Australia, reviews a large renovation and retrofitting project with an emphasis on how onsite power generation can be achieved in commercial buildings (Chapter 4.5). This project demonstrates innovation in the project management through both Energy Performance Management and the cogeneration.

Following a similar approach to the previous case, the integration of new green technologies in the form of solar thermal retrofit is discussed (Chapter 4.6). The focus is on the question of what the benefits of retrofitting solar electric are and the possibility of utilising non-low carbon energy-producing technologies.

The next study involves the development of site-based comfort standards as opposed to utilising industry standards (Chapter 4.7). The approach examines the challenges and opportunities of working with occupants' perceptions, attitudes and behaviour in working environments. It provides a mechanism to develop new site-based policy for thermal comfort.

The remaining cases move from issues concerning technology to non-technical issues. Exactly what the non-technical issues are remains unclear but it seems to mean everything that is not technical and must refer to the broader biological and ecological aspects of the parameters in the built environment. Largely it is to do with human ambitions, actions and reflections. In particular, it is focused on innovation in the quality of the work environment (Chapter 4.8).

This is followed by studies using post-occupancy evaluations to gauge the impact of renovation on workplace satisfaction and productivity (Chapter 4.9).

The final case study comprises the refurbishment and recycling of a historic Woolstore building on the Deakin University Waterfront Campus in Geelong, Victoria, Australia (Chapter 4.10). The new function for the building was to use the existing building as an office space. The project demonstrates how bioclimatic retrofitting can be carried out. In addition, it shows an evidence-based approach to office renovation works in practice where the performance benefits are within the design process through to the day-to-day management of the building.

4.2
743 ANN STREET, FORTITUDE VALLEY, QUEENSLAND

Mark Thompson

INTRODUCTION

The retrofitting of a commercial building is often undertaken to provide a building with a new lease of commercial life. In the case of 743 Ann Street, Fortitude Valley, Queensland, the building owners had a clear vision of minimising environmental degradation and creating a healthy working environment for their staff. The building owners understood that over the next 30 years the flow of their business money would result in approximately 2 per cent spending on building costs, 6 per cent spending on operational costs and 92 per cent spending on staff salaries and this determined the design approach, consistent with the principles of bioclimatic design.

At the time of purchase in 2002, the building was 15 years old. It presented a west-facing curtain glass wall, positioned 6 metres from the front boundary and was primarily accessed via unsheltered entry stairs from Ann Street (Figure 4.2.1). Designed to be air conditioned and built side boundary to side boundary, the original building had side light wells to allow natural light into the deep plan floor space. Energy and water efficiency were not important considerations at the time of the original design and access for persons of limited ability was not a mandated building regulation. The building was located on a major access road to Brisbane city and was positioned one block from a designated urban renewal centre which was favoured by the building owners due to its potential to add value and assist in the transformation of the near city area.

The building's new owners were the PMM Group who comprised urban designers, town planners and land surveyors. The retrofitting project was intended to showcase innovative sustainability measures demonstrating a commitment to managing PMM's business in a sustainable manner:

It is our belief that our building is a visible and tangible expression of who we are and where we are going. It also provides the

Figure 4.2.1 West-facing façade of the existing building prior to renovation

opportunity for our staff and our clients to share in, and to work toward, the mutually beneficial aim of promoting and practising sustainability in a profitable manner.

DESIGN CONCEPTS

The building retrofit included the redevelopment of the ground floor to add new street-level retail outlets which also allowed the addition of a lift and new atrium and entry space (Figure 4.2.2). A standard building refurbishment proposal was initially developed which included a new street appearance, a new entrance with lift access, an air-conditioning upgrade to minimum requirements, and new ceiling tiles and paint to enable commercial leasing. The PMM Group had a broader vision for the building which included a new relationship with the street, open stairways and a disabled access-compliant lift and facilities, upgraded air conditioning to premium quality, improved staff facilities, energy and water efficient features and an ecologically sustainable development (ESD) approach to inform the design process, decisions and retrofitting direction. The development encompassed 1821 square metres of net lettable area in office and retail space.

The broader PMM vision was goal-driven as the organisation wanted to demonstrate marketplace leadership via a process of constantly learning and innovating. Their stated goals included how to do more with less, how to encourage and support employees and stakeholders to achieve sustainable outcomes, how to influence and encourage governments to

Figure 4.2.2 Artist's impression of the west façade showing the new retail accommodation, atrium, and entry space

support and implement sustainable development and to create practical eco-efficient building outcomes. A multi-disciplinary and collaborative design process was undertaken to include PMM staff.

Brisbane architects, The TVS Partnership, were selected due to their previous experience in sustainable design and they facilitated various design workshops sharing their environmental brief and sustainable strategy processes. Due to the need to fast track the project to allow PMM to relocate from their existing building, a project team was formed including a project manager, builder, architect, services consultants and client representatives. A collaborative design process was undertaken with time and cost controls implemented. The Environmental Brief process established a shared vision among the numerous project stakeholders and authority approvals were achieved as quickly as possible.

Bioclimatic design issues of energy, health, well-being and sustainability were strongly encouraged by the client and were tested in value management sessions to ensure best practice solutions were realised.

Concept designs were developed to explore the three primary outcomes required by the client (Figure 4.2.3):

1. Does it look good?
2. Is it financially viable and appropriate?
3. Is it environmentally friendly at macro and micro levels?

The design team undertook various building investigations to evaluate the value of existing building attributes and to explore retrofitting opportunities. Increasing light and ventilation were primary concerns to maximise the passive design attributes of the existing building. By removing selected ceiling grid tiles, it was determined that additional natural light could be introduced into the building via the external façades; additionally, daylight

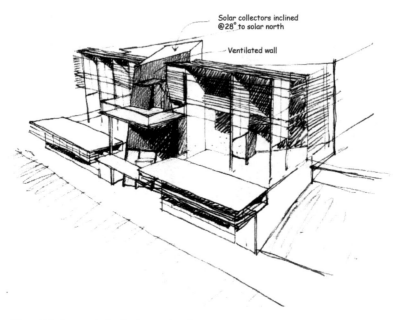

Solar collectors inclined
@28° to solar north

Ventilated wall

Figure 4.2.3 Concept design for the west façade

could also be achieved by the improvement of the existing light wells and the addition of skylights (Figure 4.2.4).

As the building floor plate was deep, the addition of natural light was considered a considerable benefit in increasing the internal ambience of the building. Early building investigations also determined that many of the existing building grid ceiling tiles were in good condition and could be reused. The design decision was made to use these existing tiles upstairs to assist in reducing the roof heat load.

The ceiling tiles to the lower level were removed to allow the thermal mass of the existing concrete floor slab to be more efficiently utilised. One of the significant environmental benefits in retrofitting is the saving of embodied energy by reusing elements of the existing building. In order to reduce waste and gain maximum benefit of the existing building materials, the following design strategies were employed. The existing building façade was retained where practical, existing internal stairways were retained, the existing ceiling grid was to be retained on Level 2 and refurbished using parts from Level 1, air-conditioning plant was refurbished and upgraded rather than being totally replaced (Figures 4.2.5 and 4.2.6). Light fittings were retained but refurbished with energy-efficient T5 fluorescent tubes.

For all new building work, embodied energy issues influenced the selection of materials. Plantation pine timber was used for internal wall studs in preference to steel studs, which is considered standard practice for office partitions. The processed energy requirement of kiln dried softwood is 3.4

(a)

(c)

(b)

Figure 4.2.4 (a and b) Enhancement of natural light though light wells; and (c) new skylights by selectively removing the ceiling tiles

mega joules per kilogram compared to galvanised mild steel which is 38 mega joules per kilogram (Lawson, 1996).

BUILDING MATERIALS ENERGY AND THE ENVIRONMENT

Demolished concrete work was crushed to produce landscaping gravel. Removed glazing was recycled into a landscape waterfall sculptural feature to the building entry courtyard. Modwood™, a recycled timber/plastic board product, was used for courtyard decking.

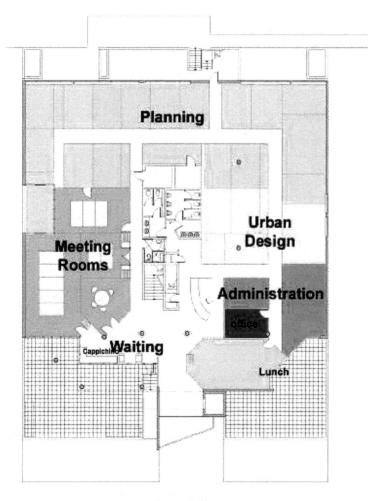

Level 2
Proposed layout 1

Figure 4.2.5 Level 2 floor plan showing the new atrium space. Level 1 car parking and retail space are below.

Grey energy, the energy expended in the transporting of materials and components from places of extraction and manufacture to the construction site was deliberately minimised by the selection of various local products. Rockcote™ low toxic paints manufactured in south-east Queensland were used extensively internally and externally on the project. Hoop pine timber from sustainably managed south-east Queensland plantation forests were selected in plywood and timber joinery.

Induced energy was also considered in the retrofit construction process. A detailed environmental management plan, certified to ISO 14000 Standard, was implemented by the builder during construction. Recycling of

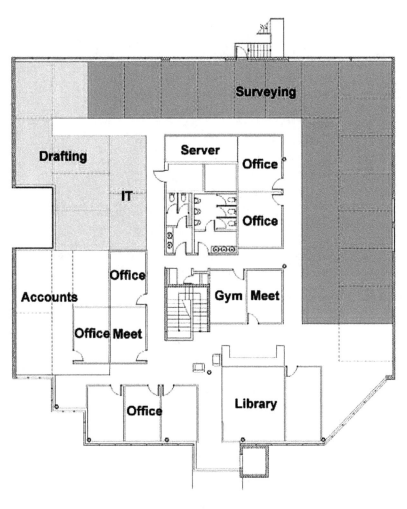

LEVEL3
Proposed layout 1

Figure 4.2.6 Level 3 floor plan showing the new atrium space

construction waste including steel, electrical and data wiring, aluminium and all cardboard packaging was undertaken by Multiplex, the principal building contractor, and imposed on all sub-contractors. Audits demonstrated that approximately 80 per cent of demolished materials were recycled.

The design process also focused on potential operational energy savings. The building retrofit included a grid-tied photovoltaic power system which has a 4.5 MW capacity to supplement the power load to the energy grid (Figure 4.2.7). Additionally, a solar hot water system was installed along with other energy-efficient appliances to reduce the operational energy load of the refurbished building.

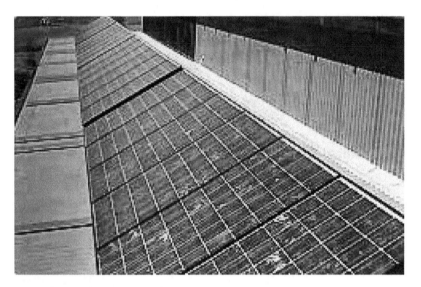

Figure 4.2.7 The building retrofit included a grid-tied photovoltaic power system which has a 4.5 MW capacity to supplement the power load to the energy grid

Energy-efficient lights were installed on split circuit lighting zones, some with dimmable controls. New electric circuits were installed on variable time switching with manual overrides with all switches clearly identified to avoid unnecessary use. These measures went well beyond the typical building practices of the day. Heating and cooling were controlled through a building management system which also manages security and timing. Air conditioning was zoned to building use and independent units were installed for tenants' after-hours and 24-hour use, to ensure the entire building was not unnecessarily air-conditioned. The basement car park ventilation is efficient via its use of a timed fresh air supply in lieu of an energy-intensive exhaust system.

Building elements which enhance the project's energy efficiency include the extensive use of sun shading, high performance solar film to reduce heat loads, daylight usage via light wells, increased window areas, insulated skylights and a natural ventilation system to the entry atrium (Figure 4.2.8). These passive system improvements to the existing building offer ongoing long-term energy-efficient benefits and value for money from a life-cycle perspective.

The retrofit program paid particular attention to the health and well-being of the future building occupants (Figure 4.2.9). Research reveals that the health of building occupants can be affected by cleaning regimes, access to daylight, thermal comfort, ventilation, exposure to chemicals and volatile organic compounds, slip resistance of materials, spaces for relaxation, air quality, acoustics, water leaks and bad drainage. Well-being is usually directly related to a project's shelter, security and comfort standards.

The building interior designers focused on products and design elements that were durable and easy to clean. Modular workstations were

Figure 4.2.8 Final elevation of the west façade to show the sun shading to the existing office façade which was retained

Figure 4.2.9 (a) Retail space and (b) new entrance foyer

chosen to allow minimal barriers to the floor to allow efficient cleaning and sub-desk air flow. Screens only extended to the floor abutting corridor areas. Interface™ carpet tiles were selected due to their low static qualities, recyclability and ability to provide flexible cleaning solutions and replacement if necessary.

Glare from daylight can adversely affect building users' productivity, and internal blinds to windows were provided to control unwanted glare sources onto desk computer screens. The predominant open office floor plan design ensures that thermal comfort is managed over office teams, with each team having its own thermal comfort zone related to the building's orientation and occupant usage. Thermostats were positioned in relation to office teams.

Extensive research was undertaken by the project's design team to minimise the use of products and materials with high off-gassing qualities. Workstations and joinery components used low formaldehyde board substrates and low odour finishes. Natural fabrics were generally chosen for furniture and screens. All paints specified for the project were non-toxic and water-based. Safety for building users was incorporated via the specification of non-slip tiles to stairs and all public areas. Linoleum flooring was used in kitchen areas to avoid resilient flooring that contained PVC, which contains chloride and is toxic if burnt. A meeting room was designed as a meditation space and staff areas have direct access to outside courtyard areas and breakout areas.

Air quality in the entry atrium has been supplemented by the addition of a biofilter (Figure 4.2.10). This internal planter has plants integrated in an activated charcoal base where electric fans draw air through the soil substrate and charcoal base. The biofilter assists in removing any traffic

Figure 4.2.10 Biofilter for improving air quality

fumes from adjacent busy Ann Street which may enter into the naturally ventilated atrium space. Prototype testing of a similar biofilter realised air quality improvements of 10–20 per cent in the removal of volatile organic compounds. Healthy plants are provided to assist air quality and are maintained regularly via a local plant hire company. PMM staff feedback on the completed project resulted in a high acceptance of the new workplace as an example of a best practice work environment.

The sustainable design approach adopted by the design team generally adopted the principle of reduce, renew and recycle throughout the project. A minimalist design approach was adopted in the retrofit of the first level. By removing the ceiling grid and exposing the underside of the concrete slab, good access was provided to all services both during construction and throughout the working life of the tenancy. Suspended service trays provided new cable routes and fixing points for lights and new fire detection services. At selected locations suspended ceiling features were integrated to provide visual relief and acoustic baffles for noisy mechanical services. This approach reduced the need for a new ceiling grid and ceiling tiles and provided the added benefit of using the concrete's thermal mass to assist in the temperature control of the first floor level. Existing air-conditioning ducts were cleaned and reused in the new mechanical design to suit the new building zones.

The existing enclosed fire stairs were renewed as the primary circulation space for the building, encouraging building users to exercise rather than rely on the lift. The stairs were modified by extending the width of the last flight to create a functional and visually appealing circulation thoroughfare. As toilet facilities were outdated, it was decided to renew fittings and fixtures to provide the latest water-efficient fixtures and provide staff with new facilities.

Numerous water-efficiency features were incorporated into the retrofit which included AAA rated shower roses, kitchen sink tap aerators, 6/3 litre dual flush toilets, waterless urinals and water tanks (Figure 4.2.11). A heat trace system on the hot water pipes makes hot water immediately available to users, saving on running cold water out of taps prior to use. Roof water was originally channelled directly into local stormwater drainage; however, the addition of an 11000-litre storage tank enabled this water to be captured and redirected to flush toilets and assist in landscape irrigation. To assist in qualifying the benefit of the water tank, water meters were installed on the mains water and staff undertook a monitoring programme to compare the water savings. At the time of the retrofit, rainwater harvesting was unusual in Brisbane, a subtropical city; however, since the retrofit south-east Queensland has experienced extensive drought conditions and water-saving measures have since become common in commercial and residential buildings.

The inclusion of a new three-storey atrium space became an important visual and climatic design feature for the retrofit (Figures 4.2.12 and 4.2.13). A thermosiphon wall was designed to assist in reducing heat

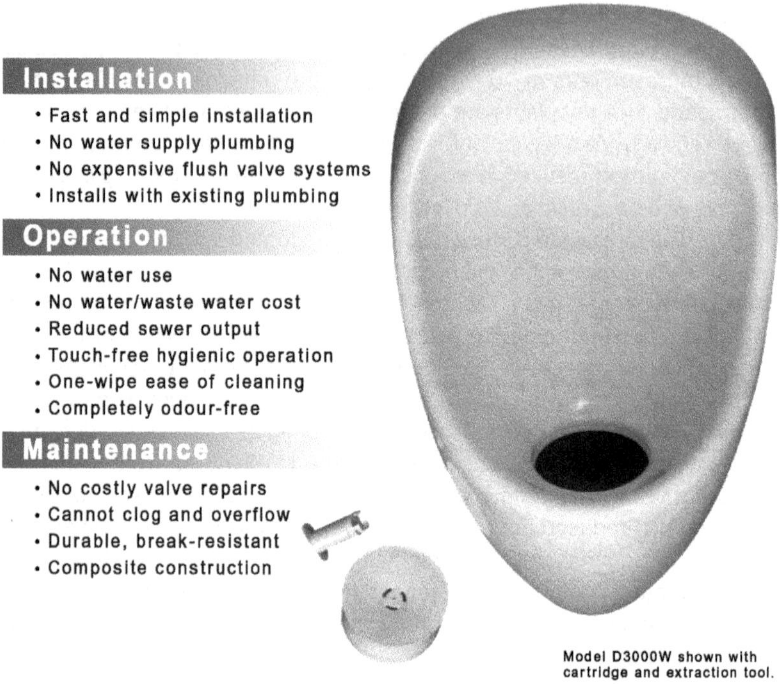

Installation

- Fast and simple installation
- No water supply plumbing
- No expensive flush valve systems
- Installs with existing plumbing

Operation

- No water use
- No water/waste water cost
- Reduced sewer output
- Touch-free hygienic operation
- One-wipe ease of cleaning
- Completely odour-free

Maintenance

- No costly valve repairs
- Cannot clog and overflow
- Durable, break-resistant
- Composite construction

Model D3000W shown with
cartridge and extraction tool.

Figure 4.2.11 Waterless urinals used for reducing water use

build-up from the western façade solar load. The vented internal atrium space allows heated air to be exhausted in summer and operable high-level louvres allow heat retention in winter. Double glazing is used to restrict both heat and noise entering the building in new façade work facing Ann Street.

Sustainability was considered at many levels for the project. It was understood that transportation contributes a major carbon emission load and the location of the building near to train and bus facilities was a strong factor in securing staff approval for the project. The basement was modified to include visitors' car spaces and bike facilities for staff and visitors. Again, the provision of 'End of Trip' facilities was unusual at the time of the retrofit and bike racks, showers and lockers were provided to encourage alternative means of transport to work. Smaller car spaces were provided to encourage more energy-efficient vehicles including hybrid and 'smart' cars. A car pooling scheme was developed by staff during a retrofitting building workshop to directly minimise their environmental impacts via transportation measures. The PMM staff were proactive in trialling electric bikes and electric scooters as viable means for inner city business commuting. The holistic sustainable retrofit process was a great stimulus to engaging staff to reconsider their individual ecological footprints. The collaborative design process was synergistic and reaped many positive project improvements. The 'green team' continues to engage staff in ongoing operational improvements through monitoring of the building's environmental performance.

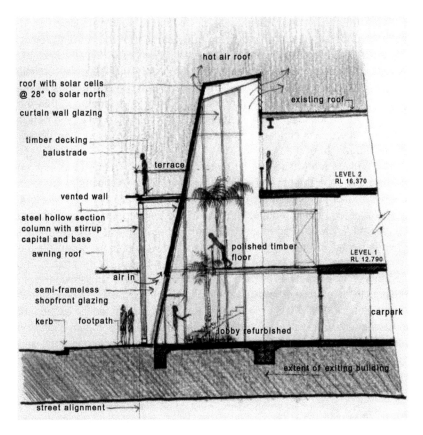

Figure 4.2.12 New atrium creates a bioclimatic chimney to avoid overheating

Figure 4.2.13 Artist's impression of the façade and the new atrium behind it

A Building Users Guide was developed to educate building occupants to understand and maintain the ESD features of the retrofit project. PMM staff have been active in assisting the ongoing ESD initiatives for the building. The staff cafeteria boasts a worm farm which receives organic waste generated on site and produces compost material for the landscaping. Staff also manage an extensive office paper recycling scheme within the building. An educational video is available in the building's foyer and details on staff energy monitoring and efficiency can be accessed via display material throughout the building.

PMM staff have hosted many tours of the building to interested government departments and industry groups to educate the marketplace on the benefits of sustainable retrofitting. By January 2005, over 1000 visitors, including international groups, had toured the building, hosted and guided by trained PMM management and staff. The Queensland Government Environmental Protection Agency produced an educational video on the retrofitted project to showcase the environmental features of the project. The success of the project has encouraged PMM to undertake similar eco-retrofit projects in its regional offices. The goal of providing a superior working environment led to PMM business growth and the company rapidly expanded, engaging new workers who were keen to experience the company's passion for practical sustainability measures. Their belief that environmental and social sustainability are directly related to financial viability was proven along with the understanding that environmental aspirations must be identified, prioritised and shared in development planning.

The project was successful in building and development industry award programmes hosted by organisations such as the Urban Design Institute of Australia, the Design Institute of Australia and the Master Building Association. The project was actively promoted by the Australian Green Development Forum as setting new standards in sustainable development.

Many lessons were learnt during the project retrofit covering three main areas: (1) the capital works period; (2) the operational phase; and (3) the management of people. As many new products and sustainable building ideas were implemented, it became essential for all stakeholders to understand the strengths and limitations of new materials and technology. Building capital costs were effectively project-managed; however, coordination of existing services and new services became critical when matching new and old equipment. Commissioning and fine-tuning of equipment take additional time when working on existing buildings and diagnostic investigations become critical to confirm the capacity and performance of existing mechanical and electrical equipment.

When using new materials, the manufacturers' specifications may need to be questioned and prototyping on site may be necessary to satisfy actual performance criteria. The recycled timber/plastic decking material required more sub-surface ventilation to control thermal expansion than both the designer and manufacturer realised before installation. The skylights were selected for their thermal and lighting characteristics and worked so well that additional glare measures were required to provide a comfortable working environment for nearby building occupants.

Facilities management personnel require access to building contractor knowledge to problem-solve certain operational issues, and unless good site records are produced, this knowledge can be lost. Building design thermal modelling is predictive and should be verified by actual building performance fully occupied as variances in modelling input can be verified. Additionally, modelling on reused equipment requires specialist skills and advice. Benchmark information is useful in ongoing building tuning and monitoring to ensure that staff understand appropriate levels of performance applicable for new and reused equipment. Recycling systems may require more detailed thought to ensure the problems are not just removed from site and recreated elsewhere.

The management of people and their expectations is important right through from designers, consultants and cost controllers through to construction workers to building occupants. Managing all the issues and complexities makes retrofitting work challenging at times and having all stakeholders working to a shared vision becomes an important project success measure.

CONCLUSION

The PMM retrofit project was progressive for its time and realised the benefits of many individuals and companies working towards common environmental outcomes. The project has inspired and influenced many building and development industry personnel to progress sustainable development concepts as the broad vision of a balanced triple bottom line approach was generally deemed appropriate and successful for future action.

ACKNOWLEDGEMENTS

Mark Thompson was the ecological architect and the TVS Partnership Architects' Director who facilitated the ecologically sustainable development strategies employed by the PMM retrofit design team.

REFERENCE

Lawson, B. (1996) *Building Materials, Energy and the Environment: Towards Ecologically Sustainable Development*, Red Hill, ACT: RAIA.

55 ST ANDREWS PLACE, MELBOURNE, AUSTRALIA: TURNING A SPARROW INTO A PEACOCK

What are the strategies for retrofitting for improved environmental performance?

Caimin McCabe

INTRODUCTION

The redevelopment of the 1960s office building at 55 St Andrews Place at Treasury Reserve in Melbourne, Victoria, in Australia represents a demonstrable example of how an existing building with poor environmental operational and functional performance can be cost effectively redeveloped to achieve best practice performance. In order to understand the steps taken to achieve this significant turnaround in environmental operational and functional performance, it is important to understand the building's history, existing issues and prevailing office quality standards.

BUILDING HISTORY

55 St Andrews Place has an interesting history:

> During the 1960s the State Government planned to build a skyscraper office tower behind the Old Treasury Building. An Architectural competition was held in 1962, in which twelve firms were invited to submit designs. The guidelines for the competition mentioned the State Government's wish for a building, which would express 'soaring wonderment'. Eleven submitted designs conformed. The twelfth, Yunken Freeman, put forward an entry by architect Barry Patten, who believed that a tower block 'would destroy Melbourne's best vista – that is – looking eastward from the top of Collins Street to the Old Treasury Building.' Patten's design was for two infill buildings of similar scale to the Old Treasury Building (constructed between 1854 and 1862) and 2 Treasury Place (constructed in 1859) which would 'stand out like brown sparrows between two peacocks of Victorian architecture.' A taller building was then to be placed facing Macarthur Street. At first the judging panel disqualified the entry, but then the decision was reconsidered. The design was finally accepted and the three pre-cast concrete panelled boxes were built with height, scale and proportioned window openings, which complemented the classical forms.
>
> (O'Neil, 2000)

Two of the buildings, 1 Treasury Place and 1 Macarthur Street, commenced construction from 1963, while 3–5 Macarthur Street (later renamed as 55 St Andrews Place) was constructed as a four-storey building in 1967 with a net lettable area of approximately 6100m^2 plus a small car park on lower ground and commenced life as the State Chemical Laboratories for the Department of Agriculture (Figure 4.3.1). The government realised the obvious issue of having a State Chemical Laboratory on the city fringe and moved it to Werribee in 1994. During the building refurbishment in 1996–97 an additional level was added as part of the fit-out for the Department of Justice. The Department occupied the building until their relocation to 121 Exhibition Street in June 2006.

OPERATIONAL AND FUNCTIONAL ISSUES

From the point of view of sustainability, the building was in poor condition (Figure 4.3.2). The energy use was high – the total electricity and gas use significantly exceeded best practice for this type of building (Figure 4.3.3) and indeed it was one of the Victorian Government Property Group's (VGPG's),

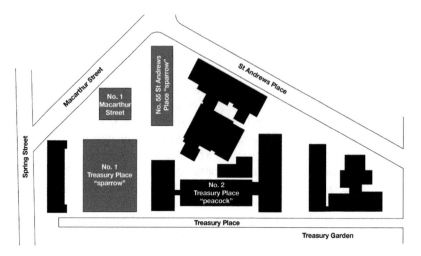

Figure 4.3.1 Location of 55 St Andrews Place to the top

Source: Clark (2006).

Figure 4.3.2 55 St Andrews Place under construction in 1967

Source: National Library of Australia.

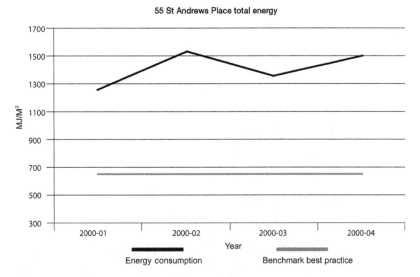

Figure 4.3.3 Historical electricity and gas consumption 2000–2004 compared to a benchmark of 'best practice' from the Property Council of Australia Energy Guidelines 2003

now the Government Services Division (GSD), worst performers. The VGPG was the State Government's Real Estate arm for office accommodation and Government properties.

The building rated about 1 star in an unofficial Green Star rating against the Green Star Office (v2) rating tool with one-third of the building's points being gained due to its proximity to Parliament railway station. The Green Star rating systems are a voluntary environmental rating system that provides an appraisal of a buildings' overall environmental performance. The tools assess a building's design response under the categories of *Management, Indoor Environmental Quality (IEQ), Energy, Transport, Water, Materials, Land Use & Ecology, Emissions* and *Innovation* to calculate its overall weighted score, which then determines its respective star or performance rating (Green Building Council of Australia, n.d.).

The building also benchmarked only a 1 star Whole Building rating under the NABERS Energy rating tool (Figure 4.3.4). NABERS (the National Australian Built Environment Rating System) is a performance-based rating system for existing buildings. NABERS rates a commercial office, hotel or residential building on the basis of its measured operational impacts on the environment. Under the rating tool a 1 star rated performance reflects a building with poor energy management or outdated systems that is consuming a lot of unnecessary energy.

While the required temperature set-points were being usually maintained within the building, from an occupant's point of view, the comfort levels were poor. This was due to a combination of building fabric and building services with issues including:

Figure 4.3.4 55 St Andrews Place prior to upgrade in 2005 (the top floor was added in 1997)

- Cold air draughts being experienced and perceived lack of fresh air.
- The outside air intake being perfectly located to collect diesel fumes from passing delivery vehicles, causing more than one building evacuation in the past.
- The heavily tinted glazing becomes very hot in summer causing localised discomfort due to radiant heat (and also cracking of a number of windows).
- The roof added in 1997 was not well insulated, causing return air to the fan assisted Variable Air Volume (VAV) boxes in the Level 4 ceiling space to reach 40°C to 50°C on extreme days.
- The addition of the VAV air-conditioning system to Level 4 starved the main air-conditioning system and was poorly designed and commissioned (e.g. the VAV boxes can run backwards if started before the main system) (Figure 4.3.5).
- There were problems with the control system (e.g. temperature sensors incorrectly positioned – or not relocated during tenancy fit outs).

Other issues which needed to be addressed included:

- The building was very dark due to the very heavily tinted windows and the artificial lighting system did not meet modern lighting practices.

Figure 4.3.5 Original outside air grille to main AHU behind and original window to lower ground floor stairwell

- There were no water conservation features in the building.
- The fire services required an upgrade.
- The carpets were nearing the end of their economic life.

In summary, 55 St Andrews Place reflected a typical 1960s building that underwent a 1990s fit-out that added another floor (and additional problems). On the plus side, the workstations were in good condition, with a number of spare parts. The walls and floors were in good condition and the solar hot water system on the roof was working well.

VGPG ACCOMMODATION STANDARDS

In 2005, VGPG issued the Victorian Government Office Accommodation Guidelines which set the scene for the new workplace as 'a living organism, as a forum, as a functional unit'. The guidelines aim to improve productivity and specify a 'green' building along with value for money. Key targets were:

- Green Star – Office Design   4 Stars or Best Practice
- Green Star – Office Interiors   4 Stars or Best Practice
- Base Building NABERS Energy   4 Stars (existing) or Strong Energy Performance

	4.5 Stars (new) or Strong Energy Performance
• Tenancy NABERS Energy	5 Stars or Best Building Performance

Key requirements include:

- engagement of a dedicated Ecological Sustainable Development (ESD) consultant;
- involvement of the department's environment manager;
- increased productivity from improved working conditions;
- access for people with a range of disabilities;
- improved water efficiency;
- material selected to minimise waste and off-gassing;
- waste – kitchen design, recycling area & construction waste improved;
- data centre design and energy consumption considered.

BUILDING UPGRADE

In mid-2005 the VGPG decided to upgrade 55 St Andrews Place as its first major government-owned office building to commence under the new accommodation guidelines. In response to this, quotations from suitably qualified ESD and Engineering Services consultants were sought to develop and implement a building improvement plan to achieve the following outcomes:

- Base Building
  - 4 star Green Star – Office Design (v2) rating
  - 4.5 star NABERS Energy Base Building rating
- Fit-out
  - 4 star Green Star – Office Interiors (v1.1) rating
  - 5 star NABERS Energy Tenancy rating

Cundall were engaged as the lead consultant as well as the Mechanical Services and ESD Consultant in September 2005 to develop the improvement plan and they engaged as sub-consultants the services of H2o Architects and Medland Metropolis (Electrical, Hydraulics & Fire Services) to assist in developing the concepts and budgets. Three strategies were developed with budget costs for each:

- *Option 1*: Measures to achieve at least 4 star Green Star and ABGR ratings.
- *Option 2*: Other measures that achieve 4 star rating and improve the health, well-being, spatial efficiency and productivity of the building.

- *Option 3*: Measures that achieve a benchmark building in fulfilling the triple bottom line (TBL) objectives of the Office Accommodation Guidelines.

In February 2006, the VGPG gave approval to proceed with Option 3 with a budget of $4.3 million allocated. Cundall were appointed as Principal Consultants to manage and prepare the design and documentation for the base building. While it might be considered unusual for the ESD Consultant to engage architects, engineers, quantity surveyors and building surveyors as sub-consultants, this was consistent with VGPG's desire to challenge conventional design processes as well as conventional design solutions.

Around the same time, the Department of Parliamentary Services was seeking new accommodation and the space requirement and location of 55 St Andrews Place proved an ideal match. The design of the tenant fit-out and base building upgrade proceeded in parallel, adopting an integrated fit-out approach to allow both to be tendered as the whole project.

DESIGN PROCESS

Very early in the design process, the team realised that an innovative and consultative approach was required. Throughout the design phase the whole project team would meet on a weekly basis for approximately two hours to review the progress made over the last week. Cundall developed an Action Plan spreadsheet which was updated at each meeting and which acted as the design brief, meeting minutes, cost plan, programme and green plan all rolled into one.

The benefits of the design approach were many: all team members had a clear understanding of the ESD elements of the building; the team were asked to contribute their own thoughts; the team members added value to the design from their own professional perspective; and the client team knew, at first hand, the issues, the cost and the resolution proposed. In essence, the whole team owned the design. The team also asked the Facility Managers of the building, Jones Lang LaSalle, to attend, which proved to be hugely advantageous, as the design had a maintenance culture embedded throughout.

UPGRADE DESIGN RESPONSE

The base building design outlined in the ESD Improvement Plan was further tested and refined and numerous changes were made. New features were added and other initiatives were deleted or adapted to either keep the project on budget or to reflect the fit-out requirements of the new tenant. Major changes included converting the ground floor into eight committee rooms and a multi-purpose room and relocating the building entry to the west elevation (its original location in 1967).

The typical approach to refurbishing poorly performing existing buildings is often to simply replace existing services, internal materials and fittings. The approach for 55 St Andrews Place was to go back to first principles and consider the same key issues as when designing a new building (Figure 4.3.6).

The core design philosophy was to do the following:

- improve daylight;
- improve comfort and air quality;
- reduce fabric loads;
- retain what we can.

The key features of the upgrade works are described as follows.

BASE BUILDING

Building fabric

The building had very heavily tinted windows which allowed minimal daylight inside the building. The building also has predominantly east- and west-facing façades. The challenge was to increase daylight and at the same time reduce solar heat loads inside the building. This was addressed, after extensive testing and simulation, by changing the punched windows to clear single

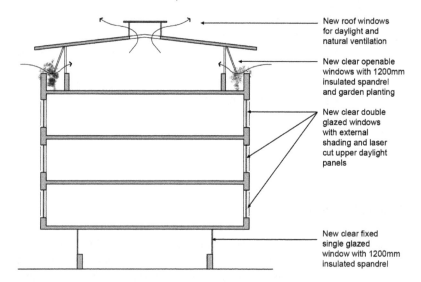

New roof windows for daylight and natural ventilation

New clear openable windows with 1200mm insulated spandrel and garden planting

New clear double glazed windows with external shading and laser cut upper daylight panels

New clear fixed single glazed window with 1200mm insulated spandrel

Figure 4.3.6 Fabric improvement concept used in original business case

Figure 4.3.8 External automated daylight guidance blinds
Source: Shade Factor.

Figure 4.3.7 Refitted windows on Level 3: clear glass and external automated daylight guidance blinds

glass and automated external venetian blinds (with individual manual override on each window) (Figures 4.3.7, 4.3.8).

On the top floor, the area of full height glazing was reduced by constructing 1200mm high insulated spandrels. On the ground floor the existing glazing was left unaltered but high performance internal blinds were added. Additional insulation was also added to the underside of the roof.

Air-conditioning system

Following detailed analysis of the existing system, the central air handling unit and over 80 per cent of all ductwork and VAV boxes were retained (also avoiding the need to remove the ceiling). The system's energy, comfort and air quality performance was improved by a combination of improving the building fabric; replacing existing four-way diffusers with swirl or induction diffusers; relocating the building's fresh air intake (to eliminate vehicle exhaust emissions); increasing the outside air quantities by over 50 per cent; cleaning the supply and return air ducts; implementing a new control strategy and re-commissioning the entire system.

A separate exhaust system for printer/photocopier rooms and tea rooms was also installed. A naturally ventilated winter garden (to be used as a meeting/resource room), accessible to all building occupants was created at the northern end of Level 4.

Lighting system

The lighting design took a different approach to standard practice. The existing inefficient lighting was replaced by a combination of a background level of lighting of 160 Lux from recessed T5 ceiling troffers and workstation-mounted task lighting to meet the additional illumination requirements. This integration of building and fit-out achieved a highly energy-efficient (6 watts per square metre) and more flexible lighting solution. In addition, an element of up-lighting was used to raise lighting in selected areas, to create visual interest within the space.

Figure 4.3.9 Naturally ventilated resource room (the high concrete upstand outside the perimeter of the top floor raises the question – why was full height glass originally installed?)

Water conservation

The roof downpipes came to a central point in the car park under the building. This led to a solution to divert the roof stormwater into new rainwater tanks located in the basement with a storage capacity of 48 ML. The collected water is then used to flush toilets and urinals in the building. Flow restrictors were also added to existing taps and new/existing showers used 7.5 L/min eco-showerheads. Water meters are also to be installed with leak detection.

Central plant

Although allowed for in the original upgrade budget by VGPG for the building, by virtue of the façade changes made which significantly reduced the external fabric loads, neither of the existing central chilled water and heating hot water plant was altered by the upgrade works. In fact the existing chillers, considered undersized prior to the design solution adopted, now have spare capacity.

Other initiatives

A variety of other initiatives have been adopted including:

- extensive commissioning requirements and commissioning agent;
- sub-metering to chilled water, gas and electrical supplies linked to a web-based reporting system;
- improving the indoor environment quality by improved material selection including low VOC paints and carpets, and low formaldehyde particleboard/MDF;
- increasing access and facilities for cyclists;
- installation of external screening plants for the top floor; and
- upgrading the Fire Indication Panel and systems.

TENANCY FIT-OUT

The tenancy fit-out aimed to deliver a similar level of environmental performance as the base building upgrade. For this work H2o Architects took the lead role in the interior design and worked with the same consultant team. Initiatives integrated within the final fit-out solution included the following.

Committee room HVAC

As the existing HVAC system could not cope with the increased occupant densities in the Ground Floor committee rooms, it was replaced with fan coil units (FCU) providing individual room control, 100 per cent outside air with heat recovery and indirect evaporative pre-cooling.

Data centre

As part of the fit-out, a data centre, sized to serve the whole of Parliamentary Services IT system throughout Victoria, needed to be accommodated. To minimise the energy uses of the data centre itself as well as the associated air-conditioning system, blade servers were used and an outside air economy cycle ventilation strategy integrated in the overall air-conditioning strategy to take advantage of both the cooler overnight air temperatures and seasonal climate variations of Melbourne.

Furniture and partitions

Over 70 per cent of the previous tenant's furniture and 50 per cent of the existing partitions were reused or refurbished for use in the new tenancy. The security counter used an existing reception counter from 1 Treasury Place. All new furniture and partitions were specified with environmental certification where possible.

Other initiatives

A variety of other initiatives have been adopted including:

- improving the indoor environment quality by improved material selection including low VOC paints and carpets, and low formaldehyde particleboard/MDF;
- installing new DDA-compliant toilets on Levels 1 and 4, adding to the existing compliant toilets;
- internal plantings to offices and breakout areas.

The combined project cost for the integrated base building upgrade and tenancy fit-out is approximately $7 million including construction costs and consultant fees (about $1,150 /m^2).

CONCLUSION

The upgrade of 55 St Andrews Place was achieved by breaking with con-ventional design processes and solutions. The ESD Consultant, Cundall, was the principal consultant and with VGPG defined the project brief and budget. Then the ESD Consultant took on the role of Principal Consultant and managed the design process with the tender coming in on budget and the construction works being completed on programme.

The design solutions are simple and, by improving the façade per-formance, the team was able to improve daylight and comfort, and reduce energy consumption, while retaining most of the original mechanical sys-tems. The fit-out and upgrade works were truly integrated and this was most apparent in the design of the lighting system, which combines ceiling lighting with task lighting.

With respect to its environmental benchmark objectives the Base Building upgrade attained its target 4 star rating, but the more onerous Office As-built rather than Design rating, and the fit-out attained its 4 star Office Interiors rating.

With respect to the target greenhouse performance or NABERS Energy ratings (NABERS, n.d.), it is not known if these were achieved as the Department of Parliamentary Services has not divulged its performance. The only feedback to date has been from the Facility Managers of the building, who have advised that it is performing well with no complaints from staff. In fact during the week in February 2010 when temperatures rose above 40°C for five consecutive days, it was surprising to find that no complaints were received from the staff within 55 St Andrews Place on the quality of comfort within their working environment.

REFERENCES

Clark, D. (2006) *55 St Andrews Place: Turning a Sparrow into a Peacock*, AIRAH Conference.
Green Building Council of Australia (GBCA) (n.d.) *Green Star Rating Systems*. Available at: www.gbca.org.au/green-star/green-star-overview/ (accessed 9 May 2012).
NABERS (National Australian Built Environment Rating System) (n.d.) *Rating Systems*. Available at: www.nabers.com.au/default.aspx (accessed 9 May 2012).
O'Neil, F. (2000) *The Treasure Reserve*, Melbourne: Heritage Victoria, Victorian Government.

4.4
BIOCLIMATIC RETROFITTING OF UNIVERSITY BUILDINGS

What issues arise in the precinct-scale retrofitting of buildings?

Margaret Liu

INTRODUCTION

The main aim of this research is to investigate the relevance of bioclimatic retrofitting in improving energy performance of existing buildings. The relevance of building scale is also included in any scope of this investigation. The study will seek to establish the relationship between bioclimatic design principles and building scale analysis. A practical outcome is a range of solution sets in building retrofitting strategies that form an integral and practical system of achieving energy efficiency for urban networks as well as individual buildings. The solution sets can also be utilised to implement green innovation and technology in a coordinated and multi-faceted manner. The research also seeks to evaluate any energy performance claims of sustainable design or innovation through post-occupancy consumption data analysis.

The methodology for this research is divided into four parts:

1 selection of case studies
2 project scope
3 data collection
4 data analysis.

SELECTION OF CASE STUDIES

The nominated cases involve an individual building, as well as its urban context. The following criteria are applied:

- The project and its impact on energy performance can be tracked and measured both before and after the retrofit works.
- Scope for quantitative energy performance analysis is presented, including the contextual impact and impact on the building as a whole.
- Relevance and compatibility to the purpose of this study can be shown.
- Clear project scope which has potential to have a measurable impact on the energy performance of a building.

Project scope

The scope of the work is defined and the design strategies are tabled based on bioclimatic principles. An applicable solution set is then translated into the specifications. It is then tracked from before to after the building works for the purpose of energy performance study.

Data collection

Relevant electricity consumption data from January 2004 to February 2009 is obtained from across three levels: from building, precinct and urban base metering systems. A five-year time frame is considered adequate to capture previous trends or other significant behaviour in electricity usage. The case studies would be ideally completed between January 2006 and January 2008, to allow sufficient time to track changes in energy performance. In instances where the retrofit work falls significantly outside of this period, a more extended period of consumption data would be required. The metering information is used to monitor the impact of retrofits and refurbishments on the electricity usage, particularly during post-occupancy. The gathering of primary data is sourced from metering software, such as the University of Sydney's UIS (Utilities Information System) monitoring software, which has real-time information on energy performance for every building and precinct on campus.

Data regarding energy costs are also gathered for the same period of January 2004 to February 2009. This is to examine the behaviour of consumption cost movement, to see whether there are any explainable differences before and after retrofit. The data from energy consumption and energy costs are then compared to investigate the relationship between monetary savings and consumption savings.

The occupants' presence and absence periods both before and after the works are also documented. This is to distinguish the energy consumption changes during and outside occupants' presence over a standard 24-hour period over 12 months.

A multitude of graphs involving daily and monthly energy usage are presented for analysis.

Data analysis

Analyses of the gathered data are performed using various graphic and mathematical software to observe any emerging trends to meet the objectives of this research.

In order to understand the impact of the retrofit on the energy performance of a building, it is necessary to compare the results both before and after the project. For the purposes of the research, the energy performance within the construction period is excluded in the analysis. The post-occupancy energy performance begins after all occupants resume full activities within the retrofitted space.

As the analyses are performed on daily, monthly and yearly gathered data, there are many potentially significant variables which are documented and considered. The following normalisation factors have been included:

- *Seasonal changes*: Energy consumption is related to the seasonal climate, as lighting and mechanical systems are utilised differently. This is expected to be reflected in the monthly data.
- *Daylight saving*: Lighting usage is not expected to be consistent due to daylight saving periods which occur every October to April. This is expected to be reflected in the monthly data.
- *Climate changes*: Although not expected to be as significant as seasonal changes, climate changes of short span of particular warmth or coolness can contribute to distortion in daily or monthly energy consumption. This is expected to be reflected in the monthly and daily data.
- *Occupancy level*: The occupancy level before and after retrofit, minor differences in occupancy level and occupants' behaviour may also influence energy consumption. Any substantial differences in the occupancy levels would deem the case studies invalid for energy performance analysis in this research.
- *Equipment usage*: All electrical equipment, such as computers, audio-visual facilities, contribute to energy consumption. The number and types of equipment used before and after the retrofit are monitored for changes.
- *Activity level*: The activity and behaviour of occupants other than occupancy rate also contribute to the energy consumption. Any significant differences in spatial usage would deem the case studies invalid for energy performance analysis in this research.
- *Academic year activities (tertiary education case studies only)*: The academic year in Australia is divided into two sessions: one from March to June, and the other from July to December. It is

expected that occupancy and activity levels would fluctuate accordingly. This is expected to be reflected in the monthly data.

The above normalisation factors require documentation and monitoring in order to conclude whether these conditions could be assumed as comparable, if not identical, for 'before' and 'after' the retrofit. In the event they are satisfactory for the purposes of this research, it is assumed that their impacts need to be taken into account in consumption differences before and after the retrofit works. However, it is noted that a combination of the accepted normalisation factors would have an accumulative effect on the energy performance such that the findings may require further investigation. The methodologies are explained further under the case studies.

THE CASE STUDIES

Metropolitan-based tertiary institutions often are presented in precincts or urban contexts. The mass building ownership, building settings, and significant land size of many city-based universities have made them appropriate departure points. Due to the size limitations of this chapter, only the case study of Darlington and Camperdown campuses of the University of Sydney is presented.

THE UNIVERSITY OF SYDNEY, CAMPERDOWN AND DARLINGTON CAMPUSES

The Camperdown and Darlington campuses cover a total area of 720,000 square metres in Sydney (Figure 4.4.1). According to recent energy audits (University of Sydney, 2008), 94 per cent of total energy use is from grid electricity with the rest being supplied by gas. The following sectors form over 90 per cent of energy consumption of an average building on campus:

- Lighting (15 per cent);
- Cooling and ventilation (27 per cent);
- Heating (22 per cent);
- Office equipment (13 per cent);
- Other equipment (teaching and research) (15 per cent).

For the purpose of this research, only three of the most energy-intensive areas are investigated: lighting, cooling/ventilation and heating, which on average, consume 64 per cent of a building's energy requirements. The three areas can be put into two categories: lighting and mechanical systems. It is also acknowledged that while existing buildings' energy performances can

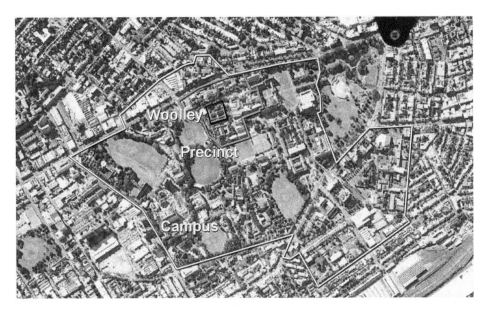

Figure 4.4.1 Aerial view of Camperdown and Darlington Campuses

vary widely from case to case, lighting and mechanical systems are generally the top two most energy-intensive sectors of a building on campus.

The retrofit project

A partial retrofit of the John Woolley Building started in June 2007 and was completed in early January 2008. Figure 4.4.1 displays the Woolley Building location within both of its precinct and urban contexts. The retrofit was to accommodate the United States Studies Centre, which is a part of the Faculty of Arts. This project was selected for the purpose of this research, due to the similarity of its attributes before and after the works. The project possessed relative consistency in the following areas before and after the retrofit, which contribute to minimise the variables in the analysis of energy consumption differences:

- purpose of usage
- occupancy density;
- occupants' behaviour;
- occupants' activity levels;
- scope of work covered significant floor area of the building, which is likely to demonstrate measurable consumption differences, if any.

The total energy consumption of the Woolley Building in 2004–09 comprises approximately 96 per cent grid electricity and 4 per cent gas. For the purpose of the measurable consumption impact of this study, only electricity data is gathered and analysed. The energy metering in the Woolley Building has a simple structure. The sole metering system of the building does present problems for the research, in that the metering is not separated into usage sectors such as lighting, cooling, heating or office equipment.

Figure 4.4.2 displays the area of the retrofit work in relation to the entire Woolley Building plan. The aim of the project was to provide a series of offices, teaching rooms, meeting rooms and administration areas. The previous space was completely demolished to its structural elements. The scope of work included a complete retrofit of mechanical, lighting and control systems, as well as a new ceiling, flooring, painting, joinery, etc. The Woolley Building is heritage-listed, hence it resulted in minimal alteration to its external façades and glazing. In terms of bioclimatic retrofit strategies, the following measures were implemented.

Active strategies

- A chilled beam system was installed to improve air cooling and ventilation. It requires less energy and mechanics to cool and recycle the air. The mechanical system involving chilled beam was the first to be installed in any buildings at the University.

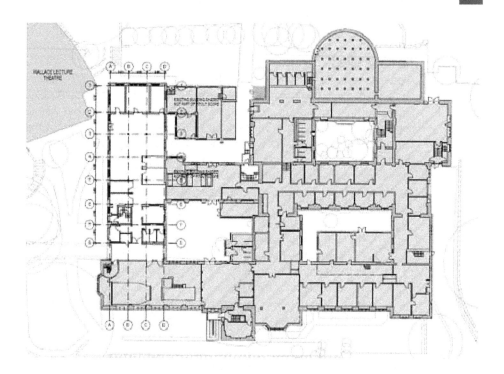

Figure 4.4.2. Plan of United Studies Centre in relation to the Woolley Building

- Energy efficiency lighting using energy-efficient compact fluorescent luminaries. Each fluorescent tube requires 30–40 per cent of the energy used by the previous lighting system, which was a conventional bulb and fluorescent setting. The lighting system is also designed to be turned off at the periphery to maximise the use of natural light and sun penetration.
- Control system with timer functions was installed to control both mechanical and lighting systems. A timer activates the systems according to when the spaces are used. For out-of-hours operations, such as week nights and weekends, it has a 2-hour manual setting to activate the systems.
- Automated lighting system allows on-demand use. Most of the lighting is automatically deactivated when not used over an extended amount of time.

Passive strategies

- Natural light is maximised by maintaining glazing openings, as well as addressing the orientation, sun penetration and the surrounding building context of the space. The design of work areas caters for the use of abundant daylight, in an attempt to reduce the requirement for artificial lighting.
- In parts of the space where air conditioning is not available, natural ventilation was encouraged through orientation and openings.

EMPIRICAL FINDINGS

Woolley Building energy performance

Daily, monthly and annual electricity consumptions for 2004–09 were gathered. An extensive series of consumption graphs were populated to demonstrate any measurable impact of a bioclimatic retrofit on the energy consumption of the building for one year. It is worth noting that for the duration of periods gathered in data, the Woolley Building underwent minimal refurbishments.

In Figure 4.4.3, the monthly usage by peak/shoulder/off-peak periods are tabled for 2004–09, based on kilowatt hours.

The energy consumption in the post-occupancy period (as marked on the consumption graph) is noticeably lower than any previous annual averages. Further analysis of the data demonstrates that during the post-occupancy period after retrofit, there was a drop in energy consumption across peak, shoulder and off-peak periods.

The reduced monthly consumption is compared to the average of the same months in the previous five years. The reduction in consumption

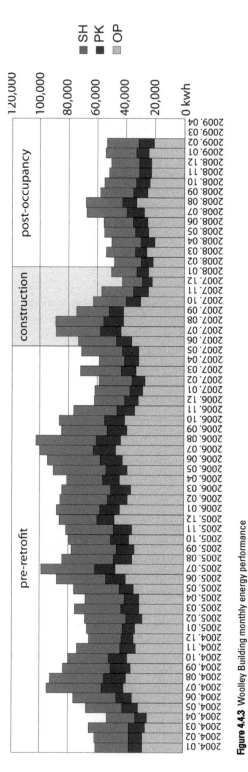

Figure 4.4.3 Woolley Building monthly energy performance

Notes: SH = Shoulder monthly usage (7am to 2pm and 8pm to 10pm working weekdays and 7am to 10pm on weekends and public holidays), PK = Peak monthly usage (2pm to 8pm on working weekdays), OP = Off-peak monthly usage (10pm to 7am every day).

during peak periods on a monthly basis has a range of 4–7 per cent reduction when compared to previous months, which averages to be around 5 per cent for 2008–09. The reduction in consumption during the shoulder period on a monthly basis shows a range of 5–11 per cent when compared to previous months, which averages to be around 8 per cent for 2008–09. The reduction in consumption during the off-peak period on a monthly basis shows a range of 9–14 per cent when compared, this averages to be around 11 per cent for 2008–09. For 2008–09, the reduction of energy consumption has been around 24 per cent when compared to previous years. In other words, for 2008–09 the Woolley Building operated on 76 per cent of total electricity used in previous years. The most significant period of consumption reduction is observed during off-peak hours. The shoulder period experienced a small reduction. The peak period was the least affected, with an observed reduction of 5 per cent. The duration of the periods is also worth noting for an average work day: off-peak and shoulder periods are both 9 hours, while peak is 6 hours. These preliminary results need further examination to justify the impact of retrofit on energy consumption during the different hours of the day.

Cost savings performance

In Figure 4.4.4, the monthly costs by peak/shoulder/off-peak periods are tabled for 2004–09, based on Australian dollars per month. The amount of energy saved in comparison with previous years is not equal to the equivalent amount of cost saved. Although cost and consumption have a direct relationship, the cost implications also involve other variables which do not precisely reflect the amount of energy saved. The accounted variables are: differences in monetary terms and in consumption, over peak, shoulder and off-peak times, the cost of penalties for exercising close to maximum demand load (as a part of the X component), changes in continuous monthly costs, rise in base unit costs, all contribute to the discrepancy between energy consumption and cost.

Precinct energy performance

Precinct energy performance is based on the local grouping of five other buildings in close proximity to the Woolley Building (see Figure 4.4.1). The United States Studies Centre has less than 10 per cent of the total floor area in the precinct, hence it was not anticipated that there would be a significant difference between pre-retrofit and post occupancy. The buildings in the same precinct simultaneously underwent retrofit works throughout 2004–09. It has been documented that no other projects in the precinct had implemented bioclimatic principles or other low-energy strategies. So far, it is not clear what the extent of impact of these other projects on precinct energy consumption is. The analysis of the precinct consumption data reveals the overall energy consumption over the investigated period as a building group.

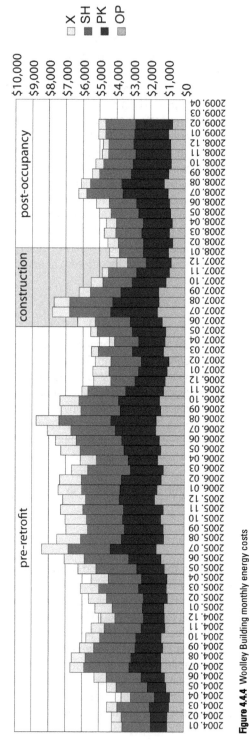

Figure 4.4.4 Woolley Building monthly energy costs

Notes: X = Continuous monthly cost, SH = Shoulder monthly cost, PK = Peak monthly cost, OP = Off-peak monthly cost.

It is revealed that at the precinct level, the average monthly post-occupancy consumption is 271,000 kWhrs (see Figure 4.4.5), which is a reduction of 3.9 per cent based on previous levels. The reduction cannot be entirely attributed to the United States Studies Centre alone, as there are variables and anomalies which need further examination. A closer examination indicates that there was a consumption decrease of around 55 per cent which sustained relative consistency between December 2004 and February 2006. This calls for the monthly consumption to be measured for more than five years, to obtain a more consistent monthly consumption average at any given time of the year. There is also the need to investigate for a valid explanation of the reduction in energy consumption during December 2004 and February 2006, and other factors that may contribute to changes in energy consumption.

Overall, at the precinct level, the consumption was reduced by around 4 per cent. However, at this preliminary stage of testing, it is not clear what the extent of impact the United States Studies Centre has had on precinct-based energy consumption.

Performance at an urban scale (campus-wide)

Energy performance at an urban scale reveals consumption patterns within the Darlington and Camperdown campuses (see Figure 4.4.6). Due to the size of the floor area of the United States Studies Centre, it was not expected to be measurable at the urban scale. The performance at the urban scale also summarises the campus-wide effect of existing retrofit works, most of which do not account for energy performance in the work scopes.

The data show a consumption increase of 32.1 per cent during 2004–09 (see Figure 4.4.6), which averages to 6.42 per cent every year. The consumption rate has been relatively stable when compared to those at the building and the precinct levels. However, if the current rate of increase is sustained and remains constant, the energy consumption is expected to double every 15 years. According to current data, the energy consumption will increase by 400 per cent by 2039, with the stipulation that the growth is linear.

DISCUSSION

At the building level, it is imperative to develop optimisation strategies that can be exercised with relative minimal cost or short payback periods in order to maximise gains in building-based energy performance. This can be done through passive strategies incorporating bioclimatic data (such as those delivered through building form, materials, openings and envelopes), active systems (such as those delivered through energy-efficient systems, automated lighting systems), or even greenery input. Each building needs to be assessed for a customised solution set which may involve one or a combination of all

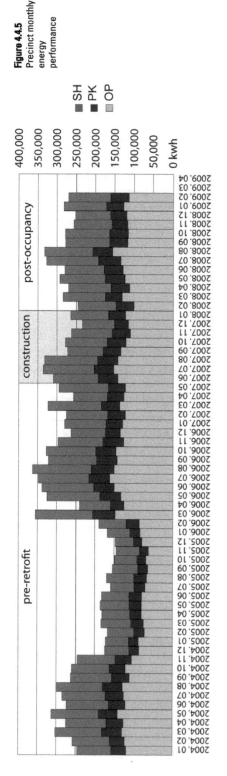

Figure 4.4.5
Precinct monthly energy performance

SH
PK
OP

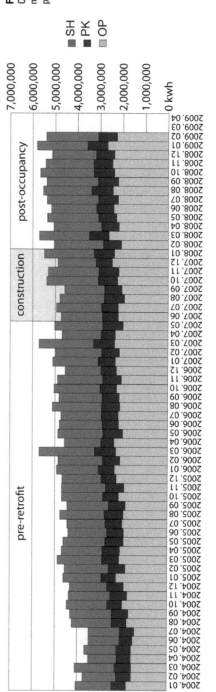

Figure 4.4.6
Campus-wide monthly energy performance

SH
PK
OP

the afore-mentioned strategies. The solution sets also need to be addressed in terms of monetary value to achieve a realistic scale of economy. A future subset of this research could look at the costs versus energy performance benefit of both passive and active strategies for existing building groups.

At the precinct level, it becomes imperative to address the issues that are relevant to this particular scale. The precinct scope shifts between the building and urban scale, and it needs normalisation between the macro (city) and the micro (building unit). The use of bioclimatic principles can benefit precinct-based solution sets for improving energy performance at that level. These include renewable energy sources and area-based retrofits. The planning of photovoltaic and other renewable energy sources need to be generated from a precinct scale to cater for future expansion. Again, the emphasis should be placed on working with bioclimatic data and relevant optimisation strategies first and foremost. At the precinct level, existing greenery with extensive canopies can act as heat sinks for adjacent buildings. In exploring precinct and urban levels, there are opportunities for incorporating bioclimatic data and renewable energy production data which would be lost in addressing these issues within the individual building envelope alone. Other strategies also need to reflect the urban-scale issues and extend these solution sets, to facilitate an integral system of energy optimisation.

At the urban level, the imperative issues are of a city planning nature. Planning for high-scale energy sourcing and distribution, such as grid and renewable sources, needs to be incorporated with sensitivity to the issues at the smaller scales. In a more general sense, the solution sets for planning need to take account of energy imperatives such as transportation, traffic and access, and other issues that are of significance at neither precinct nor building levels. It is at this macro scale where long-term objectives of urban sustainability are addressed. Any development plans or any strategic sustainable retrofitting are also anticipated to require the longest time frame to achieve at the urban level. It is envisaged that any significant macro bioclimatic strategies would take years, if not decades, to realise. As with any urban planning exercise, it is also heavily reliant on political momentum, significant monetary demands, and the collective will of stakeholders to realise any significant energy performance benefits.

COMMENTS ON CURRENT PRACTICE

The investigation into the building, precinct and campus-wide energy consumption resulted in the following comments regarding existing practices across the university:

- *Lack of strategies to improve building energy performance through retrofitting*: There has been an absence of tactics to improve existing buildings' energy performance. Current university design guidelines do not call for actions, such as those based on bioclimatic principles, that would result in improvement in energy performance. It is proposed that if such strategy were in place for every project, it is most likely there would be a visible reduction in energy consumption at the urban scale.

- *Lack of strategies to minimise off-peak expenditure*: At all three levels, the highest amount of consumption occurred during off-peak hours, i.e. 10pm–7am. This indicates there have been no effective strategies to minimise energy consumption overnight. It has been observed that a monthly average of 2,000,000 kWhr was consumed during periods when there was minimal occupant presence (see Figure 4.4.6). The off-peak consumption indicates there is a need to address lights, mechanical systems and electronic equipment being left on overnight constantly. It is understood that energy-intensive research equipment also contributes to a certain portion of the overnight consumption.
- *Lack of reinforcement and monitoring*: The University does have an environment policy which outlines its commitments and promotes the importance of a 'sustainable campus'. However, the policy has not been systematically implemented into guidelines and practical strategies that can be monitored and reinforced in everyday work processes. This is especially the case for retrofit works.
- *Lack of accountability and ownership*: There has been minimal accountability and ownership of energy performance of buildings, precincts and the campuses in general. The energy costs are not financially accountable by the buildings' occupants or faculties, but by the general university administration. There is a disconnection between the individuals who use the energy and those who pay for it.
- *A need to improve occupant education*: Previous research has indicated that occupant awareness and behaviour change can result in improved energy efficiency in building operations. It would be beneficial to investigate further occupant behaviour impact on energy performance, and how it can impact the relevant case studies.

REFERENCE

University of Sydney (2008) *Energy Savings Action Plan*. Sydney: The University of Sydney.

4.5

503 COLLINS STREET, MELBOURNE, AUSTRALIA

How can onsite power generation be achieved in commercial buildings?

Bruce Precious

INTRODUCTION

A much-loved building in Melbourne's Collins Street has undergone refurbishment to ensure it remains a key asset in its owner's property portfolio for years to come. GPT Group's 530 Collins Street building is a landmark of Melbourne's most prestigious business thoroughfare. Running through the middle of the CBD, Collins Street extends from what has become known as the 'Paris end' in the east or top end of the city, to the more recently developed Docklands business precinct to the west. It is home to a large number of well-known companies.

530 Collins Street was originally built in 1989 to a premium standard of design and finish but 20 years later it was deemed to have reached an age where refurbishment was required to dramatically improve energy efficiency and to compete with the newcomers to its west – in terms of both facilities and aesthetics.

The upgrade addressed a number of challenges, first, to ensure such an upgrade would enable the building to compete in what is still a largely unknown, low-carbon future market. Second, to respond to reported tenants who have a preference for sustainability as being an important element to their business. Third, to ensure the term 'premium grade' given to the building is maintained and that sustainability adds another dimension to the concept of 'high-quality' services. Sustainability is now synonymous with quality, and makes up a significant part of the newly defined term. We see that purely upgrading high-quality services with inefficient infrastructure is tantamount to refurbishing a building to a lower quality. Therefore, starting with a building, which already had met premium grade measure, was a significant opportunity to present to the market a true premium grade building for the future.

It is a dilemma much of Australia's existing building stock is set to face in the coming decade as the average building age nears 20 years. In detailing its options, two approaches for the building upgrade were considered: a traditional services upgrade (TSU); and an energy performance contract (EPC).

The traditional service upgrade could meet all of the performance and financial objectives set for the building; however, the risk of conducting an upgrade and falling short of the objectives was considered too great. Hence, the energy performance contract provided the best business case and highest probability of achieving a 5 star NABERS Energy rating at a known price, with guaranteed outcomes and returns for at least five years, thus enabling the building to remain a market-leading property into the foreseeable future.

DEVELOPING AN ENERGY PERFORMANCE CONTRACT (EPC)

EPCs have traditionally been applied to public buildings and infrastructure owned by government bodies wanting to reduce their risk while guaranteeing a level of performance. In the commercial world, however, traditional services upgrade agreements have been capable of delivering standard system upgrades. Where projects become more complex, such as that at 530 Collins Street, EPCs are one way to deliver the desired outcome while allocating risk to the party best able to manage it.

Some projects involving a simple services upgrade can be undertaken under a traditional specification and tender process and successfully have low risk. However, in the case of more complex projects where multiple outcomes are demanded, such as upgrading a fully operational building across multiple services requiring detailed integration of equipment, the risk needs to be carefully managed. Furthermore, based on a previous project, the EPC could provide a business model to deliver its objective of obtaining a 5 star NABERS Energy rating for 530 Collins Street, but would also deliver a 40 per cent reduction in annual greenhouse gas emissions compared to the industry average for a 2.5 star building.

The benefits to the property manager are as follows:

1   The targets are agreed.
2   Performance is covered by a guarantee and project risks are considered and allocated to the party best able to manage them.
3   The process of engagement is well defined, a competitive tender to find the right partner and the standard contracts are proven.

4   A Best Practice Guide for EPC is provided (City of Melbourne, 2011).
5   The process also allows for a number of check steps, where the parties can withdraw without undue penalty, all things considered.
6   An EPC was regarded as the fastest track to completion.

IMPLEMENTING THE EPC

A number of companies were able to deliver the upgrade under the EPC method, which included the equipment replacement, retrofitting as well as the inclusion of a cogeneration system to guarantee 5 star NABERS Energy performance. The benefits to the engineering company were working with an energy service company who could coordinate all trades, understand the technical aspects of the installation and ensure the delivery of systems to meet the design intent on program, proved a significant advantage. These types of complex upgrade projects require a unique coordination and co-operation of skills, from building modelling and system diagnostics, to risk analysis and project management. Great care is required to ensure the results can be achieved on time and to budget, and with little disruption to tenants.

For the owner, an effective EPC implementation requires close collaboration with the Energy Service Company Operator (ESCO) or contractor, as both parties are committed to the project performance outcome, which often evolves into a partnership. Most importantly, the project leaders on both sides of the partnership must be well skilled in the techniques required for the upgrade. At 530 Collins Street, the building services manager and the ESCO project manager played their parts to perfection, ensuring an outstanding result.

COMMUNITY ENGAGEMENT

The requirement for community engagement determined a complete repositioning of how the building interacted with the community. This led to the redevelopment of the building's lobby to include a new Collins Street entrance, major foyer redesign and revamped ground floor retail and dining precinct. Vacant floors were also fully refurbished for new, incoming tenants. During the refurbishment of these floors, it was possible to recycle 85 per cent of the carpet tiles left behind by the previous tenant, as well as other items. This came at an additional cost; nevertheless the opportunity was valued as an appropriate action to take as yielding significant benefits to the environment.

HVAC RETROFIT

The HVAC retrofit presented the greatest challenge during the works, and was specifically timed to coincide with the winter window to reduce any impact on comfort conditions. Following the selection of the preferred energy performance contractor, a detailed feasibility study was undertaken. This involved specialist contractors called in to assess and model the performance of combinations of replacement plant, including chillers, cooling towers, cogeneration engines, variable speed drives and other efficiency measures (Figure 4.5.1). Ultimately a fixed price was detailed to undertake the agreed works in full, with guaranteed outcomes for NABERS performance, energy (MJ and cost) savings and HVAC system performance (comfort conditions).

The scope of the works included the decommissioning, removal and replacement of three existing, large chillers, while an existing CH-4 reciprocating chiller was retained (Figure 4.5.2). The three chillers were replaced with two 2944kWr and one 1104kWr variable speed centrifugal chiller sets, with VSD motors installed on all new and existing chilled water pumps. In order to contain costs, the three new chillers, each broken down into three separate packages, were lifted on to the building's level 14 podium roof, adjacent to the plant room, on the same day. This was preceded by careful structural analysis to ensure the roof loading was not exceeded.

Figure 4.5.1 Service upgrades to floors

Figure 4.5.2 Chiller upgrade and cogeneration plant installation

RISK MANAGEMENT

The risk management strategy employed involved worst-case scenarios. This strategy was employed by all the main contractors with mitigation plans and alternative strategies developed ready for immediate implementation should they be required. For example, this process highlighted opportunities like testing high-pressure shut-off valves well before the existing chillers were to be removed. Close inspection at this stage also revealed that pipework in the plant room required new supporting structures, as the existing hangers were not taking all the pipework weight. The contractor, together with the mechanical installation team, provided a quick response to avert any programme impact. An existing pneumatic control system from the return air/smoke spill-fan pitch control mechanism was decommissioned and replaced with variable speed units to each of the return air/smoke spill fans.

The building's existing cooling towers were also decommissioned and removed, replaced by two tenant condenser water cooling towers, each selected to cool 55 litres per second and reject 1265kW. A further three common condenser water cooling towers were selected to cool 143.3 litres per second, with each dual cell rejecting 3300kW. Eight variable speed drive units were also commissioned to the towers. Lastly, an embedded, natural gas-fired cogeneration system was installed on the roof of the building (Figure 4.5.3). The one-off generator system with its acoustically designed enclosure (measuring 9m x 3m x 3m and weighing 23 tonnes) was lifted into place as three separate packages, each weighing less than 9 tonnes.

Figure 4.5.3 Cogeneration plant

Four fixed-speed pumps were installed adjacent to the enclosure for supplementary cooling of the jacket and intercooler, while a heat recovery exchanger was installed to provide for HHW to both the domestic and HHW systems. Like many experiences with cogeneration around the country, gaining the approval of the local energy retailer to synchronise the unit with the electricity grid provided a further challenge at 530 Collins Street, which took a surprising amount of determination and persistence.

ACHIEVING PERFORMANCE TARGETS

Tracking the achievement of performance targets was carried out. A control system was installed in the cogeneration unit; a new building management system (BMS) has been able to provide a high level of analysis on energy and water meter data. This gives detailed information right down to tracking the hourly performance of the building and major sub-systems. The utility also provides full tracking of the building's energy performance via an automated, rolling NABERS Energy rating generated to ensure the building is operating at the targeted efficiency.

As cogeneration has been a critical strategy in delivering a 5 star NABERS Energy outcome, the performance of the system is monitored daily. The project reached practical completion on 1 December 2009 (the start of summer), with monitoring and verification commencing a month later. CitiPower approval of grid synchronisation was achieved on 1 April 2010, and signalled commencement of operation for the cogeneration plant.

As a result of the building upgrade to 5 star NABERS Energy performance, greenhouse emissions have been reduced by 4,700 tonnes CO_2/annum, while 4.3GWh of grid electricity has been saved annually. This translates to a yearly energy cost reduction of A$360,000.

CONCLUSION

The significance of the building upgrade, and the benefits of improved services, lower energy bills and greatly reduced greenhouse emissions were seen as necessary to ensure the building was attractive to the widest range of tenants. Enhancing the premium grade office by sustainable retrofitting takes the building to a higher level of performance demanded by some tenants, hence making the building available to a wider potential market in the premium grade sector.

The EPC provided the fastest track to upgrading services and seeing performance improvements; however, it remains to be seen if the traditional service upgrade could have delivered the same quality in the same time frame.

ACKNOWLEDGEMENTS

This chapter was based on an article developed for *Ecolibrium, The Official Journal of AIHRA*, 9(11), December 2011.

REFERENCE

City of Melbourne (2011) *Energy Performance Contracts.* Available at: http://www.melbourne.vic.gov.au/1200buildings/what/Pages/EnergyPerformanceContract.aspx (accessed 21 December 2011).

4.6
SOLAR RETROFIT

What are the benefits of retrofitting solar-producing technologies?

Brett Pollard

According to the United Nations, the energy used in buildings is responsible for more than a third of global greenhouse gas emissions (Yamamoto and Graham, 2009). In response to the need to reduce the greenhouse gas (GHG) emissions related to building energy use, architects and engineers are focusing on improving the energy efficiency of the buildings they design and renovate. Increasingly, they are also seeking to incorporate renewable energy systems such as photovoltaic panels and wind turbines into their projects. However, the technical and economic feasibility of incorporating these systems into buildings is not widely understood by most architects and their clients.

This chapter is a summary of a detailed research project undertaken by the author into the feasibility of using a grid connected photovoltaic (PV) system to decrease the GHG emissions of an Australian case study commercial building. This research project is particularly relevant for Australia because over 90 per cent of Australia's electricity is generated using fossil fuels, primarily coal. This is despite Australia having one of the best solar energy resources in the world.

BACKGROUND AND CONTEXT

In 2010, the Australian federal government confirmed its intention to help drive the commercial building sector to reduce its share of greenhouse gas emissions by introducing the Commercial Building Disclosure (CBD) scheme. This scheme requires building owners to disclose the energy efficiency rating of all commercial office buildings and tenancy spaces of 2,000 square metres or more at the time of sale or lease. The energy efficiency rating is based upon the National Australian Built Environment Rating System (NABERS) Energy rating scheme.

NABERS ENERGY

The Australian Greenhouse Building Rating, as NABERS Energy was previously called, was first released in 1999 and is an operational energy-efficiency rating tool somewhat similar to the United States Energy Star for Buildings rating system. NABERS Energy rates existing commercial buildings based on the GHG emissions associated with their actual energy use. NABERS Energy is part of a suite of performance-based rating tools that also include NABERS Water and NABERS Waste.

Information including energy consumption from the previous 12 months of operation, floor area, hours of operation, number of occupants and energy sources is used to determine a NABERS Energy star rating. A maximum of 6 stars can be awarded, with 2.5–3 stars representing average building performance and 6 stars representing excellent performance. NABERS can be used to rate a whole building, base building and/or tenancy spaces within a building. The Property Council of Australia estimates that about 40 per cent of the 21 million square metres of commercial office space in Australia currently has a NABERS Energy rating.

NABERS Energy allows renewable energy generated and used within a building to be discounted against a building's overall energy use. However, NABERS does not permit the reduction of a building's total energy consumption figure by the energy exported from the building when calculating the rating (DECC, 2008).

CASE STUDY RESEARCH PROJECT DESCRIPTION

The primary research objective of the project was to determine the technical and economical feasibility of using a grid connected PV system to increase the NABERS Energy rating (i.e. decrease the GHG emissions) of a case study commercial office building. To help direct the research project, a number of specific research questions were developed: (1) determining the optimal arrangement of the PV panels; (2) examining the impact of future economic conditions; and (3) investigating alternative strategies for improving the NABERS Energy rating. This enabled a detailed research methodology to be developed for the project (see Figure 4.6.1). All costs throughout this chapter are given in Australian dollars ($).

CASE STUDY BUILDING

The case study is a commercial office building located approximately 15 kilometres west of Sydney's central business district. The building was completed in 2009 and is six storeys in height with two levels of below ground

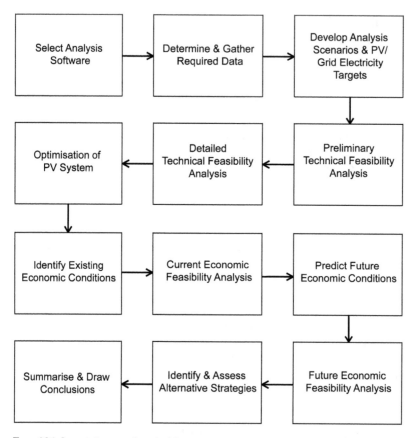

Figure 4.6.1 Case study research methodology

car parking (Figure 4.6.2). There are retail facilities on the ground floor, five floors of offices and plant space on the roof. The Net Lettable Area (NLA) of the office space is approximately 5,770 square metres and the gross floor area is approximately 11,800 square metres, including the car parking. The building is rectangular in shape, measuring approximately 70 metres by 20 metres. There are no adjacent structures shading the building and future development will not shade the northern façade or roof.

The building's air-conditioning system is a centralised variable air volume (VAV) system with high efficiency chillers. A gas-fired boiler system is used for space heating and the domestic hot water. The lighting for the offices is high efficiency T5 fluorescent light fittings. The building has been designed to achieve a 4.5 star NABERS Energy rating for the base building component.

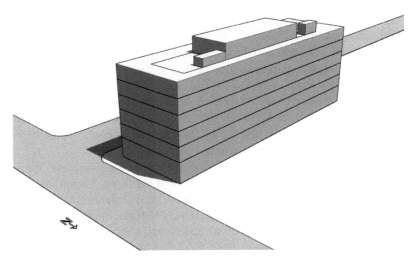

Figure 4.6.2 Case study building

RENEWABLE ENERGY FEASIBILITY SOFTWARE PROGRAMS

There are a number of software programs currently available to assist with determining the feasibility of renewable energy systems. In this study the HOMER program (US National Renewable Energy Laboratory) was used for the technical feasibility modelling and RETScreen (Natural Resources Canada) was used for the economic feasibility modelling. Both programs use information on available solar radiation levels, PV panel performance, tilt and orientation to calculate how much electricity can be supplied from the PV system and how much additional grid supplied electricity is required to meet the building's energy demand.

In addition, a number of other software programs were used during the course of the research project. These programs were:

- eQUEST – used to simulate the building's energy consumption and load profile;
- NABERS Office Rating Calculator – used to determine the NABERS Energy rating for the building;
- Sunny Design – used to select and size of inverters for the PV system;
- Panel Shading – used in the optimisation of the orientation and spacing of the PV panels.

GRID CONNECTED PV SYSTEM

The two primary components of a grid connected PV system are the photo-voltaic panels and the inverter. In basic terms, the PV panels convert solar energy into DC electricity and the inverter converts this into AC electricity for use within the building or distribution to the grid (see Figure 4.6.3).

The PV panel selected for this study was the BP 4175 mono-crystalline PV module produced by BP Solar. It was selected because it has a relatively high output per panel (175W), a good level of efficiency (13.9 per cent) and both RETScreen and Sunny Design have its performance infor-mation in their inbuilt product database.

The inverters were from the Sunny range by SMA Australia. These inverters were selected simply on the basis that the Sunny Design program used to size the inverters only gives details on Sunny inverters.

In terms of embodied energy, the US National Renewable Energy Laboratory (NREL) has estimated that multicrystalline PV systems are able to pay back their embodied energy within four years of operation (NREL, 2004). This time frame include the PV modules, support framing and the balance of the system.

TECHNICAL FEASIBILITY

The technical feasibility assessment of using PV panels to increase the NABERS Energy rating of the case study building involved gathering relevant solar and building energy use information, setting energy targets, and under-taking an initial feasibility analysis followed by a more detailed feasibility analysis. As NABERS Energy is able to rate different categories of spaces within a building (i.e. tenancy, base building and whole building), it was decided to test the feasibility of using PV systems to increase the NABERS Energy rating for each of these categories.

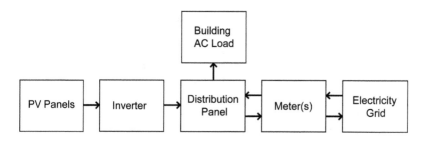

Figure 4.6.3 Typical grid connected PV system

SOLAR AND BUILDING ENERGY INFORMATION

The available solar resource information was obtained for Sydney using the data sources provided in HOMER and RETScreen (Figure 4.6.4). Both programs use solar data provided by the National Aeronautics and Space Administration (NASA).

As detailed building energy consumption and load profile information were not readily available, simulation of the building's energy use was undertaken using eQUEST. The results of the simulation were checked using the NABERS Energy online and reverse calculators to confirm that the simulated results achieved the 4.5 star NABERS Energy rating. The base building and tenancy energy consumption figures were extrapolated from the whole building figures and then rechecked using the NABERS calculators (Table 4.6.1).

ELECTRICITY TARGETS

The PV production and grid supplied electricity targets for each of the NABERS categories (Table 4.6.2) were developed by calculating the maximum annual grid supplied electricity consumption allowed by NABERS Energy for the building to achieve a 5 star rating and then subtracting this from the simulated annual electricity consumption. The difference represented the minimum

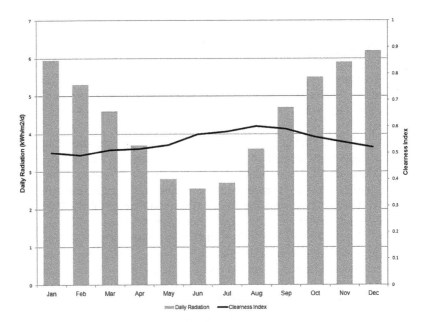

Figure 4.6.4 Annual horizontal solar radiation level for Sydney

Source: HOMER

Table 4.6.1 Case study building simulated energy consumption and GHG emissions

NABERS category	Electricity (kWh/yr)	Gas (GJ/yr)	NABERS rating	GHG emissions (kgCO$_2$e/yr)
Tenancy	586,894	–	4.5	551,680
Base building	509,129	299.357	4.5	497,706
Whole building	1,095,989	299.357	4.5	1,049,355

Note: NABERS Energy GHG emission factors for electricity and gas in NSW are 0.94 kgCO$_2$e/kWh and 83.4 kgCO$_2$e/GJ respectively.

Table 4.6.2 Case study building energy targets

NABERS category	PV production (kWh/yr)	Grid electricity (kWh/yr)	Gas (GJ/yr)	NABERS rating
Tenancy	50,229	536,665	–	5
Base building	62,127	447,002	299.357	5
Whole building	109,933	986,056	299.357	5

amount of electricity the PV system is required to produce annually. In order to cope with possible fluctuations in annual PV output, the minimum PV electricity production targets were increased by 10 per cent. The gas consumption of the building, which was not targeted in this study, has been shown for clarity.

PRELIMINARY TECHNICAL FEASIBILITY ANALYSIS

The first step of the analysis involved determining the area of PV panels required to achieve each of the PV electricity production targets and then comparing these areas to the available roof and façade area of the case study building.

The required output capacity of each PV system was determined by using an electricity production figure of 1382 kWh per kW per year (Watt, 2007) (Table 4.6.3). The required area and number of PV panels were then calculated using the rated power output and dimensions of the selected solar panel (175W and 1.28 square metres). A de-rating factor for the PV panel power output was not used in this preliminary exercise.

The area of the roof of the building is approximately 1,250 square metres; however, a large proportion is taken up by plant room, cooling towers, lift overruns and expansion space for future plant. The residual roof space, although reasonably large in total area, is narrow, discontinuous and subject to potential overshadowing by the plant room and cooling towers. However, the roof of the plant room (321 square metres) and an area above the lift overrun and future plant space (151 square metres) offered potential unshaded space for locating PV panels (Figure 4.6.5).

Table 4.6.3 Preliminary PV system sizing

NABERS category	PV production (kWh/yr)	PV capacity (kW)	PV area (m²)	No. of PV panels
Tenancy	50,229	36.3	265.8	208
Base building	62,127	45	328.8	257
Whole building	109,933	79.5	581.8	455

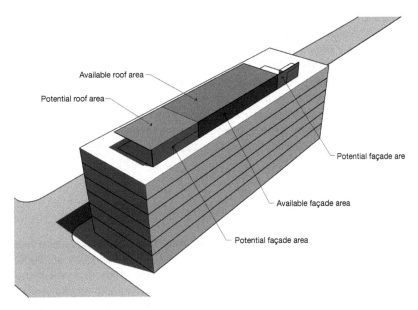

Figure 4.6.5 Roof and façade areas available for PV panels

The northern façade of the plant room offered approximately 105 square metres of blank north-facing façade while the area in front of the future expansion space and cooling towers offered another 84 square metres of potential vertical space to mount PV panels. The main northern façade of the building has approximately 250 square metres of continuous blank façade, but was not considered suitable due to potential maintenance access, and sufficient space had already been identified on the roof. It was assumed that sufficient space could be found within the existing plant rooms for inverters and associated equipment (Table 4.6.4).

DETAILED TECHNICAL FEASIBILITY ANALYSIS

A more detailed assessment of the identified areas was undertaken using the dimensions of the selected PV panel (1.6m x 0.8m) to determine the precise

Table 4.6.4 Roof and façade areas available for PV panels

Location	Available area (m²)	Potential area (m²)	Total (m²)
Roof (sloping)	321	151	472
Façade (vertical)	105	84	189
Total	426	235	661

area and number of PV panels that could be installed in each of the identified areas. The available areas were assessed first and if insufficient space was available, the potential areas were then assessed. Once the number of PV panels was determined, the capacity (kW) of each PV system was calculated. The Sunny Design software program was then used to size the inverter(s) for each system and this information, together with the PV panel orientation and tilt angle, were entered into HOMER. The results were then compared to the PV and grid electricity targets to determine the technical feasibility of each system. The technically feasible systems are shown in Table 4.6.5.

OPTIMISATION ANALYSIS

The optimal orientation of PV panels for maximum annual electricity production in the southern hemisphere is due north (Prasad and Corkish, 2006) while the optimal tilt angle is equal to the latitude of location where the panels are installed (Gong and Kulkarni, 2005). The minimum recommended tilt angle for PV panels is about 10 degrees (Prasad and Corkish, 2006) which allows for self-cleaning of the panels from rainfall to remove dust and grime that would otherwise decrease the output of the panels. The reduction in the performance of vertically mounted panels can be up to 30 per cent when compared to those tilted to the angle of latitude (Eiffert and Kiss, 2000).

Another important consideration is the self-shading of panels, which can lead to a dramatic reduction in PV output (Prasad and Corkish, 2006).

Table 4.6.5 Technically feasible PV systems from detailed feasibility analysis

NABERS category	No. of PV panels	PV capacity (kW)	Tilt & azimuth of PV panels	PV production (kWh/yr)
Tenancy	210 (s)	36.8 (s)	2° x 180° (s)	51,571
Base building	222 (s)	38.9 (s)	2° x 180° (s)	65,956
	64 (v)	11.2 (v)	90° x 180° (s)	
Whole building	330 (s)	57.8 (s)	2° x 180° (s)	104,986
	134 (v)	23 (v)	90° x 180° (s)	

Note: (s) indicates sloping PV panels and (v) indicates vertical PV panels.

There are a number of methods to calculate the required spacing to avoid self-shading. In this study the Panel Shading program was used to test a wide range of panel tilt angles and spacing. Once the optimal PV panel arrangements were determined, they were assessed using Sunny Design and HOMER to confirm that the PV system output and resultant grid electricity consumption would meet the targets.

The results indicated that while only the tenancy and base building PV systems achieved the PV output target, all the systems were considered technically feasible as they kept the case study building's grid electricity consumption below the maximum allowed to achieve a 5 star NABERS Energy rating (Table 4.6.6).

GRID ELECTRICITY AND GREENHOUSE GAS EMISSION REDUCTION

The optimised PV systems achieved between 8.4 per cent and 11.6 per cent reduction in grid electricity consumption for the case study building with only a relatively small amount of PV generated electricity being exported to the grid (Table 4.6.7). In terms of GHG emission reductions, the percentage reduction was slightly smaller than the reduction in grid electricity consumption for the Base Building and Whole Building categories due to the inclusion of the emissions associated with gas consumption (Table 4.6.8).

Table 4.6.6 Optimised PV systems

NABERS category	No. of PV panels	PV capacity (kW)	Tilt & azimuth of PV panels	PV production (kWh/yr)
Tenancy	148 (s)	25.9 (s)	25° x 180° (s)	50,276
	46 (v)	8.1 (v)	70° x 180° (s)	
Base building	185 (s)	32.4 (s)	10° x 180° (s)	66,276
	82 (v)	14.4 (v)	70° x 180° (s)	
Whole building	330 (s)	57.8 (s)	10° x 180° (s)	106,031
	101 (v)	19.5 (v)	85° x 180° (s)	

Note: (s) indicates sloping PV panels and (v) indicates vertical PV panels.

Table 4.6.7 Case study building grid electricity consumption with and without PV

NABERS category	Grid electricity without PV (kWh/yr)	Grid electricity with PV (kWh/yr)	Reduction in grid electricity (%)	PV electricity exported (kWh/yr)
Tenancy	586,894	537,336	8.4%	718
Base building	509,528	450,300	11.6%	7,048
Whole building	1,095,989	995,558	9.2%	5,600

Table 4.6.8 Case study building GHG emissions with and without PV

NABERS category	GHG emissions without PV (kgCO$_2$e/yr)	GHG emissions with PV (kgCO$_2$e/yr)	Reduction in GHG emissions (kgCO$_2$e/yr)	Reduction in GHG emissions (%)
Tenancy	551,680	505,096	46,584	8.4%
Base building	497,706	442,407	55,299	11.1%
Whole building	1,049,355	954,949	94,406	9.0%

Note: Emissions figures include gas consumption but exclude PV electricity exported to the grid. NABERS Energy GHG emission factors for electricity and gas in NSW are 0.94 kgCO$_2$e/kWh and 83.4 kgCO$_2$e/GJ respectively.

Consistent with NABERS methodology, the GHG reduction associated with the PV generated electricity exported to the grid was also not included in the GHG emissions reductions for the case study building. However, the exported PV electricity would be available to other users of the electricity grid and assist very slightly in lowering overall GHG emissions.

ECONOMIC FEASIBILITY

Understanding the economic feasibility of PV systems is extremely important as PV systems must be not only technically viable but also economically viable if they are to be considered by building owners and operators as practical methods of increasing the NABERS Energy rating of a building. The economic feasibility indicators used in the assessment were: simple and equity payback, Net Present Value (NPV) and Internal Rate of Return (IRR).

PV SYSTEM COSTS

A report to the International Energy Agency (IEA) (Watt *et al.*, 2011) indicates that the typical installed (capital) cost for Australian grid connected commercial building PV systems over 10kW in 2010 was between $6,000 and $7,000/kW. This excludes goods and services tax (GST) as well as operation and maintenance costs. A base capital cost of $6,500/kW was used for the economic feasibility modelling of each system. This cost was assumed to include basic support structure, inverters and wiring. Annual operating and maintenance costs of approximately $20/kW were also included in the modelling.

SUBSIDIES, REBATES AND FEED IN TARIFFS

The Australian Federal Government offers a range of subsidies and grants for renewable energy systems. The incentives identified as being applicable to PV systems were: Small-scale Technology Certificates, Solar Credits and the Green Building Fund (Table 4.6.9).

Small-scale Technology Certificates (STCs) are a market-based financial mechanism that can be sold or traded. The price of STCs fluctuates depending upon demand and currently the price varies between $20 and $40 per STC. A range of renewable energy and solar hot water systems, including PV systems under 100kW capacity and with a total output of less than 250MWh per year, are eligible for STCs. The method for calculating the number of STCs for eligible PV systems is based on a set or deemed rate of annual electricity production and the capacity of the PV system.

Alongside STCs, the Federal Government has the Solar Credit (SC) scheme which offers an additional incentive for small-scale PV systems. Businesses as well as households and community groups are able to receive additional STCs for PV systems up to 1.5kW in capacity by multiplying the number of STCs that can be claimed. Systems larger than 1.5 kW are eligible but only receive the multiplier for the first 1.5kW and then receive STCs at the standard rate for the remainder of the system. When the scheme was first introduced, a multiplier of 5 was used but this was scaled back to 3 on 1 July 2011. The scheme phases out completely on 30 June 2013.

The Federal Government offers a range of other subsidies and grants to assist business reduce greenhouse gas emissions. One of them, the Green Building Fund (GBF) provides grants to the owners of existing commercial buildings for projects that reduce a base building's energy consumption. The GBF will fund up to 50 per cent of an eligible project's costs to a total grant of $500,000. The GBF is currently closed to new applications.

There is a range of gross and net feed in tariffs currently in operation across Australia. In January 2010, the New South Wales State Government introduced a gross Feed in Tariff (FiT) of $0.60/kWh for PV systems up to 10kW in capacity. However, due to an overwhelming number of applications, the FiT was reduced to $0.20/kWh in October 2010 for all new applications. The FiT was completely closed to all new applications in April 2011. Currently,

Table 4.6.9 Impact of subsidies and grants on the PV system capital cost ($)

NABERS category	Base capital cost	Capital cost after STC	Capital cost after STC & SC	Capital cost after STC SC & GBF
Tenancy	221,000	206,920	190,320	190,320*
Base building	304,200	284,800	262,920	131,460
Whole building	502,450	470,410	435,890	217,945

Note: *GBF is not available for tenancy upgrade projects.

no government FiT exists for new systems in New South Wales although some electricity retailers are offering net metering for smaller-scale systems. Given that all the case study PV systems are larger than 10 kW, the impact of a FiT was not included in the current economic feasibility modelling.

CURRENT ECONOMIC FEASIBILITY

RETScreen is better equipped than HOMER to perform economic analysis as it offers a range of economic indicators such as payback periods, NPV and IRR. As RETScreen is able to assess the cost of reducing greenhouse (GHG) emissions in $ per tonne of CO_2e, this was also included in the modelling.

The following economic parameters were used as the basis of the economic feasibility modelling:

- inflation rate: 3 per cent/year
- average electricity rate (tenancy): AUD$0.1602/kWh
- average electricity rate (base building): AUD$0.1419/kWh
- average electricity rate (whole building): AUD$0.1340/kWh
- electricity cost increase: 3 per cent/year (as per inflation)
- discount rate: 7 per cent
- project life: 25 years
- GHG emission factor: 1.063 kg CO_2e/kWh of grid supplied electricity.

The results of the modelling for each of the technically feasible PV systems are shown in Tables 4.6.10, 4.6.11 and 4.6.12.

The modelling indicated that none of the systems could be considered economically viable, as all but one of the scenarios modelled failed to deliver a positive NPV or an acceptable IRR. In the case of the one scenario that did deliver a positive NPV, the simple and equity payback periods were greater than 15 and 12 years, respectively. These payback periods could possibly be acceptable to a public sector building owner but were unlikely to

Table 4.6.10 Tenancy PV system economic feasibility modelling

Rebate/Grant	Simple payback (yr)	Equity payback (yr)	NPV ($)	IRR (%)	GHG reduct cost ($/t$CO_2$e)
No rebates	29.9	21.2	−104,137	1.6	167
STC	28	20.2	−90,057	2.1	145
SC + STC	25.8	18.9	−73,457	2.8	118
GBF* (incl. SC + STC)	NA	NA	NA	NA	NA

Note: *GBF is not available for tenancy upgrade projects.

Table 4.6.11 Base building PV system economic feasibility modelling

Rebate/Grant	Simple payback (yr)	Equity payback (yr)	NPV ($)	IRR (%)	GHG reduct cost ($/tCO$_2$e)
No rebates	36.5	24.5	−172,515	0.2	210
STC	34.2	23.4	−153,115	0.7	186
SC + STC	31.6	22.1	−131,235	1.2	160
GBF (incl. SC + STC)	15.8	12.8	225	7	0

Table 4.6.12 Whole building PV system economic feasibility modelling

Rebate/Grant	Simple payback (yr)	Equity payback (yr)	NPV ($)	IRR (%)	GHG reduct cost ($/tCO$_2$e)
No rebates	39.4	>25	−300,984	−0.3	228
STC	36.9	24.7	−268,944	0.1	204
SC + STC	34.2	23.4	−234,424	0.6	178
GBF (incl. SC + STC)	17.1	13.7	−16,479	6.3	12

be acceptable to a commercial building owner. The modelling also indicated that the majority of the systems and scenarios delivered greenhouse gas reduction at high cost.

FUTURE ECONOMIC FEASIBILITY

As the effective service life of PV panels is usually considered to be a minimum of 25 years, it is highly likely that a range of different economic conditions and factors will operate during the lifespan of a PV system. While it is not possible to predict all the different future economic conditions, there are several factors that are likely to impact in the short to medium term. These factors are:

- rising electricity costs;
- introduction of an emission trading scheme (ETS) or carbon tax;
- feed in tariffs (FiT).

RETScreen was used to model the impacts of a range of grid electricity cost increases above the rate of inflation. The increases could be caused by rising supply and infrastructure costs, as well the impact of the introduction of an ETS/carbon tax. In addition, the impact of a range of different FITs was also modelled alone and in combination with the rising electricity costs.

The results of the future economic modelling indicated that an annual 3 per cent electricity price increase above the rate of inflation would cause only a moderate improvement in the economic indicators, but would not be sufficient to make the systems commercially viable. An annual 7 per cent above inflation electricity price increase created positive NPVs and IRRs above 7.0 per cent for all the technically feasible PV systems. However, even at this rate of increase, the equity payback periods for all the systems remained quite considerable at between 11.3 and 12.4 years.

A range of gross FiTs were modelled and these had a more significant impact on the financial viability of the systems, especially when combined with the above inflation electricity price increases. For example, a gross FiT of $0.20/kWh operating for 15 years in combination with 3 per cent annual above inflation electricity price increase reduced the equity payback period to between 8.8 and 10.9 years and IRRs ranging from 8.2 to 11.4 per cent.

ALTERNATIVE STRATEGIES FOR IMPROVING NABERS ENERGY RATINGS

Apart from using renewable energy systems such as PV, there are a number of alternative strategies that could be used to reduce energy consumption and/or decrease greenhouse gas emissions that would increase the NABERS rating of the case study building to 5 stars. These strategies include:

- use of a chilled beams HVAC system;
- inclusion of a gas fired cogeneration/tri-generation plant;
- purchase of Accredited Green Power;
- improved building management.

Of the identified strategies examined in this study, the two most cost-effective and easily implemented were the purchase of Accredited Green Power (AGP) and improved building management. As NABERS Energy assesses a building on its greenhouse gas emissions, the purchase of AGP is permitted under NABERS to increase the NABERS Energy rating of a building. Under the CBD scheme, a building owner must disclose how much AGP is purchased for a building and the building's NABERS rating without AGP. The annual cost of purchasing sufficient AGP to increase the case study building rating to 5 Stars would be approximately AUD$8,800 for the Tenancy space, AUD$7,600 for the Base Building or AUD$16,400 for the Whole Building.

In regard to improved building management, a report released by the Warren Centre (WCAC, 2009) delivered some quite surprising findings pertaining to the NABERS Energy ratings of buildings. The report found that the buildings included in its research survey operated at a 1.3 star higher level where the management of the building was undertaken by an in-house building manager rather than contracted out to a third party. There were other similar findings in regard to energy reporting, staff knowledge and incentives.

The implications of these findings for the case study building are probably more related to the building achieving its 4.5 star rating rather than gaining an extra 0.5 star, but it does show how important the human factor is in maximising building energy efficiency.

Consideration of the use of passive, low energy design techniques such as the use of natural ventilation, redesign of the building envelope, exposure of thermal mass, and redefinition of the air-conditioning design comfort criteria was outside the scope of this study.

CONCLUSION

In this study it was found that it is technically feasible to increase the NABERS Energy rating of an office building using PV panels, and that it is possible to increase the rating for all of the rating categories within NABERS. Approximately 10 per cent reductions in grid electricity consumption and associated GHG emissions are possible using PV panels located on the available roof area. It is not possible to predict whether this could be achieved on other projects with any certainty as each building project will have a unique set of circumstances such as orientation, energy use load profile and suitable roof area. It would be difficult to achieve substantially greater improvements unless there was an extremely large amount of suitable roof space available and the building had a relatively low floor area to roof area ratio.

In regard to the optimal arrangement of PV panels, it was concluded that a PV system with panels facing directly north and having a tilt angle close to 35 degrees would deliver the maximum amount of electricity annually. This is in accordance with the literature. However, achieving this arrangement is not always possible due to various physical and planning constraints.

In regard to the economic feasibility of using PVs to increase the NABERS rating of an office building and decrease grid electricity consumption and GHG emissions within the current energy prices and subsidies regime, the answer for the case study building is no. Even when high levels of available subsidies were applied, the PV systems analysed failed to deliver positive returns within a time frame that would be acceptable to the vast majority of commercial building owners.

In order to make the PV systems for the case study building economically viable, it is necessary for very high levels of subsidies to be provided or the base capital cost of PV to be substantially reduced. One way to achieve this is through a combination of rebates, grants and an uncapped Gross FiT. If this approach were applied across the commercial building sector, it would certainly increase the use of PV systems, but it would also create a great cost burden on governments and electricity users who would have to accept higher electricity bills to cross-subsidise the FiT.

It appears from the modelling of the impact of electricity price rises undertaken for this study that modest increases in prices will not make the PV systems analysed economically feasible. Dramatic and sustained increases in electricity prices would be required to make the PV systems financially viable. Such increases, if applied to the broader society, would certainly sharpen the focus on reducing energy consumption as well as lead to the further development of renewable technologies.

In summary, this research project determined that while it is technically feasible to use PV systems to increase the NABERS rating and decrease GHG emissions of the case study building, it is not currently or likely to be economically feasible to do so for the foreseeable future without very high levels of subsidies or substantial reductions in the cost of PV systems. On this final point, it is hoped that the growth of PV production, particularly in China, as well as the continued development of lower cost PV technologies will continue to drive down the cost of PV systems so these renewable energy systems can be more widely used to reduce GHG emissions across the built environment.

REFERENCES

DECC (NSW Dept of Environment & Climate Change) (2008) *NABERS Energy: Guide to Building Energy Estimation*. Sydney: DECC.

Eiffert, P. and Kiss, G.J. (2000) *Building Integrated Photovoltaic Designs for Commercial and Institutional Structures: A Sourcebook for Architects*. Golden, VA: US Department of Energy: National Renewable Energy Laboratory.

Gong, X. and Kulkarni, M. (2005) 'Design optimization of a large scale rooftop photovoltaic system', *Solar Energy*, 78: 362—74.

NREL (National Renewable Energy Laboratory) (2004) *PV FAQs: What is the Energy Payback for PV?*, Washington, DC: US Department of Energy, Office of Energy Efficiency and Renewable Energy. Available at: www.nrel.gov/docs/fy04osti/35489.pdf (accessed June 2009).

Prasad, D. and Corkish, R. (2006) 'Integrated solar photovoltaics for buildings', *Journal of Green Building*, 1(2): 63–76.

Watt, M., Passey, R. and Johnston, W. (2011) *Australian PV Survey Report 2010: PV in Australia 2010*, prepared for IEA Cooperative Programme on PV Power Systems by Australian PV Association, New South Wales

Watt, M. (2007) *National Survey Report of PV Power Applications in Australia 2006*, Co-Operative Programme on Photovoltaic Power Systems, St. Ursen, Switzerland: International Energy Agency.

WCAC (The Warren Centre for Advanced Engineering) (2009) *Low Energy High Rise: Suite of Initiatives Report*, Sydney: The Warren Centre, University of Sydney.

Yamamoto, J. and Graham, P. (eds) (2009) *Buildings and Climate Change: Summary for Decision Makers*, United Nations Environment Programme: Sustainable Buildings and Climate Initiative, Paris: United Nations.

4.7

BENEFITS AND IMPACTS OF ADJUSTING COOLING SET-POINTS IN BRISBANE

How do office workers respond?

Wendy Miller, Rosemary Kennedy and Susan Loh

It has been accepted industry practice in the design of commercial and institutional buildings in Australia that internal spaces be conditioned to between 21°C and 24°C, using international standards for thermal comfort endorsed by ASHRAE 55-1992 and ISO 7730. While ASHRAE 55-2004 now encompasses an Adaptive Comfort Standard, the building services industry continues to adopt the approach that assumes that 'comfort' is universal, that thermal variation outside the band is undesirable and that occupants of buildings want neutral, dry, still air. The Heating Ventilation and Air Conditioning (HVAC) systems in the buildings which were the subject of this case study, and many similar existing commercial buildings, embody the methodology of universal comfort standards.

A search of the literature regarding climate control for indoor comfort reveals the intricate relationship between occupants' perceptions of thermal comfort and the provision of that comfort via the buildings' HVAC systems. Because of the biological and cultural diversity among human occupants, it is difficult to apply a universal air-conditioning temperature setting that meets all occupants' perceptions of internal thermal comfort

(Brager and de Dear, 2003). These perceptions of thermal comfort are strongly influenced by social norms and cultural influences (Brager and de Dear, 2003; Chappells and Shove, 2005; Peterson and Williams, 2006). Building designers struggle with balancing the occupants' desire for individual local control to improve their satisfaction with indoor thermal comfort levels, and the need to maintain central control of the systems in order to run efficiently (Bordass, 1990, 2001; De Dear, 2004; Leaman and Bordass, 2005). Climatically responsive design integrating architectural and mechanical elements from the outset of the design process, would likely offer building occupants better thermal comfort (Leaman and Bordass, 2005).

The aim of the research was to investigate the social, environmental, economic and human comfort implications associated with adjusting commercial office air-conditioning temperatures to levels which aligned more closely to the relatively benign subtropical climate of Brisbane. The purpose was to qualify whether occupants of commercial buildings in a warm, humid, sub-tropical location would tolerate changes to the generally accepted industry standards for thermal comfort in order to reduce energy use associated with HVAC systems. The project methodology was simply to implement changes to the physical environment through increasing the air-conditioned temperature set-points to 25°C in two buildings for the four summer months, December 2006 to March 2007, and to collect the perceived physiological and psychological responses to these physical changes in order to ascertain whether such a change was acceptable. Data was also gathered regarding any economic and environmental impacts in order to ascertain whether such a change was viable.

This chapter describes how office workers modified their attitudes towards thermal comfort and what behavioural adjustments they made. The savings in energy usage made as a consequence of the altered set-point provides evidence to support the notion that a 1° or 2° C change in the set-point reduces energy consumption significantly (1°C warmer may reduce energy used by 10 per cent). Further benefits accruing to the owners of existing buildings using this approach are demonstrable savings in greenhouse gas emissions, water and significant cost savings.

CASE STUDY: METHOD

The research was undertaken as a pilot project and predominantly involved staff and buildings of the Faculty of Built Environment and Engineering (BEE) at Queensland University of Technology (QUT) in Brisbane, Australia, during summer 2006–07. Brisbane (latitude 27.5°S) has a subtropical climate characterised by hot, humid summers with mild, dry winters. QUT operates three campuses with over 3,000 staff and 38,000 students (2007 figures). The main campus at Gardens Point in Brisbane's central business district (CBD) is characterised by buildings dating from early twentieth century to early twenty-first century. The buildings on this campus offer a wide range of fully conditioned, partially conditioned and non-conditioned spaces, with a trend towards full conditioning of all spaces as building infrastructure and finances allow. Two buildings, utilising the same HVAC chiller plant, were selected for alteration of the cooling set-point (Table 4.7.1). Four control buildings were also selected.

INTERNAL AND EXTERNAL TEMPERATURES

Brisbane's long-term climate data, from the Bureau of Meteorology (BOM), was compared with daily weather data for the period December 2006 through to March 2007. Temperature and humidity sensors were placed in five 'problematic' offices, selected to provide quantitative data to compare with building management system (BMS) performance parameters and occupants' perceptions. The sensors, in place for between 28 and 56 days, were set to record temperature and relative humidity every ten minutes. The mean, maximum and minimum were analysed for 9am and 3pm to enable comparison with BOM data. The cooling set-point for buildings A and D was raised from 23°C to 25°C on 11 December 2006, one week after notifying staff that the thermostat would be adjusted. Building D remained on this set-point until the first week in April. No other HVAC parameters were adjusted. Building A was changed to 24°C on 24 January 2007 due to a high number of staff complaints (formal and informal). The number of HVAC-related

Table 4.7.1 Case study building specifications

Building parameters	Building A	Building D
Construction date	Circa 1919	1999
Total floor area	2197m^2	6205m^2
Useable levels	3	5
General building condition	Fair (average condition, services functional but require attention; backlog maintenance work exists)	Fair–Good (superficial wear and tear and minor defects)
AC services / Functionality assessment	Fully conditioned (retrofitted) Functionality: barely adequate	Fully conditioned Functionality: barely adequate
Type/Function	Offices/administration 84% Lecture/seminar rooms 16%	Offices/administration 80% Lecture rooms 17% Computer labs 3%

Source: QUT's (2006) Condition Audit of Built Assets.

complaints received by the Facilities Management call centre during the study period was compared with the previous summer.

OCCUPANTS' SURVEY AND FOCUS GROUP

Occupants' physiological and psychological responses to the indoor environment were monitored through online surveys which were sent fortnightly during the study period to all BEE staff in the two affected buildings and the four control buildings. Some 106 staff participated, with a total of 273 responses. Further qualitative information was gathered through staff emails to project members and through a one-hour focus group at the conclusion of the study period. Responses were analysed demographically (age, gender, building, work location within building, time of completion of survey) and statistically. In preparation for future research, the survey also included questions that enabled initial exploration of the relationship between levels of comfort and the nature of recent activity levels, clothing levels and mode of transport to work. Respondents were also surveyed about the extent of air-conditioning use in their homes and cars in order to ascertain whether this had any bearing on their responses to the indoor climate at work.

PERFORMANCE OUTCOMES MODELLING

Because neither building had independent energy meters by which to verify actual savings attributable to the project, a firm of consultant engineers was contracted by QUT Facilities Management (FM) to undertake a series of energy simulations for three common occupancy types (lecture theatre, computer laboratory and office) and temperature set-points, to determine energy, water and greenhouse gas emissions from each variation. As-built drawings of the buildings were provided to the consultants who also undertook a site visit to confirm the locations of relevant shading and glazing. Table 4.7.2 summarises the parameters of each occupancy type. The Building Code of Australia 2006 Class 5 schedules were assumed for the occupancy hours of the office and lecture theatre while it was assumed the computer laboratory would be utilised 24 hours/day, 7 days/week. Electricity loads for each of the spaces were nominated by the consultant to reflect typical university usage. The consultants used the Beaver program to perform the modelling.

Each space was assumed to have a single air-conditioning unit with its own appropriately sized chiller unit (with a cooling efficiency (COP) of 5) to handle the space's expected cooling needs. Weather data and public holidays for Brisbane were taken into account. Each occupancy type was modelled on three different summer/winter set-points: current (23°C/21°C); summer

Table 4.7.2 Modelling parameters for selected occupancy types

Space type	Occupancy hours	Size	Proportions of total electricity use (%)		
			Lighting	Equipment	HVAC
Lecture theatre	7am–6pm, 5 days/week	296m^2	26	6.6	67.5
Office	7am–6pm, 5 days/week	329m^2	25.2	21.2	52.6
Computer laboratory	24 hours, 7 days/week	329m^2	16.6	33.1	50.3

efficiency (25°C /21°C); and winter efficiency (23°C /20°C). The results of this modelling were applied to the usage areas of the case study buildings.

RESULTS

With the exception of higher than normal temperatures for March, the summer of 2006–07 did not present any weather extremes that would need to be taken into account for the purposes of this project.

Thermal comfort

In explorations of differences in comfort levels, chi-square tests and an ANOVA demonstrated that:

- There was no association between comfort levels of respondents and the buildings (all six) in which they were located. This suggests that manipulating the thermostat settings in two of the buildings did not significantly affect participants' perceptions of comfort.
- 39 per cent of respondents found the thermal environment of their work space to be unacceptable.
- 49 per cent of respondents wanted more air movement.

There was no significant association between general comfort levels and age or gender; general comfort levels of respondents and the use of air conditioning in the car or home; general comfort levels and clothing levels worn.

In comparison with call centre-logged complaints for the previous summer, the total number of complaints was slightly lower for Building A and slightly higher for Building D. The thermal comfort complaints, for both summers, related to the same rooms, which were described by occupants as 'too hot', 'too cold', 'too stuffy' or 'no ventilation'. Analysis of the temperature data recorded for four of these workspaces in Building A showed the mean maximum temperature for each of the offices was outside of the operating parameters of the HVAC system (25°C) by 1.3–6.9°C. Close examination of the Building Management System (BMS) by Facilities Management in January 2007 revealed that raising the set-point (to 25°C) had unmasked pre-existing sensor calibration errors and control algorithm errors in Building A. The AC system was subsequently re-commissioned, allowing the AC system to perform to its design parameters, and there was a subsequent drop in the number of complaints from occupants.

QUANTITATIVE IMPACTS

Modelled savings, derived from the consulting engineers' report, are presented in Table 4.7.3. For all occupancy types, raising the summer thermostat setting by 2°C would result in savings in end use energy, associated electricity costs, primary energy, greenhouse gas emissions, and water use. If the altered set-points were adopted in the design of new HVAC systems for these spaces, capital savings would also be realised due to the smaller chiller plant size required to meet the new performance requirements.

DISCUSSION

Economic and environmental benefits of technical changes

The modelled data reinforced the expectations of the project proponents that increasing the set-point by 2°Celsius would result in decreased electricity

Table 4.7.3 Indicative savings for different occupancy types attributable to increased cooling set-point

Room type and HVAC thermostat cooling set-point	End use energy MWh/y	Primary energy MWh/y	CO_2 emissions Tonnes/y	Water usage L/day	Water usage L/y	Electricity costs $/y*	Chiller plant capacity
Lecture room 23°C	32.51	101.58	34.13	33	8125	2,600	76
Lecture room 25°C	30.55	95.47	32.08	29	7150	2,444	67
Office 23°C	35.86	112.05	37.65	20	4908	2,868	46
Office 25°C	33.39	104.33	35.05	18	4485	2,671	42
Computer lab 23°C	133.50	417.19	140.18	50	17892	10,680	64
Computer lab 25°C	111.52	348.51	117.10	47	16529	8,922	59

Note: * For the purposes of this study, energy consumed was costed at 7.6c/kWhr, representing a long-term bulk purchase agreement price. It is not reflective of current prices.

consumption, resulting in annual savings of 78 tonnes of carbon emissions and 14 Kl water, for these two buildings alone. The demonstrated emissions and cost reductions achievable would be quite significant if extrapolated across all buildings on all campuses of the university. Additional capital and operational savings could be realised through the 8–12 per cent reduction in chiller plant capacity and through the reduction in demand charges due to a reduced peak demand (calculated at a 70Kw peak demand reduction from office occupancy in the two buildings). While these outcomes could have significant financial and environmental benefits for the university, they also impact on regional electricity generation, transmission and distribution. The outcomes of this study suggest that both significant environmental benefits (reduced carbon emissions due to reduced demand) and economic benefits (reduced need for infrastructure investment to meet peak and total demand) accrue from management of air conditioning by end users.

OCCUPANT KNOWLEDGE OF THERMAL CONTROLS

Analysis of the type of complaints made by workers about thermal comfort revealed that people were not always aware of how the AC in their areas operated, for example, whether the AC was triggered by a manual switch, a sensor or a BMS timer. There was little perception of how long it might take some spaces to respond to AC operational commands. AC operational and response times were particularly important for staff arriving early in the morning expecting immediate thermal relief (depending on their commuting habits and the prevailing outdoor temperature and humidity). This would suggest that there is a disconnect between the knowledge, understanding and expectations of building occupants, and the design and operational assumptions and practices of HVAC systems designers and facilities managers.

SOCIAL IMPACTS

The pilot study highlighted the salience of managing the social impacts of instigating energy-saving measures such as increasing the HVAC thermostat setting during the summer months. The social research and feedback components of the project enabled the change to be managed effectively. Hence, it is assumed that an effective change management strategy is as important for success as the technical FM dimension. Several conclusions can be drawn from the social dimensions of this research:

- The 'comfort' survey administered to occupants acted as an effective change management tool that provided occupants with a valuable avenue for feedback. It also enabled the project team to monitor general levels of comfort and identify areas where intervention may be required. As a research tool, the small sample size and self-select sampling method mean that the social survey results cannot be extrapolated beyond this sample.

- The project itself acted as an effective means of awareness-raising and resulted in occupants identifying specific sources of discomfort and, in some instances, taking steps to address the problem. It also highlighted the value of consultation with occupants prior to making changes that affect their thermal environment.

- The commissioning and operation of a BMS must involve the occupants and some measure of whether the aim of occupant comfort is being achieved (as opposed to whether the HVAC system is performing to its engineering design parameters). This includes implementation of a FM process whereby occupants' complaints are accepted as legitimate at face value, and holistic investigations are conducted to identify the source of the problem. Constant cooperative communication between FM and occupants is essential in achieving the desired indoor comfort and the proper use of the building functions.

- Occupants were highly amenable to people in management positions taking a leadership role in encouraging a corporate dress code that responds appropriately to local climate.

- The project appeared to act as a catalyst for staff to express their ideas of additional programmes that could be implemented at the university to reduce its greenhouse gas emissions. These unsolicited suggestions encompassed areas such as a sustainability audit of the university's policies and processes, greening the vehicle fleet and introducing a carbon offset programme. This would seem to suggest that a change management process needs to consider how to encourage and utilise the momentum gained from an initial environmental improvement programme (for example, energy and greenhouse reduction through HVAC controls) to implement a culture and practice of continual improvement in environmental performance in other areas.

FURTHER OPPORTUNITIES FOR ENERGY SAVINGS

While air conditioning accounted for 50–67 per cent of the energy use of each occupancy type, the economic and environmental benefits could be magnified through addressing the lighting and equipment energy use. The enhanced awareness of at least some of the participating staff led to involvement in other energy-saving behaviours, such as switching off lights and computers when not in use, and closing doors to non-conditioned spaces. A well-orchestrated environmental programme could capitalise on the enhanced awareness generated through one activity to encourage sustainable behaviours in other areas. Corporate procurement practices also play a critical role in energy reductions: energy efficiency should be a core criterion for all purchasing decisions (in addition to other environmental considerations such as toxicity and life-cycle costs).

The use of energy-efficient appliances and fittings in turn has positive benefits for the internal thermal comfort, contributing significantly less heat gain that would otherwise need to be managed through either HVAC systems or natural ventilation strategies. A dramatic reduction in internal heat loads from building equipment would in turn reduce the operational costs and physical stresses on air-conditioning plant (for example, in dealing with hot spots created in rooms with a large number of computers) and could potentially affect the size of HVAC plant required in the first instance (thereby providing capital cost savings).

CONCLUSION

Air conditioning accounts for a significant proportion of commercial buildings' energy consumption. Energy and greenhouse gas reduction strategies in existing commercial buildings have primarily focused on physical engineering solutions, such as fine tuning or upgrading existing air-conditioning units, to meet existing engineering practices that define thermal comfort. The pilot project conducted on the QUT campus aimed at qualifying whether building occupiers would tolerate changes in the generally accepted industry standards for office temperature.

The four-month pilot project, which increased the air-conditioning temperature set-points by 2°C during a sub-tropical summer, revealed that accepted industry norms regarding system design for indoor climate in the workplace need to be revisited in order to accomplish a global goal of energy-efficient, low carbon buildings. Occupants' responses to the adjustment of the HVAC thermostat cooling set-points in this case study showed that building occupants can be meaningfully engaged in a change management process that delivers financial and environmental savings as well as occupant comfort.

This case study found that the usual approach to providing comfort via the standard engineering process can be challenged by responding to cultural and social issues in building and HVAC systems design.

REFERENCES

Bordass, B. (1990) 'The balance between central and local control systems', paper presented at the Environmental Quality 90, British Gas HQ, Solihull, England.

Bordass, B. (2001) 'Flying blind: everything you wanted to know about energy in commercial buildings but were afraid to ask', *Energy Efficiency Advice Services to Oxfordshire*. Available at: www.usablebuildings.co.uk (accessed 6 June 2007).

Brager, G. and de Dear, R. (2003) 'Historical and cultural influences on comfort expectations', in R. Cole and R. Lorch (eds) *Buildings, Culture & Environment: Informing Local & Global Practices*, Oxford: Blackwell, pp. 177–201.

Chappells, H. and Shove, E. (2005) 'Debating the future of comfort: environmental sustainability, energy consumption and the indoor environment', *Building Research & Information*, 33(1): 32–40.

De Dear, R. (2004) 'Thermal comfort in practice', *Indoor Air*, 14(Supplement 7): 32–9.

Leaman, A. and Bordass, B. (2005) 'Productivity in buildings: the "killer" variables', *Ecolibrium by AIRAH, Australian Institute of Refrigeration Air Conditioning and Heating*.

Peterson, E. and Williams, N. (2006) 'New air conditioning design temperatures for Queensland, Australia', *Ecolibrium by AIRAH, Australian Institute of Refrigeration Air Conditioning and Heating*.

4.8
DELIVERING ENERGY-EFFICIENT BUILDINGS

The Low Energy High Rise (LEHR) Project

Alexandra McKenna

INTRODUCTION

The Low Energy High Rise (LEHR) Project commenced in 2005 seeking to identify if high-rise buildings could offer significant energy efficiency opportunities. A number of studies indicated that significant energy savings could be made but it was apparent that barriers were preventing the uptake of energy efficiency initiatives. Further, the National Built Environment Rating System (NABERS) indicated a significant spread of building performance from zero to five stars which did not appear to correlate across the same building types. The LEHR project involved a number of stages.

STAGE 1: BACKGROUND RESEARCH

The first stage of LEHR included industry work groups and an empirical survey to identify potential action items – with a strong but not exclusive focus on non-technical actions – that can improve energy efficiency in commercial office towers with a literature review which indicated the barriers to energy efficiency uptake warranted further investigation. A comprehensive survey was conducted across all levels of the building management spectrum to assess attitudes to energy efficiency initiatives. All responses were assessed against NABERS Energy and Water ratings for the base building to determine issues of statistical significance. In a world-first result, the study has been able to demonstrate clear statistical evidence that certain factors really do correlate with efficient outcomes. Key factors that correlate with improved performance include:

- new buildings, and buildings with economy cycles;
- regular incremental efficiency upgrades and the elimination of older and unserviceable technologies;
- reporting NABERS performance to tenants and the public, 'ownership' of efficiency by operators and maintenance staff/contractors, strong management leadership with regard to efficiency, common objectives and agendas throughout the management chain and/or retention of efficiency savings in budgets;
- training in energy efficiency and building managers who are not overly conservative with respect to energy efficiency technologies.

STAGE 2: TARGET AREAS

Stage 2 of the project identified key 'Target Areas' related to the findings in Stage 1 and developed materials that would guide building operators towards greater energy efficiency.

The statistical study methodology

At the heart of this study lies a large and complex survey methodology which was composed of three separate surveys:

- a base building survey, covering the technologies and management of the building;
- a tenant survey, covering the interactions between the base building and the tenant;
- a managers survey, covering the knowledge, attitudes, authorities and responsibilities of the building, property and asset managers associated with the building.

Well over a hundred questions were covered within these surveys, which were distributed to 189 buildings, 188 tenancies and 296 managers. Satisfactory responses were received from 127 buildings, 102 tenancies and 173 managers. However, the need to cross-correlate base building data with tenancy and manager data meant that the cross-survey analyses were based on 67 base building and tenancy sets, 93 base building and manager sets and 53 base building, tenancy and manager sets.

The hypothesis testing processes were based on statistical testing of the following types of proposition:

- Do the answers to an individual survey question correlate with the NABERS Energy or Water rating?
- Do buildings with NABERS Energy or Water ratings of 3 or above have statistically significant different responses to an individual question than those with ratings below 3 stars?
- Do the answers to logically related aggregates of individual survey questions correlate with the NABERS Energy or Water rating?
- Do buildings with NABERS Energy or Water ratings of 3 or above have statistically significant different responses to logically related aggregates of individual survey questions than those with ratings below 3 stars?

In addition, some information was derived directly from managers' responses to what they considered to be barriers or facilitators for efficiency decisions.

The data sample

The resultant data sample was tested for a number of factors to establish to what extent biases existed. The following factors were noted:

- The sample was biased to the 'top end of town', being premium, A and B grade buildings, with lower grades being largely absent from the sample.
- The state/city distribution was a reasonable reflection of the national distribution of office buildings.
- The sample had an average NABERS Energy Base Building performance of 2.87 stars and a median performance of 3.25 stars, both marginally higher than the population average.
- The sample has an average NABERS Energy Whole Building performance of 2.96 stars and a median performance of 2.91 stars. Again both figures are marginally better than the population average.

These factors indicate that the statistical results can only be considered appropriate for the upper end of the market. However, in many cases it is reasonable to postulate extrapolations to the broader market and indeed to some other building types.

Basic results

From the hundreds of tests and hypotheses undertaken, the following results were identified (Table 4.8.1).

These results provide an insight into factors affecting building performance but are of limited assistance in informing actions and decisions relating to efficiency. As a result, a range of action-oriented findings have

Table 4.8.1 Statistically significant relationships identified

Measure	NABERS Energy impact	Measure summary
Economy cycle	0.6 stars	Buildings with Economy cycles outperform those without
Building technology	1.4 stars	Buildings with current good practice façade and services technology perform better
Management	1.3 stars	Buildings where management is at least partially in-sourced perform better
	0.9 stars	Buildings where building, asset and portfolio manager all feel able to affect efficiency perform better
Disclosure	0.5 stars	Buildings that disclose their NABERS performance to tenants perform better
Incentives and penalties	0.4 stars	Buildings that provide efficiency penalties/incentives to maintenance contractors perform better.
Training and skills	0.5 stars	Buildings where there is an efficiency training programme perform better
	1.3 stars	Buildings where the manager reports a higher level of energy efficiency knowledge perform better
Incremental improvement	0.6 stars	Buildings where incremental investments have been made in efficiency perform better than those where no such investment has occurred

Note: Due to cross-correlation between factors, the individual results are not additive.

been developed based on these key findings combined with some of the other results from the study and a degree of extrapolation. These action-oriented findings have been developed in the second stage of the project through a Low Energy High Rise Handbook.

Low Energy High Rise Handbook

Stage 2 commenced with a short list of potential energy efficiency-related actions grouped into related areas or 'Target Areas' based on the empirical findings in Stage 1. Each target area outlined a short overview and a description of the potential outputs that could be developed from the trial process. The potential outputs were numerous and it was expected they would be rationalised and consolidated to more specific strategies through the Stage 2 process. These target areas were initially presented at workshops held in Brisbane, Sydney and Melbourne with industry professions representing all areas of the building operation process. Groups at these events worked together on a target area further refining the considerations and nominated themselves for specific target areas of interest. Target Area Working Groups were established from these lists under a Working Group Leader from the Gold Sponsor group.

The target areas

Target areas identified in Stage 1 of the LEHR project fell into five key areas of building management: Management, Monitoring and Reporting, Tenants, Technical and Training.

All target areas were presented in the same format:

1 How focusing on the target area can assist in delivering greater efficiency.
2 What the research tells us, either by referencing the empirical evidence identified in Stage 1 or, where appropriate, the anecdotal indications.
3 Strategies on managing the target area to deliver greater efficiency.

Each target area had a working group established for the purposes of developing materials. These materials or ideas were communicated to five implementation sites or test buildings for relevance and benefit throughout the development process.

Interestingly, in working with the implementation sites it was evident a number of simple strategies had been overlooked. For example, discussing with one site how to incorporate energy efficiency in managing the tenant fit-out process, it was revealed they did not have a tenant fit-out manual and sought advice on how to construct one. Similarly, in identifying the building specifications, such as load, capacity and air-conditioning pro-vision, a number of building operators were not collating this information in a single process. Further, this information was not fully integrated into the leasing process, potentially creating a risk that lease offerings could be made in excess of the building capabilities. These discoveries led to the incor-poration of a far broader range of guidance materials than what was initially envisaged for a number of the target areas. Further investigation may determine whether this was unique to the project or whether it indicates a systemic problem in the built environment of overlooking the critical first steps in a process.

The LEHR process sought to avoid duplicating information already available so where relevant, reference is made to other sources. For exam-ple, where AIRAH (Australian Institute of Refrigeration, Air-conditioning and Heating) had training courses relating to energy efficiency, links to these are provided, with the LEHR materials remaining focused on how to establish a training strategy. The elements incorporated in each target area are outlined as follows.

Management

Stage 1 demonstrated that organisations where all parties in the building management chain are engaged perform better. Further, those organisations which had internalised management structures and which were encouraged to explore less proven technology also performed better. The management materials provide guidance to encourage:

- organisation-wide engagement;
- energy efficiency policy development;
- strategic planning;
- benchmarking assets and developing improvement roadmaps;
- engaging with experts;
- developing a strong implementation team;
- establishing a monitoring and reporting strategy;
- developing Key Performance Indicators;
- establishing a policy to disclose performance.

Monitoring and reporting

Stage 1 demonstrated buildings with better technology performed better although there was limited empirical evidence to support the suspected importance of good monitoring systems and public reporting.

The monitoring and reporting materials were developed in a sequential manner to assist building operators who did not have any benchmarking data on their buildings' performance through to guidance on how to establish a comprehensive sub-metering and monitoring system. The monitoring and reporting materials covered:

- identifying and documenting utility meters;
- establishing a building log book;
- benchmarking performance using utility bills;
- establishing a baseline;
- establishing a monitoring strategy;
- designing and implementing a sub-metering system;
- understanding building loads;
- understanding consumption profiles;
- commissioning a sub-metering system;
- engaging with tenants;
- reporting efficiency performance publicly.

Tenants

The Tenant Target Area was initially focused on green leases and management of the fit-out process, however, green leases have evolved considerably over the past 12 months. As a result, this target area focused more on the management of the fit-out process. Although there was no empirical evidence to support this, anecdotal evidence suggested that failure to manage the tenant fit-out process and ensure compliance with the building's service capability presented a major risk to energy efficiency. The working group quickly established a consensus view that materials to help guide the tenant fit-out process would not only be very effective but were often not available. The tenant materials covered the following considerations:

- establishing the building's capability and service constraints by developing a building services summary;
- developing a tenant fit-out manual;
- incorporating energy efficiency considerations in the tenant fitout manual.

Technical

The Technical Target Area was developed to assist building operators in how to manage the technical aspects of a building through a simple three-step approach. This process was incorporated into a series of checklists which covered the three key technical areas of a building which are critical in managing energy efficiency:

- basic energy efficiency checklist;
- basic control functionality checklist;
- building management system graphics screens.

Training

Stage 1 established a strong correlation between organisations which had an energy efficiency training programme in place or managers who felt they understood energy efficiency and a higher NABERS Energy rating. The Training Target Area provides context around establishing training requirements rather than specific direction as to what training should be undertaken through the following:

- establishing the level of general building services knowledge;
- identifying the understanding of energy management;
- identifying potential courses available.

FUTURE INITIATIVES

The Low Energy High Rise methodology is not specific to high-rise buildings. In fact, the materials make almost no mention of high-rise buildings and the next stage for the project will be to refine the materials into an approach that is applicable for all building types. This will be developed and delivered through a web-based set of materials to maintain an efficient process.

To date, the Low Energy High Rise results and methodology have been incorporated into a Green Star Operating Green Buildings course, indicating a convergence in the language and approach required to improve energy efficiency. Further, an MOU between the Green Building Council and NABERS has established a clearer language around the performance of efficient buildings which is expected to evolve through the Green Star Performance Tool. This is an exciting development that will assist building operators in distilling a more unified message about what makes a green building and how to deliver greater operational efficiency.

CONCLUSION

The Low Energy High Rise project has conclusively demonstrated that building operational factors are critical in delivering energy efficiency, translating into a 30 per cent improvement in sector performance which in turn translates into a 1.2 per cent reduction in Australia's national greenhouse emissions. Further delivery and testing of the Low Energy High Rise materials will refine the approach, lead to the development of practical case studies and significantly improve building operational efficiency. This may also lead to greater recognition of the value of building operators.

4.9

REFURBISHMENT FOR CARBON REDUCTION AND OCCUPANT COMFORT: INSIGHTS FROM THE POST-OCCUPANCY EVALUATION OF THREE OFFICE BUILDINGS

What is the feedback from occupants in developing the case for retrofitting?

Leena Thomas

INTRODUCTION

As with most developed economies, existing buildings comprise 98 per cent of Australia's building stock and account for the bulk of greenhouse gas emissions from this sector. It has been argued elsewhere (Thomas, 2010) that the questions surrounding the refurbishment of existing buildings are critical in determining 'how to address the issue of the ageing building stock and seize the opportunities presented to future-proof these buildings in the context of climate change'.

This chapter elicits lessons for future redevelopment of existing buildings that can be gleaned from post-occupancy studies of three office refurbishment projects in Australia. The three buildings selected for discussion in this chapter have been noted for their attention to environmentally sustainable design and represent different scales at which refurbishment can occur within existing buildings. These buildings range from a single floor refurbishment for a tenancy within a 16-storey building (Green Building Council of Australia head office or GreenHouse) in the Sydney CBD; a major refurbishment across eight floors for Stockland Pty Ltd (Stockhome) in Sydney; and a total building refurbishment at 40 Albert Street (head office for Szencorp) in Melbourne on the CBD fringe.

The varied scale of these projects provides insights as to appropriate design strategies and bioclimatic approaches that were implemented given the opportunities and constraints for each project. The post-occupancy studies provide useful information to close the loop on performance by evaluating consequent outcomes for energy performance, carbon reduction and occupant satisfaction in the study buildings.

Through these studies, this chapter argues for a strategic approach to building refurbishment that recognises the nexus between indoor environmental quality, building performance and energy efficiency and occupant satisfaction and productivity.

STUDY APPROACH

The three projects were evaluated using a post-occupancy evaluation (POE) approach, which is widely accepted for its ability to provide vital feedback regarding a building's performance in use. Understanding building performance is vital, given that the indoor environmental quality of a building affects occupants' well-being and productivity (Wyon, 2004; Vischer, 2007; Leaman et al., 2007) as well as consequent impacts on energy performance.

The study approach included a review of project information, site visits, interviews with key stakeholders (architects, owner developer, environmental consultants, building managers) and a survey of building occupants using the Building Use Studies (BUS) Workplace Questionnaire. The

BUS survey has been used to evaluate more than 95 buildings in Australia and over 500 buildings in over 17 countries internationally including the seminal PROBE studies between 1995 and 2002.[1] For the three buildings discussed here, a pre and post occupancy approach was adopted where occupants were first surveyed in their previous accommodation before relocating to their refurbished location.

This approach allowed for evaluation of the individual buildings against the BUS Australian and/or International benchmarks and with respect to experience in the previous accommodation. In contrast to estimations for thermal comfort that may be obtained by applying calculation methods such as PMV (Predicted Mean Vote, Fanger, 1970) to monitor temperature data, the survey data and open-ended comments provide a direct and rich description of users' experience and assessment of temperature, air, noise, lighting and comfort overall. Occupant feedback is sought for over 63 variables covering environmental comfort, building operation, user control, design as well as perceived productivity and perceived health. The approach of assessing perceived productivity and health in the BUS survey has been argued to provide a consistent indicator for comparison across buildings while overcoming problems of wide variance associated with context-specific measures such as 'sick days off' or 'efficiency of keyboard strokes' (Leaman and Bordass, 1999).

This chapter refers to the Australian benchmark dataset, which comprises other Australian buildings that have been surveyed using the BUS methodology. In addition to office buildings, the dataset includes workplaces in a range of educational and institutional buildings. It must be noted that the datasets represent buildings and tenancies that would perform better than the norm (Leaman *et al.*, 2007). Approximately 50 per cent of the datasets comprise recently completed buildings designed with a green or sustainable design intent, and several of the occupant studies in conventional buildings are also prompted by owners and designers, partly to gauge the impact of positive interventions.

In this chapter, only a summary of the workplace survey results are provided. Figure 4.9.3, Figure 4.9.10 and Figure 4.9.17 show the overall performance based on 12 summary variables covering environmental comfort and workplace satisfaction. The mean value from the survey responses for each variable is compared with the scale midpoint and mean value of the buildings in the Australian benchmark dataset together with its upper and lower 95 per cent confidence intervals, and the scale midpoint. This creates the criteria for the variables as follows:

- Squares represent mean values significantly better or higher than both benchmark and scale midpoint (a good score).
- Circles represent mean values that are not significantly different from benchmark and scale midpoint (a typical score).
- Diamonds represent mean values significantly worse or lower than benchmark and scale midpoint (a poor score).

The benchmarks and their upper and lower 95 per cent confidence interval are represented by three short lines above the slider scale of each variable. All variables use a 1 to 7 scale with the exception of perceived productivity, which is assessed on a 9-point scale.

Two overall indices are also calculated: a comfort index, based on the overall comfort, lighting, noise, temperature, and air quality scores; and a satisfaction index based on the design, needs, health, and productivity scores. The average of these is termed the summary index[2] for the building and provides a way of characterising the overall performance of the building.

All the three projects as designed and completed have been rated under the Green Star[3] rating system which is the prevailing Australian tool for evaluating environmental design and construction of buildings. Further, the operational energy performance and consequent CO_2 equivalent emissions reported here are based on the official ratings achieved under the NABERS[4] protocol drawing on actual utility bill data for one year.

The key design interventions and outcomes for occupant evaluation and energy performance for each building are discussed below.

CASE STUDY 1: A SINGLE FLOOR TENANCY IN AN EXISTING BUILDING

The GreenHouse
Client: Green Building Council of Australia
Fit-Out Design: Bligh Voller Nield (BVN Architecture)
Environmental design: Lincolne Scott
Year of completion: 2009
Certified ESD Rating: 5 Star Green Star – Office Interiors v1.1 (2009)
Certified NABERS Rating: 5 Star Energy Rating – Tenancy

Design strategies

The refurbishment covered 804 m² of net lettable area on the 15th floor. The floor plate had the advantage of providing a column-free space of approximately 26m x 42m with a central core, and was designed to accommodate 55 workstations with a range of meeting rooms and other office amenities. The refurbishment occurred within the constraints of an existing shell with no possibility for changes to the external façade via window to wall ratios, shading devices or the like. As evident in the plan (Figure 4.9.1), the predominant glazed façades are oriented to the east and west. The extensive floor-to-ceiling glazing to the east retains unobstructed views to Hyde Park, while the west-facing windows (window-to-wall ratio approximately 62 per cent) are partially shaded by buildings opposite.

The design for the refurbishment integrated ambitions for minimising material by doing away with a traditional suspended ceiling, and reusing and recycling furniture and other equipment (Figure 4.9.2). The layout

Figure 4.9.1
Floor plan of
GreenHouse at
Elizabeth Street

Source: BVN
Architecture.

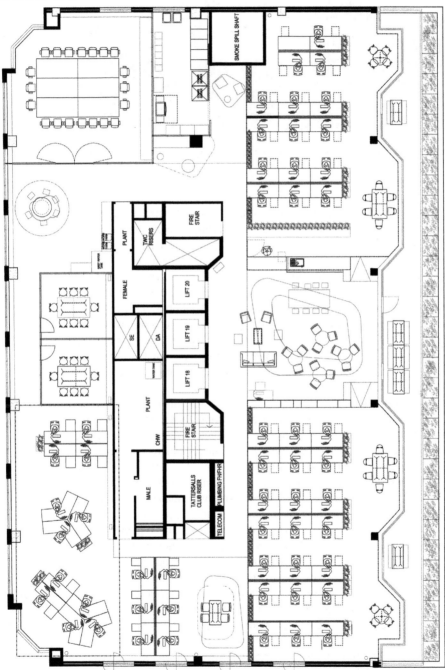

Figure 4.9.2 Raised floor to workstations, light shelves on east-facing glazing, exposed ceiling, and recycled and repurposed furniture at GreenHouse

Source: BVN Architecture.

was organised so that 90 per cent of workstations were located no more than 8m from the glazing and had access to views. A key design intervention for the space was the introduction of light shelves on the inner face of the window bays along the east designed to redistribute daylight further into the space coupled with an integrated electric lighting system. The electric lighting for the main workstation areas comprised a direct and indirect ambient lighting (70/30) system with supplementary task lighting and provision for user control at the individual workstations. In the absence of external shading devices, fully automated internal blinds with less than 5 per cent visible light transmittance are installed on all windows to control glare and limit heat gain. The operation of the blinds has been set on a timer with provision for manual override.

In a departure from many tenancy refurbishments that are limited to changes in layout, lighting and office equipment, this tenancy was also redesigned to incorporate a displacement ventilation system via floor vents in a raised floor east of the core where the bulk of the workstations are located and side wall diffusers for remaining areas. The floor vents enable individual control, where occupants can rotate them to direct the flow of air either towards or away from them. Displacement ventilation systems are noted for their potential to reduce cross-contamination with return air, allowing better air distribution at slow velocity. They enable a higher supply air temperature around 18°C which increases opportunities for free cooling when compared to the lower temperatures (13–14°C) required for conventional overhead systems. Outside air is provided at a 50 per cent improvement over AS 1668 requirements and additional moves to improve indoor environmental quality extended to careful selection of material for low toxicity and VOC content. Other aspects of interior fit-out and ESD initiatives for water, waste and materials can be found in the GBCA case study and certified Green Star Assessment documentation available on the organisation website (GBCA, 2009).

Occupant evaluation

Before moving into the current premises in 2008, Green Building Council of Australia's (GBCA) offices were located at Pitt Street, Sydney. A pre-occupancy evaluation of occupant experience in the Pitt Street premises was undertaken prior to relocation. That study indicated that employees were hindered in their ability to work due to unwanted interruptions and noise, lack of space, poor storage, lack of meeting rooms, stuffy and hot conditions in summer and an inability to control their work environment. The ratings placed the building in the bottom 10 per cent of buildings in the Australian bench-mark dataset.

The post-occupancy study at the GreenHouse was conducted after 18 months of occupancy in the new premises. Some 28 staff participated in the BUS post-occupancy survey, resulting in a response rate of 88 per cent. As seen in the summary chart shown in Figure 4.9.3, occupants at the GreenHouse rated the tenancy significantly better than the scale midpoint and benchmarks for 10 of the 12 key variables evaluating comfort and satisfaction.

Occupants at GreenHouse rated their overall comfort highly, with just 6 per cent dissatisfaction and a mean score in the top 10 per cent of the dataset. Significantly they also rated an increase in perceived health (3 per cent dissatisfaction) and perceived productivity. The mean rating for per-ceived productivity was +14 per cent. In addition, open-ended comments revealed that a number of occupants rated natural light and access to views

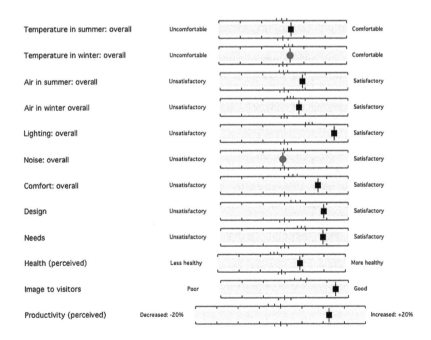

Figure 4.9.3 GreenHouse: summary chart, Australian benchmark, 2009

among the top aspects of things that work well and have a positive impact on their ability to do their work. Other positive aspects were good quality IT and meeting rooms and access to outdoor spaces and break-out spaces in the tenancy. Unacceptable noise and distractions were consistently nominated as the primary issues that hindered occupants' ability to work, with a small percentage (11 per cent) listing concerns with air conditioning and temperature issues. The results for the key aspects of the indoor environment are discussed briefly.

The GreenHouse achieved occupant ratings for air overall in summer and winter which were better than the Australian benchmark. These ratings were accompanied by a high level of satisfaction with air freshness. Significantly, dissatisfaction with air freshness dropped from 40–70 per cent at the Pitt Street tenancy to 17–22 per cent at GreenHouse.

Occupants' ratings for temperature overall were better than Australian benchmarks for summer; however as seen in Figure 4.9.3, the mean score for winter in the building is considered typical for buildings in the dataset. The results for winter overall are a consequence of more occupants rating the office environment on the colder side of neutral, accompanied by concerns for supply air close to occupants' feet being cold and uncomfortable. The latter experience is partly explained by a faulty thermostat that has since been fixed.

The average rating for lighting overall was significantly better than Australian benchmarks. Further to the positive feedback on natural light mentioned above, feedback on detailed aspects showed glare concerns were minimal (14 per cent dissatisfaction). Employees were positive with all aspects of the artificial lighting system that was installed, with a dissatisfaction of just 3 per cent.

Although employees were appreciative of the open plan in delivering access to views, there were some concerns with noise and interruptions resulting in the typical score for noise overall. However, it is worth noting here that dissatisfaction with noise had halved from 73 per cent at Pitt Street to 37 per cent at GreenHouse.

Occupants also registered a high level of satisfaction with design, image and workplace needs which were remarkably improved at the GreenHouse when compared to their Pitt Street accommodation. Figure 4.9.4 shows the overall performance for the GreenHouse for the three derived indices – comfort, satisfaction and summary index – in relation to the buildings in the Australian benchmark dataset for 2009. As seen, GreenHouse was ranked at the 96th percentile for overall building performance (i.e. Summary Index) and is a significant improvement from pre-occupancy ratings at GBCA in Pitt Street which was ranked at the 9th percentile. The Comfort Index for GreenHouse was ranked at the 84th percentile while the Satisfaction Index was ranked at the 99th percentile of the Australian benchmark.

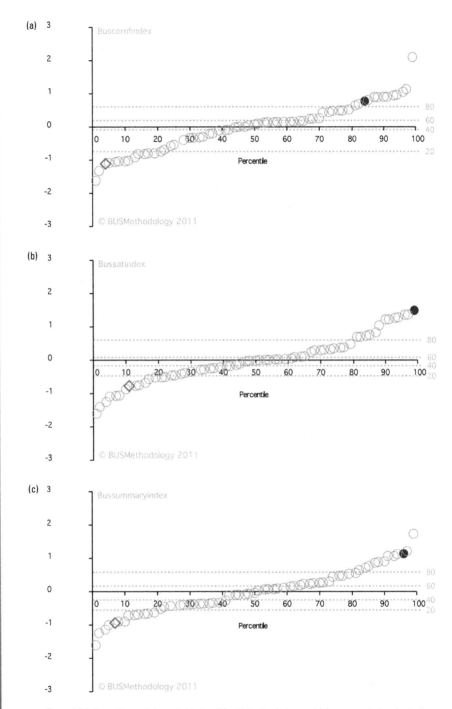

Figure 4.9.4 GreenHouse: (a) comfort index; (b) satisfaction index; and (c) summary index, Australian benchmark, 2009

Energy performance

In the absence of supplementary air conditioning to the meeting rooms, the major end uses in the tenancy were lighting and equipment power use. Energy use for the end users was separately metered and monitored on a monthly basis to ensure the tenancy met its energy targets. Substantial reductions in lighting energy are made possible through user-controlled supplementary task lighting for workstations that can be dimmed for comfort and switched off when workstations are unoccupied and the use of energy-efficient high frequency ballasts on all lighting fixtures and occupant sensors in meeting rooms. Additional savings arise from the ambient lighting systems that require a lower general lighting level (no more than 220 lx) when compared to uniform lighting systems often designed to provide 400 lx requirements across a whole floor plate. Energy use for light and power in a tenancy is also affected by user behaviour, and in this case a highly motivated and committed tenant group ensures energy-efficient practices are adhered to.

The results for energy consumption reflect the efforts that have been put in. GreenHouse achieved a 5 star NABERS Energy Office Tenancy rating with normalised carbon dioxide emissions of 57 kg CO_2e/m^2/year (213 MJ/m^2/year) for the billing period 15 October 2008 to 14 October 2009. This includes purchased 'Green Power'[5] generated from renewable resources at 10.2 per cent of the total electricity consumption. It should be noted that even without purchase of green power the tenancy would have still achieved a 5 star NABERS Energy rating.

CASE STUDY 2: A WHOLE BUILDING REFURBISHMENT

40 Albert Road, Melbourne Australia
Client: 40 Albert Road Commercial P/L a subsidiary of Szencorp Group of Companies
Design team: SJB Architects and Interiors
ESD Consultants: Connell Mott MacDonald and Energy Conservation Systems (ECS)
Year of completion: 2005
Certified ESD Rating: Green Star: 6 Stars (Green Star Office Design v1 Certification)
Certified NABERS Rating: 5 Star Energy Rating – Whole Building since 2007

Design strategies

This 1200m^2 office building was originally built in 1987 and refurbished between 2004 and 2005 (Figures 4.9.5, 4.9.6). The challenge for the design team was to convert a building performing well below industry average into a high performance building with ambitious client targets for sustainability and energy efficiency. The building was constrained by its narrow floor plate (10m x 55m) bounded by buildings to its north and south, with opportunities for daylight and views restricted to its 10m east (front) and west (rear) façades which in turn raised concerns for solar control. On the other hand, structural bracing via rigid connections between the floor plates and long north and south precast walls made architectural interventions to the service core and lifts on its northern boundary possible without undue stress on structural integrity. The plan and section depicted in Figure 4.9.7 and Figure 4.9.8 show aspects of the bioclimatic approach that was adopted for the refurbishment.

The revised environmental control strategy for the building integrated a mixed mode of operation with openable windows on the east and west ends for natural ventilation when ambient conditions permit. The main stairwell (see Figure 4.9.8 and Figure 4.9.9) was remodelled to serve as a light well and thermal stack designed to draw air out of the office floors. The office spaces were designed to operate in a 19–25°C temperature range during occupied hours, with a building management system (BMS) controlling the switch over from passive to air-conditioned mode. The air-conditioning system comprised a Variable Refrigerant Volume (VRV) three-pipe gas heat pump system consisting of two outdoor units powered by natural gas engines and 21 indoor fan coil units in the ceiling. The natural gas engines,

Figure 4.9.5 40 Albert Road: street view before refurbishment

Figure 4.9.6 40 Albert Road: street view after refurbishment

trialled for the first time in Australia, were selected for their capacity to deliver lower CO_2 emissions when compared to brown-coal fired grid supply prevalent in the State of Victoria.

The existing suspended ceilings were removed to expose existing concrete floors which were integrated with a night purge cycle to 'pre-cool' the mass and stabilise internal temperatures. The existing concrete façade with windows to the east was replaced with full height glazing with external shading via a perforated screen mesh to increase daylight penetration to the open plan office floor, while also enhancing the street presence of the building. High performance clear Low-E double glazing was retrofitted throughout, to maximise light transmission and minimise solar heat gain.

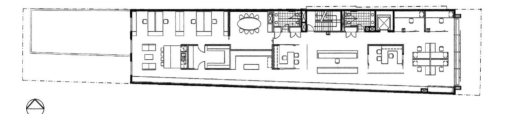

Figure 4.9.7 40 Albert Road: office floor plan

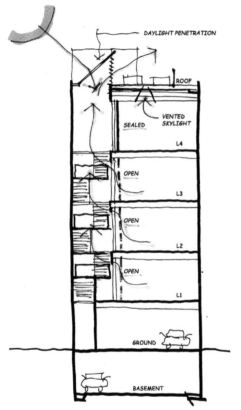

Figure 4.9.8 40 Albert Road: section through stairwell

Figure 4.9.9 40 Albert Road: view of sky-lit stairwell adjoining office floor

The refurbishment incorporated low toxicity materials and finishes throughout and the interior layout was designed to ensure the design intent for view corridors and air flow was not compromised. A proprietary managed lighting system with integrated occupancy sensors was developed for the project by the team at ECS. The occupancy sensors were linked to switching and dimming controls for lighting as well as the BMS to switch fan coil units off when an area became vacant.

With a holistic view to sustainability, other strategies integrated include crystalline PV panels and amorphous PV pergola and solar hot water, grey water recycling, rain water harvesting, waterless urinals and other water efficiency features, waste management, cycle parking with showers and change-rooms in the basement and inclusion of outdoor terrace gardens for staff use. More details of the integrated design process and environmental outcomes for water, waste, and energy can be found in Thomas and Vandenberg (2007).

Occupant evaluation

A post-occupancy study incorporating the BUS survey of 40 Albert Road was undertaken 12 months after occupancy (Encompass Sustainability, 2006). Some 92 per cent of staff (24 respondents) participated in the survey, and the results were compared with a pre-occupancy BUS survey completed in Szencorp's previous tenancy in a conventional building on St Kilda Road in Melbourne. A subsequent post-occupancy study was also undertaken (Encompass Sustainability, 2009) as part of the organisation's commitment to gaining routine feedback, which yielded a response rate of 88 per cent (34 respondents) of staff employed in 2009.

As seen in the summary chart in Figure 4.9.10, occupants in the 2006 study at 40 Albert Road rated the tenancy significantly better than the scale midpoint and 2006 benchmarks for 11 of the 12 key variables evaluating comfort and satisfaction.

Occupants at 40 Albert Road rated their productivity as increased by 10 per cent as a consequence of the environmental conditions in the building. This was matched by a high rating for overall comfort with only 12 per cent dissatisfaction. However, feedback to individual environmental variables suggested that some aspects could be improved.

Occupants were very positive about lighting overall and natural light; however, staff identified concerns with glare and heat coming through the west-facing window which has no external shading. Although 40 per cent of occupants expressed some dissatisfaction with glare, the rating was typical for buildings in the Australian dataset. Noise was not a major concern and this could be attributed to the lower level of occupancy at the building despite its open plan arrangement.

The overall ratings for temperature were generally positive with occupants' ratings for temperature overall in both summer and winter better than the Australian benchmark and discomfort levels at 10 per cent for summer and 30 per cent for winter. Temperature variation in winter caused initially by incorrect set-points meant that staff rated the building on the colder side of neutral, and air was also perceived as too dry.

On the other hand, occupants returned high ratings for air freshness with almost no dissatisfaction either in summer or winter. Interestingly their satisfaction with air quality was despite large spikes in introduced dust when windows were open and when recorded levels of VOCs, carbon monoxide and carbon dioxide mirrored outside levels.

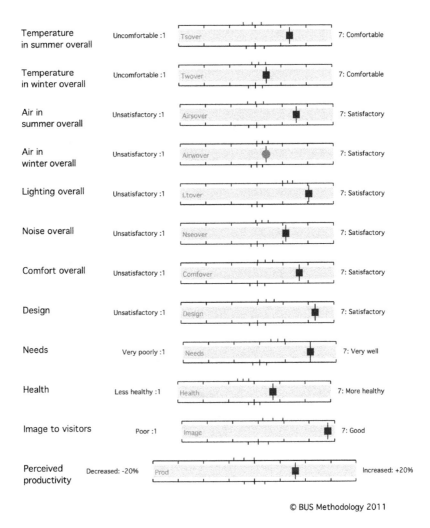

Temperature in summer overall — Uncomfortable :1 — Tsover — 7: Comfortable

Temperature in winter overall — Uncomfortable :1 — Twover — 7: Comfortable

Air in summer overall — Unsatisfactory :1 — Airsover — 7: Satisfactory

Air in winter overall — Unsatisfactory :1 — Airwover — 7: Satisfactory

Lighting overall — Unsatisfactory :1 — Ltover — 7: Satisfactory

Noise overall — Unsatisfactory :1 — Nseover — 7: Satisfactory

Comfort overall — Unsatisfactory :1 — Comfover — 7: Satisfactory

Design — Unsatisfactory :1 — Design — 7: Satisfactory

Needs — Very poorly :1 — Needs — 7: Very well

Health — Less healthy :1 — Health — 7: More healthy

Image to visitors — Poor :1 — Image — 7: Good

Perceived productivity — Decreased: -20% — Prod — Increased: +20%

© BUS Methodology 2011

Figure 4.9.10 40 Albert Road: summary chart, Australian benchmark, 2006

With temperature, humidity, daylight and lighting levels, air quality and occupation all monitored by the BMS, and each zone with its own metering system, and thermostat display, there is a pro-active approach to environmental control. Although users perceived a low level of personal control over their environment, this had little impact on their ratings for overall comfort.

The user response at the refurbished building was a significant improvement to occupants' experience in the earlier accommodation. 40 Albert Road was rated highly for how the facilities met workplace needs, design, image to visitors, and space in the building. Occupants also rated the building above benchmarks for perceived health, with only 4 per cent rating themselves as less healthy inside the building. Figure 4.9.11 shows the

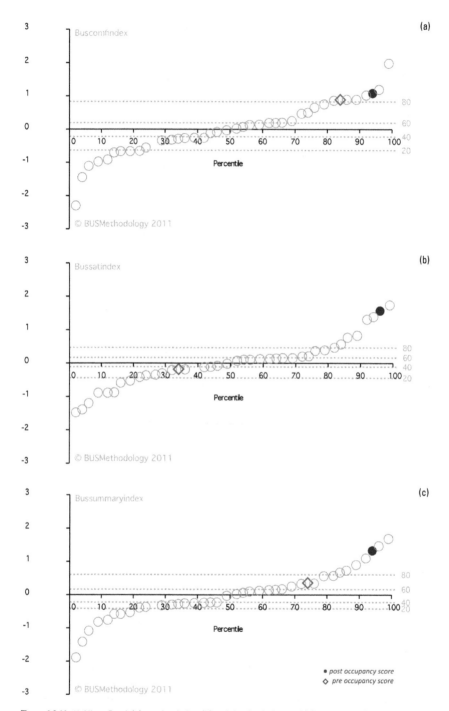

Figure 4.9.11 40 Albert Road: (a) comfort index; (b) satisfaction index; and (c) summary index, Australian benchmark, 2006

overall performance for 40 Albert Road for the three derived indices – comfort, satisfaction and summary index – in relation to the buildings in the Australian benchmark dataset for 2006.

The results for 2009 were consistently good for the building with overall performance indicator – the summary index is now in the top 4 per cent of Australian buildings (see Figure 4.9.12). Although there is a high degree of occupant satisfaction, the 2009 study highlighted ongoing issues with the building's internal temperature and concerns for feeling too cold.

Energy performance

In the first 15 months of occupation, 40 Albert Road experienced teething problems with the gas-fired heat pumps and achieved lower than expected outputs from the more unconventional strategies such as onsite power generation. Nevertheless, following a programme of monitoring to optimise performance, the building was rated at a 5 star level (NABERS Energy) for the whole building (base + tenancy) with normalised carbon dioxide emission of 168 kg CO_2e/m^2/year (701 MJ/m^2/year) for the period April 2006 to April 2007. This rating was achieved without inclusion of purchased accredited Green Power or carbon credits to offset CO_2 emissions on site. Through continued attention to building management, the building has continued to maintain its 5 star rating, with its most recent NABERS Energy rating (2010) being 162 kg CO_2e/m^2/year (602 MJ/m^2/year) also achieved with 0 per cent Green Power.

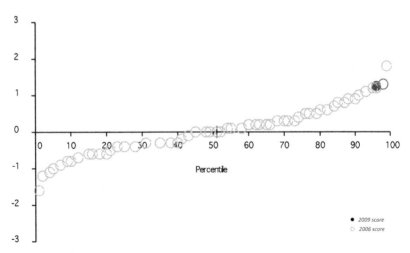

Figure 4.9.12 40 Albert Road: summary index showing 2006 and 2009 performance against the Australian benchmark, 2009

CASE STUDY 3: A LARGE-SCALE REFURBISHMENT WITHIN AN EXISTING BUILDING

Stockhome, 133 Castlereagh Street, Sydney
Client: Stockland Pty Ltd
Architects: BVN Architecture
ESD Consultants: Arup
Year of completion: 2007 Certified ESD Rating: 6 Star Green Star –
    Office Interiors v1.1 Certified Rating
Certified NABERS Rating: 5 Star Energy Rating – Tenancy since 2008

Design strategies

Stockhome was designed as the head office for a property developer and building portfolio owner company to accommodate over 600 employees. It was designed to be a showcase for sustainability with clear goals for sustainability under the Green Star rating system as well as energy efficiency targets under the NABERS scheme. A detailed analysis of the design strategies, performance and occupant satisfaction for this building can be found in Thomas (2010). What follows is a brief description in order to review outcomes and implications for large-scale refurbishments.

The building selected for the new head office was representative of many buildings from the 1980s. It was a sealed, centrally air-conditioned building with an octagonal deep floor plate, a large central core with no other vertical connection between floors and had a poor environmental performance. A major constraint in the development process was that the refurbishment across the allocated eight floors in a 31-storey tower had to occur while other floors in the building continued to be occupied (Figure 4.9.13). This prevented radical alteration to the air-conditioning system and also meant that changes to the façade were limited to minimal glazing upgrades. While the design team adopted a multi-disciplinary integrated approach, a number of change management and employee consultation initiatives were used to engage users in the design process and support their transition from a cellular office arrangement to one that was totally open plan (Figure 4.9.14).

Notwithstanding the climate-rejecting shell, two climate responsive strategies were integrated as part of the major refurbishment. The first was the creation of an open stair and void or light well spanning the eight-floor tenancy through an incision on the eastern side of the floor plates (Figure 4.9.15). This became the main circulation hub connecting departments across the floors, linking a number of formal and informal breakout spaces while simultaneously drawing daylight and access to views into the deep floor plate (Figure 4.9.16).

In a second move, a floating set-point in sympathy with outdoor conditions was adopted to extend the potential for free cooling via the economy cycle, thereby reducing energy for space conditioning. Although the

Figure 4.9.13 External view of 133 Castlereagh Street

Source: BVN Architecture.

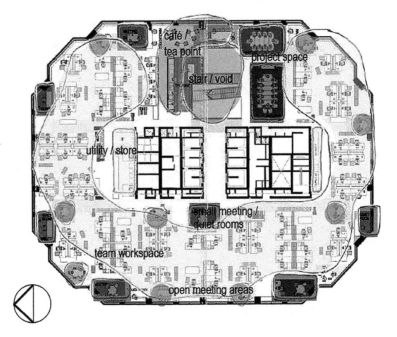

Figure 4.9.14 Stockhome: plan showing internal layout of workplace

Source: BVN Architecture.

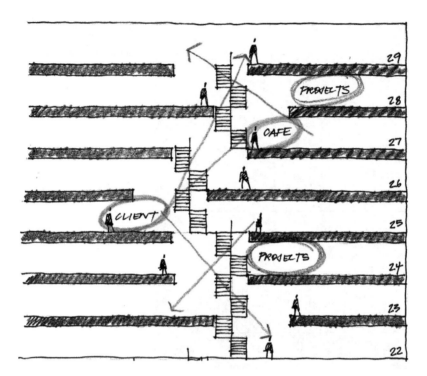

Figure 4.9.15 Stockhome: conceptual section
Source: BVN Architecture.

Figure 4.9.16 Stockhome: internal view
Source: photo by author.

potential range of 19.5–24.5°C does not push the boundaries greatly, this is seen as the first step to challenging the notion of a constant thermal environmental in an air-conditioned setting. During the refurbishment, variable speed drives and high efficiency VAV boxes were installed in the building to improve the efficiency of the air-conditioning system

Other strategies included retrofitting high performance glazing to the void to allow higher levels of daylight while continuing to manage heat gain, energy-efficient lighting, occupant sensors and operable blinds. Low VOC emission products were used throughout and fresh air at 15 L/s, twice the minimum requirement of the Australian Standard, was provided to the office workstations. A total of 60 bicycle parking spots with showers were provided. The tenancy also integrated a tri-generation power which was estimated to deliver 20 per cent reduction in CO_2 emissions.

Occupant evaluation

In addition to a user-responsive approach to the daily operation of the building that was adopted, feedback from occupants was sought through structured post-occupancy evaluation. Following a pre-occupancy survey of the organisation at their previous location in the CBD, the internet-based survey was administered 15 months after relocation and yielded a 40 per cent response rate comprising 238 respondents.

The results for Stockhome are impressive, especially given the size and scale of the tenancy compared to many boutique green building developments that are generally much smaller. On average, occupants rated Stockhome higher or better than the Australian benchmark for 11 of the 12 main study variables (Figure 4.9.17).

Interestingly, the survey results show that the floating set-point did not appear to have any significant negative impact on occupant ratings for satisfaction with temperature overall, while the open-ended comments indicated more concerns of feeling 'too cold' both in winter and in summer. The efforts to increase fresh air rates, and maximise access to views and access to daylight were ratified in the positive ratings for air quality and lighting. Users were also appreciative of their ability to control glare with the blinds that had been provided.

Users commented that the open plan layout increased collaboration and the breakout spaces alongside the light well worked well to provide formal and informal meeting spaces. Although the level of dissatisfaction with noise (34 per cent) and concerns of unwanted interruptions from colleagues in the workplace observed at Stockhome is typical of buildings in the dataset, there have been some efforts to manage these issues through glass screens separating the breakout spaces and the introduction of acoustic treads to central stairs.

The positive feedback for air, temperature, and lighting overall, and the efforts to manage problems as they arose were backed up by a high rating for overall comfort (5 per cent dissatisfied) and a positive rating overall

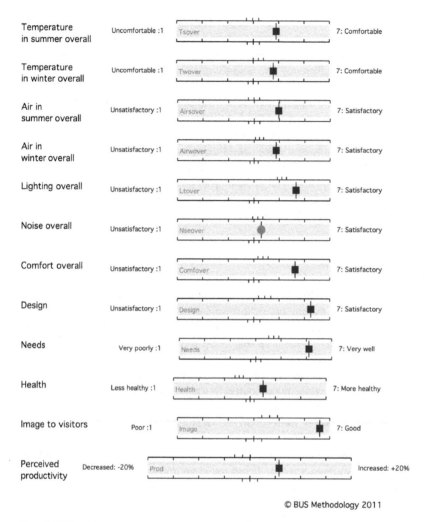

Temperature in summer overall	Uncomfortable :1 — Tsover	7: Comfortable
Temperature in winter overall	Uncomfortable :1 — Twover	7: Comfortable
Air in summer overall	Unsatisfactory :1 — Airsover	7: Satisfactory
Air in winter overall	Unsatisfactory :1 — Airwover	7: Satisfactory
Lighting overall	Unsatisfactory :1 — Ltover	7: Satisfactory
Noise overall	Unsatisfactory :1 — Nseover	7: Satisfactory
Comfort overall	Unsatisfactory :1 — Comfover	7: Satisfactory
Design	Unsatisfactory :1 — Design	7: Satisfactory
Needs	Very poorly :1 — Needs	7: Very well
Health	Less healthy :1 — Health	7: More healthy
Image to visitors	Poor :1 — Image	7: Good
Perceived productivity	Decreased: -20% — Prod	Increased: +20%

© BUS Methodology 2011

Figure 4.9.17 Stockhome: summary chart, Australian benchmark, 2008

for perceived productivity. The study revealed that a perceived productivity loss of 2 per cent on average at the previous location had been converted into a gain of 7 per cent at Stockhome, a change of 9 per cent all told.

Figure 4.9.18 shows the overall performance for Stockhome for the three derived indices (comfort, satisfaction and summary) in relation to the buildings in the Australian benchmark dataset for 2008, while Figure 4.9.19 shows the summary index in relation to buildings in the international benchmark dataset.

As seen, the overall summary index for occupant satisfaction using the Building Use Studies survey placed in the top decile of the Australian building dataset. With the benchmarks for the International dataset being more stringent, the building lies in the top quintile of this dataset. Clearly the

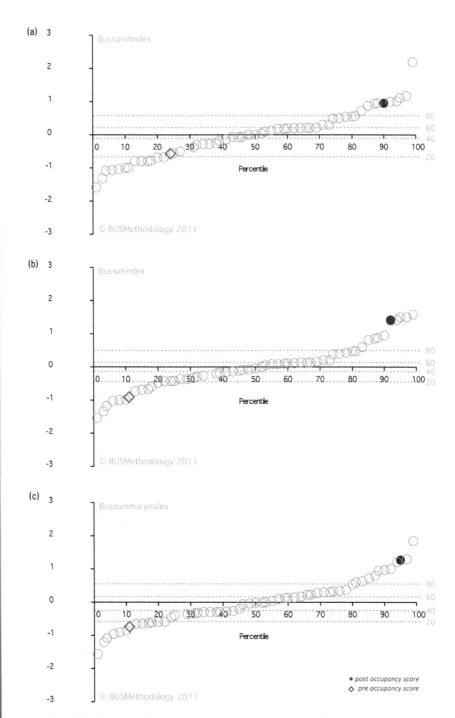

Figure 4.9.18 Stockhome: (a) comfort index; (b) satisfaction index; and (c) summary index, Australian benchmark, 2008

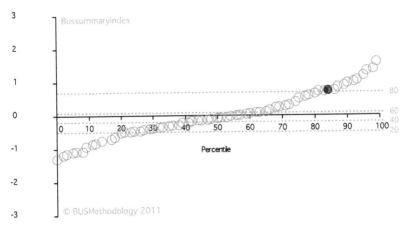

Figure 4.9.19 Stockhome: summary index, international benchmark, 2008

overall performance is also a significant improvement from the experience in the previous tenancy which was ranked in the bottom decile of the Australian dataset.

Energy performance

In addition to the energy-saving measures as part of the refurbishment, initiatives such as monthly competitions between floors, awareness campaigns, and sustainability champions on each floor were used to get occupants engaged during the operation of the building. These complemented a programme of monitoring and energy audits to ensure the building performed to its full potential.

In its first officially monitored rating, Stockhome achieved the highest level or 5 star NABERS Energy Office Tenancy rating with normalised carbon dioxide emissions of 61 kg $CO_2e/m^2/year$ (356 MJ/m²/year) for the period 1 October 2007 to 30 September 2008. This included 18.75 per cent purchased Green Power that was used to make up for the tri-generation system that was not implemented in the first year.

In a tri-generation system, power needs are supplied through electricity generated from an onsite gas turbine which emits lower carbon dioxide emissions in comparison to coal-fired electricity. Additionally, the waste heat can be used for cooling via an absorption chiller and to generate hot water for space heating. The tri-generation system implemented since 2009 at the Stockland building accounts for 10862 GJ of gas to generate the electricity for tenant lighting, equipment and supplementary air conditioning. As seen in the latest rating, the tenancy continues to maintain its 5 star NABERS Energy performance with normalised carbon dioxide emissions of 57 kg $CO_2e/m^2/year$ (1121 MJ/m²/year) for the period 1 August 2009 to 31 July 2010 even in the absence of purchased Green Power.

A STRATEGIC APPROACH TO BUILDING REFURBISHMENT

In response to concerns of global warming and climate change and its impact on the environment, we have seen the recent development of a number of green buildings. Designed with a 'clean slate', these are noted for their success in integrating a number of innovative design features to reduce their environmental impact. Nevertheless the bulk of the building stock comprises existing buildings which return a very poor environmental performance. Such buildings are generally characterised by a climate-rejecting approach and remain constrained by their deep floor plates, sealed windows, clean unshaded façades and insensitive use of glass which forces dependence on air conditioning and artificial lighting.

An incremental approach towards achieving efficiency in such buildings would entail the introduction of better building and energy management systems, increasing user awareness and improving the efficiency of lighting and air-conditioning systems as discussed elsewhere in this book. All of these aspects are included in the three study buildings discussed above, although each of them goes further as discussed below.

A contrasting approach to the incremental approach is to adopt a strategic approach from a triple bottom line perspective. This approach recognises opportunities for occupant comfort and well-being in tandem with carbon reduction and energy efficiency. It is in this context, that the outcomes of the three buildings discussed here are most valuable. A common thread in their development is an integrated approach to questions of sustainability throughout the design, development and operation phases of the building. Client commitment to real outcomes, establishment of tangible environmental goals and user participation in the design process play a critical role in their success. (See also Thomas and Hall, 2004.)

A bioclimatic and user-responsive approach to refurbishment calls for the ability to realign the building to changing user needs while enhancing indoor environmental quality. 40 Albert Road provides insights of what can be achieved through a complete overhaul of a building even when it has the worst orientation in a very tight site. As seen here, shading, introducing daylight and reinstating a more climate responsive approach to environmental control while moving from former cellular arrangements to open plan layouts are the first steps even before more technology-based solutions for renewable energy and reduced emissions are introduced.

When considering the limitations of the existing shell, it is worth noting that key interventions which have maximum impact are those which serve more than one function – such as the introduction of a light well via the stair at two different scales as seen at 40 Albert Road and Stockhome. Apart from changing the dynamic circulation in the building, and drawing natural light into the building, at 40 Albert Road it becomes the conduit for introducing a mixed mode of operation, whereas at Stockhome it allows for break-out spaces that provide relief in a deep plan layout enhancing user experience. Even on a single floor plate without the possibility of external shading, the

GreenHouse design shows opportunities for integrating light shelves to distribute natural light further into the floor plate in conjunction with an integrated system of task and ambient lighting that offers user control.

Within the constraints of sealed air-conditioned buildings and partial refurbishments as seen in the GreenHouse and Stockhome, it is clear that users respond positively to increased rates of fresh air and efforts to reduce VOCs. Significantly, in all cases, it is worth noting that users rate overall comfort more highly than their ratings for individual aspects such as noise, temperature, air quality and lighting, with the positive features reinforcing one another. Coupled with their positive responses to natural light, access to views and fresh air, this confirms our studies (see Leaman *et al.*, 2007) that show that people actually like the bioclimatic features included in green buildings and as a consequence are sometimes forgiving of minor shortcomings. On the other hand, it is important to note that buildings that do not perform well from the users' perspective run the risk of greater dissatisfaction and become expensive to operate and fix. Increased monitoring and feedback become vital tools for user-responsive building management during commissioning and operation.

The 5 star NABERS Energy ratings for GreenHouse and Stockhome tenancies and the 40 Albert Road building demonstrate actual reductions of at least 58 per cent of greenhouse gas emissions compared to the industry average. These reflect the value of the efforts towards energy efficiency that have been put in, and demonstrate the potential for carbon reduction within the existing stock. Even in the absence of calculations for embodied energy for the study buildings, the carbon-reduction impact arising from refurbishment in comparison with rebuilding is well accepted (see Cole and Sattary, Chapter 3.9 in this volume; Bullen, 2007). Significantly in each of these cases, the refurbishment of existing commercial buildings extends the life-cycle of the building to preserve resources and ensure a reduction in carbon over the life-cycle of the building.

The positive occupant feedback seen in this study confirms the potential for strategic refurbishments that enhance workplace quality. In each of the three instances, occupants reported high satisfaction with overall comfort that was matched with an improvement in their perceived productivity. From an organisational perspective, in a scenario where staff-related costs often comprise more than 80 per cent of the operational budget, it is the benefit to the occupant well-being and comfort and consequent impact to worker productivity that is valued more highly than energy savings *per se*.

Clearly the strategic approach to building refurbishment outlined in this chapter offers positive outcomes at all levels and provides useful insights for retrofitting the vast portfolio of existing buildings in our cities.

ACKNOWLEDGEMENTS

The author acknowledges Green Building Council of Australia, Szencorp Group and Stockland Pty Ltd for facilitating the extended post-occupancy studies at GreenHouse, 40 Albert Road, and Stockhome respectively. The detailed POE studies would not have been possible without the generous assistance and participation of key personnel in each of these organisations, as well as BVN Architecture (GreenHouse and Stockhome), SJB Architecture (40 Albert Road), Arup, ECS and Connell Wagner. Each of the POE studies included the use of the BUS questionnaire under licence, and the feedback and support received from Adrian Leaman of Building Use Studies, UK, during this project is acknowledged. The Stockhome study was undertaken by the author for Stockland Pty Ltd, and the BUS study of GreenHouse was completed with the assistance of Monica Vandenberg of Encompass Sustainability. The author acknowledges Monica Vandenberg, who also undertook the BUS studies for 40 Albert Road and co-authored the detailed case studies for 40 Albert Road with the author, for her support.

NOTES

1   Twenty-three Probe study papers were published in *Building Services Journal* between September 1995 and October 2002.
2   An overall summary index provides one way of benchmarking building performance in relation to other buildings in the dataset. It is derived as the average of the comfort index and the satisfaction index, where the comfort index is the average of the z-scores of the variables for overall comfort, lighting, noise, temperature, and air quality, while the satisfaction index is the average of the z-scores for design, needs, health, and productivity.
3   Green Star is a national, voluntary environmental rating system that evaluates the environmental design and construction of buildings. The Office and Office Interiors tools attributes and inclusions for Management, Indoor Environmental Quality, Energy, Transport, Land Use, Emissions, Materials, Water and Innovation. A 5 Star Green Star Certified Rating recognises Australian excellence while the top 6 Star Green Star Certified Rating recognises World Leadership
4   National Australian Building Environmental Rating Scheme (NABERS), NABERS Office: Developed in 1998 as Australian Building Greenhouse Rating (ABGR), the NABERS energy rating system is a protocol for rating actual energy performance and CO_2 equivalent emissions post occupancy. It is currently administered by the New South Wales Government on behalf of commonwealth, state, and territory governments. In Australia, a building that achieves the top level of a 5 star NABERS Energy rating delivers 58 per cent CO_2 reductions in comparison with the industry average of 2.5 stars. Separate benchmarked ratings are available for Tenancy (tenant light and power, and any supplementary air conditioning), Base Building (common areas and central services including air conditioning and hot water) and Whole Building (Tenancy + Base Building).
5   Green Power refers to electricity generated from renewable resources and purchased by energy providers on behalf of consumers who sign up for the scheme through a government accreditation programme for renewable energy.

REFERENCES

Birkeland, J. (2008) *Positive Development from Vicious Circles to Virtuous Cycles through Built Environment Design*, London: Earthscan.
Building Use Studies (n.d.) available at: http://www.usablebuildings.co.uk (accessed 21 June 2011).
Bullen, P. (2007) 'Adaptive reuse and sustainability of commercial buildings', *Facilities*, 25(1/2): 20–31.
Encompass Sustainability (2006) *Post Occupancy Evaluation: 40 Albert Road Site Report*, Melbourne: Encompass Sustainability.
Encompass Sustainability (2009) *Post Occupancy Evaluation: Site Summary Report, 40 Albert Road*, Melbourne: Encompass Sustainability.
Fanger, P.O. (1970) *Thermal Comfort, Analysis and Applications in Environmental Engineering*, Copenhagen: Danish Technical Press.
GBCA (2009) 'The GreenHouse: It is easy being green', available at: http://www.gbca.org.au/uploads/132/2436/greenhouse%20case%20study_It%20is%20easy%20being%20green_270809.pdf (accessed 28 April 2011).
Green Star (n.d.) Green Star Rating Tools, available at: http://www.gbca.org.au/green-star/rating-tools/ (accessed on 21 June 2011)
Leaman, A. and Bordass,W. (1999) 'Productivity in buildings: the "killer" variables', *Building Research & Information*, 27(1): 4–19.
Leaman, A., Thomas, L.E. and Vandenberg, M. (2007) '"Green" buildings: what Australian users are saying', *EcoLibrium*, 6(10): 22–30.
National Australian Building Environmental Rating Scheme (NABERS), NABERS Office available at: http://www.nabers.com.au/office.aspx (accessed 26 June 2011).
Thomas, L.E. (2010) 'Evaluating design strategies, performance and occupant satisfaction: a low carbon office refurbishment', *Building Research & Information*, 38(6): 610–24.
Thomas, L. and Hall, M.R. (2004) 'Implementing ESD in architectural practice – an investigation of effective design strategies and environmental outcomes', in M.H. deWit (ed.) *Proceedings of the 21st PLEA (Passive and Low Energy Architecture) Conference*, Technische Universiteit Eindhoven, Eindhoven, the Netherlands, Vol. 1, pp. 415–20.
Thomas, L. and Vandenberg M. (2007) *40 Albert Road, South Melbourne: Designing for Sustainable Outcomes: A Review of Design Strategies, Building Performance and Users' Perspectives*, BEDP Environment Design Guide (CAS45, May 2007), Melbourne: Royal Australian Institute of Architects.
Vischer, J.C. (2007) 'The effects of the physical environment on job performance: towards a theoretical model of workspace stress', *Stress and Health*, 23(3): 175–84.
Wyon, D.P. (2004) 'The effects of indoor air quality on performance and productivity', *Indoor Air*, 14(7): 92–101.

4.10
THE DEAKIN UNIVERSITY WATERFRONT CAMPUS: CALLISTA OFFICES

How can an evidence-based approach to office
renovation work in practice?

Mark B. Luther

INTRODUCTION

This project showcases the refurbishment and recycling of a historic
Woolstore building on the Deakin University Waterfront Campus in Geelong
Victoria. At the inception of the Deakin Callista Offices project, daylight and
energy simulation studies were performed for the 1000m² predominantly
east-facing and vacant space (Figure 4.10.1). Simulation results served the
architect and consultant as an initial benchmarking of the building environmental performance retrofitting potential. The following is a brief summary of
the findings using the Mobile Architecture and Built Environment Laboratory
(MABEL) testing facility as well as the 'before and after' retrofitting results.
This case study serves as an example of what owners and management can
accomplish for their buildings through rigorous worthwhile investigation and
research.

At the initial project stage, the client specifically sought advice on
the clerestory south (non-equatorial facing) 'saw tooth' structure for its daylighting retrofitting capabilities. Several studies applied computational software to optimise the number of clerestory windows and their location. One

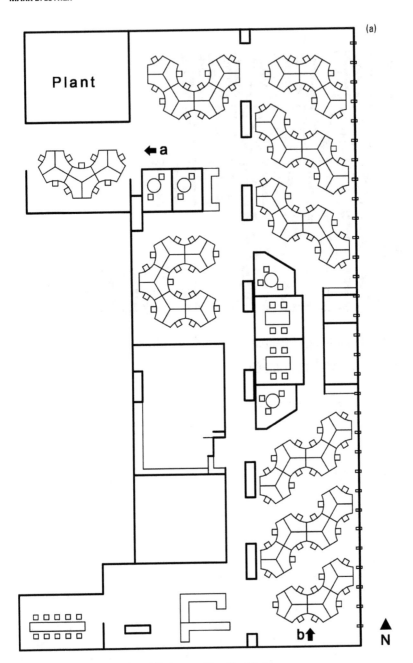

(a)

Plant

←a

b↑

N

Figure 4.10.1 Floor plan with (a and b) pictures of the office interior

(b)

(c)

Figure 4.10.1 Continued

of several daylighting conditioning configurations is illustrated in Figure 4.10.2. This demonstrates that under a clear sky condition, 250 lux is achieved at the working plane for a 'general office' lighting. Note that only every other clerestory applied a glazing installation. This study potentially saved the client 60 per cent in capital costs over a full clerestory glazing retrofit and also considered the benefit of the east façade (see Figure 4.10.3).

The clerestory glazing system was also modelled and designed for its thermal as well as visible light contribution. A profile of the thermal as well as visible glazing properties is shown in Figure 4.10.4. The idea is to fabricate a 'cool daylighting' double glazed system. Such a glazing system would also yield a high luminous efficacy: $Ke = Tv / SC$ while keeping the heat out, as discussed by Lyons (2001). Luminous efficacies greater than $Ke = 1.2$ are desired. The advice on daylighting with regard to glazing selection, area and location as well as control was accepted and proved to be a success, while the information on HVAC sizing, solar shading, and heating, ventilating and air-conditioning concepts were somewhat ignored.

After the first year of occupation, and several instances of discomfort complaints, the management decided to engage Kodo Pacific to conduct an occupant PROBE survey as well as the Mobile Architecture and Built Environment Laboratory (MABEL, Deakin University) to investigate the interior environmental performance. Figure 4.10.5 is an overall result of the Kodo Pacific occupant survey where 50 per cent indicates the average of the database on surveyed Australian buildings. It is acknowledged that 'user control' and 'building comfort' are lower than average for this building.

The MABEL facility measured thermal comfort of the office space around the time of the Kodo Pacific survey and confirmed (objectively) that here were thermal comfort problems especially at workplaces near the building perimeter.

This initial investigation indicated that the occupants were mainly concerned about their thermal discomfort. Shortly thereafter, a company introducing a new HVAC control technology approached MABEL. They wanted to test run the performance and capabilities of their product, claiming superior HVAC control, air quality conditioning and energy savings. This new system offered a demand-controlled and slightly interior–exterior positively pressurised room condition. The proposed system also permitted all the existing ductwork to remain in place. However, additional capital, introducing variable speed drive fans and proper dampeners on the supply and return ducting needed to be installed. This new system provides an economiser cycle, introducing variable proportions of fresh air intake as required by the 'demand control' sensors located within the space. In this system, temperature, CO_2, humidity and pressure differential all define the 'demand' for fresh air intake and energy-efficient operation. This concept takes advantage of the external air conditions and often utilises 50–100 per cent outside air to condition the space which is very possible for the Melbourne, Australia, climate. Additional options are a 'floating' thermostat, where external mean daily temperatures are considered and the interior set-point is raised accordingly.

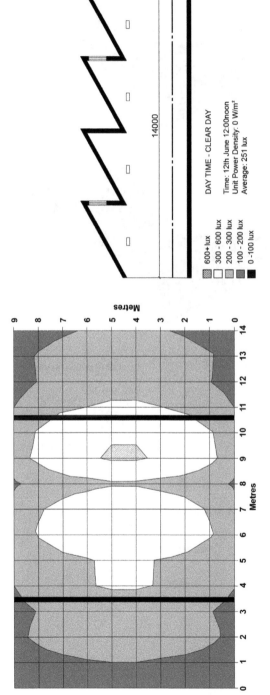

Figure 4.10.2 Winter clear sky daylighting working-plane illuminance result from clerestory

Figure 4.10.3
(a) A full clerestory glazing retrofit; (b) over a selective optimised requirement

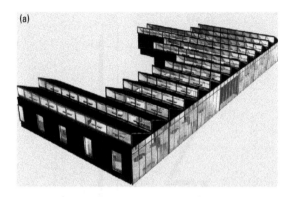

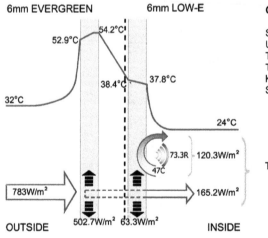

6mm EVERGREEN   6mm LOW-E

52.9°C 54.2°C

38.4°C 37.8°C

32°C

24°C

73.3R 120.3W/m²

47C

783W/m²   165.2W/m²

502.7W/m² 63.3W/m²

OUTSIDE   INSIDE

Overall Glazing Results:

S.C. = 0.39
U-value = 2.21 W/m². K
Tvis = 0.53
Tsol = 0.21
Ke = Tvis/SC ≥ 1.2
Specified glass:Tvis/S.C ≥ 1.5

TOTAL HEAT GAIN
285.5W/m²

Figure 4.10.4 Proposed clerestory glazing design

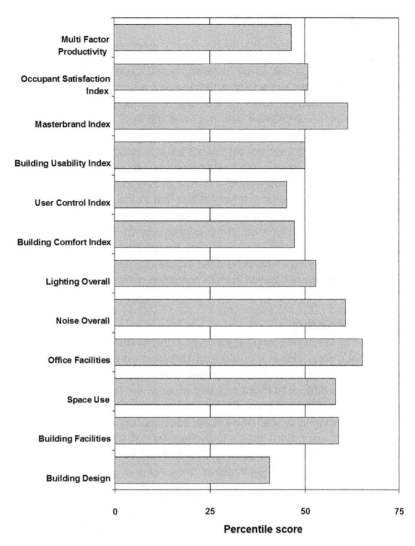

Figure 4.10.5 Overall result of the Kodo Pacific occupant survey

In a sense, the new HVAC control offers a 'bioclimatic-controlled' system, being mechanically accomplished through the recognition of external wind forces on the building (pressurisation balance), the implementation of an economiser cycle, the air quality (CO_2 levels) within the room and the response of a set-point temperature according to mean daily temperatures.

PROJECT MEASUREMENT OVERVIEW

An important and major part of the project was to measure the energy use and consumption for the newly installed control system over that of the original one. Given the circumstance that the new system could not be completely implemented because of the central chiller plant serving additional building zones, a compromise was met, and the new system utilised the existing generated chilled water.

Nevertheless, substantial savings are indicated from this study. The primary savings would be in the total energy costs to produce chilled water for cooling purposes which were found to be of the order of 60 per cent less during summer conditions. Further savings are in the variable fan speed drives compared to the original constant drive fans. Here again, almost 50 per cent savings occurred from that of the original system. The other important realisation is that several of the existing air handling units (AHUs), conventionally used, were turned off and not required to maintain comfort conditions during the newly installed system controlled periods (Figure 4.10.6).

Energy analysis summary

As a result of this initial change in the HVAC control retrofitting of this space, a comparison was made in annual energy consumption between the original and the new system. Due to a 'lumped' metering approach within the university, it is difficult to make an exact assessment of the energy breakdown between HVAC, lighting and other plug loads. In lieu of this, an assumption is made where a 'typical' office HVAC consumption is considered on the order of 65 per cent of the total energy load. Table 4.10.1 provides an estimate, based upon measurement, of two different years of occupancy. In 2004, the Callista offices were newly occupied and operated under the original system. In 2006, the new demand-controlled system was fully implemented. Note that the only change made between these two years was in the HVAC control system. It is estimated that there would be about a 70 per cent total energy reduction from the HVAC operational energy alone (Table 4.10.1). Given the fact that there are four less air handling units in operation, this is believed to be a reasonable estimate.

These findings merit further investigation for newly installed building systems where up to a 40 per cent reduction capital expenditure through reduced peak load sizing and capital equipment (eliminating unnecessary AHUs and ductwork) could take place.

Further investigation

One of the planned investigations by MABEL was the claimed increase in comfort for the east façade of the building. Such was investigated with the façade analysis meter as well as the thermal imaging camera (see Figures 4.10.7 and 4.10.8).

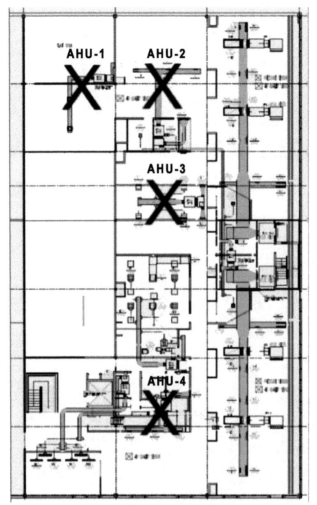

Air Handling Unit Layout
(Marked units shut down)

Figure 4.10.6 Mechanical AHU layout: indicating those units not in use with the newly controlled system

Table 4.10.1 A comparison between the annual energy consumption of the original and the new HVAC control system

Year	Description	Total energy cost ($)	HVAC operation ($)	Lighting & plug loads ($)
2004	Original design	27,315	17,755	9,660
2006	Bauer design	14,774	5,114	9,660
		SAVINGS:	**12,641**	

Figure 4.10.7 Façade analysis meter

Figure 4.10.8
Exterior façade thermal imaging

A façade analysis meter was constructed using a Bruel & Kjaer 1221 Comfort Meter. This instrument includes an interior glass surface temperature as well as a radiant asymmetry probe. The radiant asymmetry indicates the net radiant heat gain (as well as loss) between the east façade and the interior space. For this case, the façade meter indicated extraordinary differences in the comfort between the original and newly conditioned office. These results are presented in Figure 4.10.9, beginning with the results of the radiant asymmetry probe and glass surface temperature, followed by the calculated Predicted Percentage Dissatisfied (PPD) and operative temperatures (Figure 4.10.10).

The radiant asymmetry (indicated by shading) is shown for both demand control and original conditioned spaces. It is interesting to note that the radiant component reaches the same intensity on both days. However, in the forthcoming comfort 'vote' – predicted percentage dissatisfied (PPD) – there is an indication of better comfort under the demand controlled condition. Also, it is noted that the glass surface temperature does not rise as high as under the original conditioning system (about a 5°C difference).

The above two graphs indicate the comfort achieved at a 0.8 meter distance from the east façade. Although the day is slightly different with respect to external temperature, there remains a significant difference between the two levels of comfort.

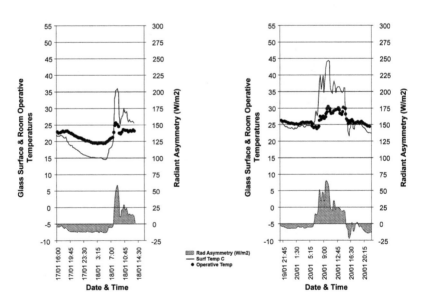

Figure 4.10.9 Glass surface and operative temperature with radiant asymmetry for the demand control and original HVAC system

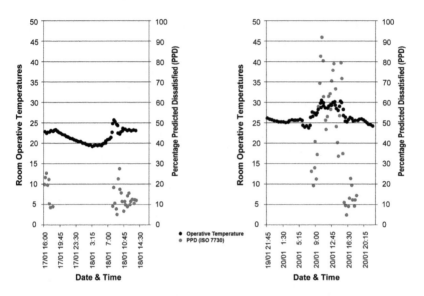

Figure 4.10.10 The predicted percentage dissatisfied as well as the operative temperatures

Recent façade developments

For several years ongoing, the management and occupants of this building still remain unsatisfied with the thermal conditioning near the façade perimeter on extremely hot days. As a result several new glazing systems and an external shading device are being researched. The bar graph in Figure 4.10.11 indicates the actual measured results of radiant asymmetry taken at 0.6m from the interior glass surface. Note that the external shading device performs the best out of all the systems in terms of blocking radiation gains to the space.

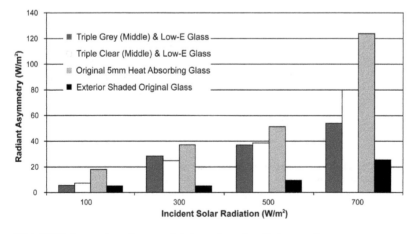

Figure 4.10.11 A comparison of measured glazing systems via radiant asymmetry

A calculation of each of the glazing systems was performed using the Vision 3.0 software program. The total inward heat gain together with the solar transmittance for the original glazed unit over the entire east façade area under a possible 700 W/m2 incident solar radiation would yield a 45kW load for that hour. For the exact same façade with the proposed external shading system this total load for the specific hour would be reduced to 7.5 kW. Such implications would have an enormous impact on the sizing of HVAC equipment for this space. The reductions of these façade loads together with those already implemented through the retrofitting control of the HVAC system would substantially improve the overall energy efficiency, greenhouse gas emissions and operational cost of this office, while improving comfort.

CONCLUSION

There is room for improvement in the common-sense application of reducing basic solar loads to our office buildings as well as introducing the use of effective daylight and its control. However, new control technologies that appear to comply with our passive design and bioclimatic principles can now also be introduced. Such was the case with the introduction of a new demand control HVAC system as demonstrated here. This system has allowed for natural ventilation conditioning, a response to external temperature, humidity and wind speeds and internal set-point as well as offering night-time purging strategies to be implemented and controlled. Further studies on the manner in which we design our curtain wall systems are also important to building control optimisation. Together, a reduction of 50 per cent or more in energy savings over a conventional 'business as usual' approach can be easily obtained.

REFERENCE

Lyons, P. (2001) 'The energy impact of windows in building design', *BDP Environment Design Guide*, PRO 3, Melbourne: The Australian Institute of Architects.

4.11
SUMMARY
Francis Barram and Nathan Groenhout

There are two main lessons to be learned from the projects. The PMM case study (Chapter 4.2) shows the significance of adaptive reuse of exiting buildings as an economic and environmental strategy for sustainable retrofitting. While the core building function appears to remain, additional retail space is added and the construction is used to address problems with a largely obsolescent mirror glass façade. The project also points to the importance of facilities management personnel and their role in the retrofittng process. These needs are:

- Access to building contractor knowledge to problem-solve operational issues.
- Good site records should be produced and as-built drawings of the project to provide knowledge security.
- Building design thermal modelling is predictive and should be verified by actual building performance fully occupied as variances in modelling input can be verified.
- Information from additionally modelling on the benefits of upgraded and reused equipment requires specialist skills and advice.
- Benchmark information is useful in ongoing building tuning and monitoring to ensure that staff understand appropriate levels of performance applicable to new and reused equipment.
- Recycling systems may require more detailed thought to ensure the problems are not just removed from site and recreated elsewhere.
- The management of people and their expectations is important right through from designers, consultants and cost controllers to construction workers to building occupants. Managing all the issues and complexities makes retrofitting work challenging at times and having all stakeholders working to a shared vision becomes an important project success measure.

55 St Andrews Place's retrofitting process (Chapter 4.3) followed a more typical approach to environmental retrofitting which is to upgrade the

environmental performance of the building to improve comfort and reduce energy consumption. The design solutions are technically straightforward:

- Improving the façade performance.
- Improving daylight and comfort and reducing energy consumption.
- Retaining most of the original mechanical systems.

This was achieved through the following:

- Integrated design process, using design performance.
- Synergies in the solution set for the design of the façade, the lighting and internal environmental quality.
- Use of building simulation modelling in the design process to achieve environmental rating objectives. The Base Building upgrade attained its target 4 star rating, but the more onerous Office As-built rather than Design rating, and the fit-out attained its 4 star Office Interiors rating.

However, it is clear that the issues of sustainable retrofitting can be seen to operate at both the building and precinct level with large numbers coming under the portfolio of one organisation. Universities and other education institutions are examples (Chapter 4.4). The progressive rise in energy costs and the environmental imperative on these organisations to address environmental issues provide a rationale for progressive improvement of the building stock. Current areas for improvement which could equally apply to other large organisations are:

- Develop a strategic and tactical approach to improve existing buildings' energy performance.
- Develop design guidelines that are action-oriented; where performance falls below baselines, principles should be triggered which can apply, such as those based on bioclimatic principles.
- Focus on the consumption pattern; for example, the largest amount of consumption in this case study occurred during off-peak hours, i.e. 10pm–7am. This indicates there were no effective strategies to minimise energy consumption overnight. It has been observed that a monthly average of 2,000,000 KW/h is consumed during periods when there is minimal occupant presence.
- Improve accountability and ownership of energy performance of buildings, precincts and the campuses in general. The energy costs are not financially accountable by building occupants or faculties, but by the general university administration.
- Address the disconnection between the individuals who use the energy and those who pay for it. This requires improvement

in *education:* previous research has indicated that occupant awareness and behaviour change can result in improved energy efficiency in building operations.

Hence the message from this is that for retrofitting to operate effectively, it needs both the technical fix and policy frameworks to effectively manage this process of change. The studies have conclusions which focus on the technical fix.

The case study on 503 Collins Street (Chapter 4.5) focuses on the innovation in the energy contracting process and the supply of energy. Cogeneration is seen as a way of diversifying the source of energy other than the grid supplied to a building and to use less polluting fuel sources. Lessons learned were:

- Cogeneration can be a critical strategy in delivering a 5 star NABERS Energy outcome. The building upgraded to 5 star NABERS Energy performance; greenhouse emissions have been reduced by 4,700 tonnes CO_2/annum, while 4.3GWh of grid electricity has been saved annually, a yearly energy cost reduction of A$360,000.
- The building upgrade, and the benefits of improved services, lower energy bills and much-reduced greenhouse emissions were seen as necessary to ensure a building is attractive to the widest range of tenants. Enhancing the premium grade office by sustainable retrofitting takes the building to a higher level of performance demanded by some tenants, hence making the building available to a wider potential market in the premium grade sector.
- The EPC provided the fastest track to upgrading services and seeing performance improvements; however, it remains to be seen if the traditional service upgrade (TDU) could have delivered the same quality in the same time frame.

As well as cogeneration, it is also possible to use other green technologies such as photovoltaic panels (Chapter 4.6). Cost per Watt of these systems is decreasing, hence improving the economic feasibility of the systems. In addition it is technically feasible to increase the performance rating using tools such as a NABERS Energy rating of an office building. The benefits are:

- Approximately 10 per cent reduction in grid electricity consumption and associated GHG emissions is possible using PV panels located on the available roof area.
- This is dependent on orientation, energy use load profile and suitable roof area, with a relatively low floor area to roof area ratio.

- Optimal arrangement of PV panels directly facing equator and having a tilt angle close to 35 degrees would deliver the maximum amount of electricity annually (this may vary with latitude).
- This is consistent with the literature. However, achieving this arrangement is not always possible due to various physical and planning constraints.
- Economic feasibility of using PVs to increase the NABERS rating of an office building and decrease grid electricity consumption and GHG emissions within the current energy prices and subsidies regime for the case study building still is weak.
- Even when high levels of available subsidies and grants were applied, the PV systems analysed failed to deliver positive returns within a time frame that would be acceptable to the vast majority of commercial building owners.
- To make the PV systems for the case study building economically viable, it is necessary for very high levels of subsidies to be provided or the base capital cost of PV to be substantially reduced. One way to achieve this is through a combination of rebates, grants and an uncapped Gross FiT. If this approach was applied across the commercial building sector, it would certainly increase the use of PV systems, but it would also create a great cost burden on governments and electricity users who would have to accept higher electricity bills to cross-subsidise the FiT.
- The growth of PV production, particularly in China, as well as continued development of lower cost PV technologies will rapidly drive the cost of PV systems down so these renewable energy systems can be more widely used to reduce GHG emissions across the built environment.

Conclusions from the next study concern the policy for managing human systems for retrofitting. A pilot project conducted at the Queensland University of Technology, on its Urban Technology campus, aimed at qualifying tolerance of building occupiers to changes in the generally accepted industry standards for office temperature (Chapter 4.7). The findings were that:

- Adjustment of the HVAC thermostat cooling set-points resulted in meaningful engagement by the occupants in a change management process that delivers financial and environmental savings as well as occupant comfort.
- The usual approach to providing comfort via the standard engineering process can be challenged by harnessing the building occupants to assist with managing the HVAC systems.

A further study by The Low Energy High Rise project (Chapter 4.8) has conclusively demonstrated:

- Building operational factors are critical in delivering energy efficiency, translating into a 30 per cent improvement in sector performance, which in turn translates into a 1.2 per cent reduction in Australia's national greenhouse emissions.
- Further delivery and testing of the Low Energy High Rise materials will refine the approach, lead to the development of practical case studies and significantly improve building operational efficiency. This may also lead to greater recognition of the value of building operators.

Further post-occupancy studies provide more depth to the way occupants behave and give insights into some of the complexity of renovation and retrofitting buildings and the benefits to organisations. These are:

- Where staff-related costs often comprise more than 80 per cent of the operational budget, it is the benefit to the occupant's well-being and comfort and consequent impact to worker productivity that are valued more highly than energy savings *per se*.
- The strategic approach to building refurbishment outlined in this chapter offers positive outcomes at all levels and provides useful insights for retrofitting the vast portfolio of existing buildings in our cities.

The argument which underlies this study is that positive occupant feedback is seen in the case of refurbishments, that this enhances workplace quality, creating high satisfaction. Workplace quality includes a range of factors but central to this core construct is the overall comfort of the occupants.

In the final case study we find the identification of the use of occupant control as a central determinant of workplace quality (Chapter 4.10). This study examines an adaptive reuse project. This places the user in control of the environmental aspects of the building as an important determinant if not a proxy for comfort. So rather than designing a space for comfort conditions, the control system drives both the passive and active systems either automatically or manually adjusted by the occupants. In this case the new office space was designed to do the following:

- Reduce basic solar loads through improved envelope performance.
- Apply the use of effective daylight.
- Apply new control technologies that appear to comply with our passive design and bioclimatic principles which allowed for natural ventilation conditioning, a response to external temperature, humidity and wind speeds and internal set-point as well as offering night-time purging strategies to be implemented and controlled.

- Create a reduction of 50 per cent or more in energy savings over a conventional 'business as usual' approach, which can be easily obtained.

INDEX

Locators in **bold** are to diagrams, photographs, plans and tables

Milton Keynes UK
Ingram Content Group UK Ltd.
UKHW021915071024
449327UK00022B/1669

9 780367 576677